REASON AND THE SEARCH FOR KNOWLEDGE

BOSTON STUDIES IN THE PHILOSOPHY OF SCIENCE

EDITED BY ROBERT S. COHEN AND MARX W. WARTOFSKY

DUDLEY SHAPERE

*Department of Philosophy, University of Maryland*

# REASON AND THE SEARCH FOR KNOWLEDGE

## *Investigations in the Philosophy of Science*

D. REIDEL PUBLISHING COMPANY

A MEMBER OF THE KLUWER ACADEMIC PUBLISHERS GROUP

DORDRECHT / BOSTON / LANCASTER

**Library of Congress Cataloging in Publication Data**

Shapere, Dudley.
Reason and the search for knowledge.

(Boston studies in the philosophy of science ; v. 78)
Bibliography: p.
Includes index.
1. Science—Philosophy. I. Title. II. Series.
Q174.B67 vol. 78 [Q175] 001'.01s [001.4] 83–11182
ISBN 90–277–1551–3

---

Published by D. Reidel Publishing Company,
P.O. Box 17, 3300 AA Dordrecht, Holland.

Sold and distributed in the U.S.A. and Canada
by Kluwer Academic Publishers,
190 Old Derby Street, Hingham, MA 02043, U.S.A.

In all other countries, sold and distributed
by Kluwer Academic Publishers Group,
P.O. Box 322, 3300 AH Dordrecht, Holland.

2-0390-200 ts



Printed in The Netherlands

# CONTENTS

# EDITORIAL PREFACE

An impressive characteristic of Dudley Shapere's studies in the philosophy of the sciences has been his dogged reasonableness. He sorts things out, with logical care and mastery of the materials, and with an epistemological curiosity for the historical happenings which is both critical and respectful. Science changes, and the philosopher had better not link philosophical standards too tightly to either the latest orthodox or the provocative upstart in scientific fashions; and yet, as critic, the philosopher must not only master the sciences but also explicate *their* meanings, not those of a cognitive never-never land. Neither dreamer nor pedant, Professor Shapere has been able to practice the modern empiricist's exercises with the sober and stimulating results shown in this volume: he sees that he can be faithful to philosophical analysis, engage in the boldest 'rational reconstruction' of theories and experimental measurements, and faithful too, empirically faithful we may say, to both the direct super-highways and the winding pathways of conceptual evolutions and metaphysical revolutions. Not least, Shapere listens! To Einstein and Galileo of course, but to the workings of the engineers and the scientific apprentices too, and to the various philosophers, now and of old, who have also worked to make sense of what has been learned and how that has happened and where we might go wrong. We think that Shapere's title for this selection of his essays is a description of himself as it is of his work and his book. Can more be asked of a philosopher?

*October 1983*

ROBERT S. COHEN
*Center for Philosophy & History of Science,*
*Boston University*

MARX W. WARTOFSKY
*Department of Philosophy,*
*Baruch College, The City University of New York*

# PREFACE

The essays collected in this volume were written over a period of nearly three decades, from "Philosophy and the Analysis of Language," originally a term paper for Morton White's course on Analytic Philosophy at Harvard in 1953 or 1954, to "Modern Science and the Philosophical Tradition," which has been presented as a public lecture on various occasions during the last two or three years. In spite of the time spanned, however, the papers present a unified perspective on the nature of science, and of human knowledge in general. Although some exceptions will be found, the changes that have taken place in my viewpoint turn out to have consisted, for the most part, in a gradual broadening and deepening of perspective rather than in major shifts of doctrine. But not all of the viewpoint which I have developed is represented in the papers included here; other aspects of it will be found in recent publications, and still more will appear in future work. In order that the ideas found in the present papers may be placed within that larger framework, I have tried to give a rather extensive, though still all too sketchy, outline of it in the Introduction to this volume.

It is impossible to name all the students, colleagues, and friends who have contributed to the development of the ideas presented here. Some people must, however, be given special thanks: Robert Cohen, for encouraging me to publish this collection; Morton White, for constant help and enlightening discussions; and Hannah Hardgrave Shapere, for her valuable suggestions about the work, and for her support while I was putting it together. The University of Maryland generously provided me with a Faculty Research Grant to pursue these investigations, and the National Science Foundation has also supported them on several occasions. I also wish to express my special gratitude to the Institute for Advanced Study, Princeton, New Jersey, for providing the time and the environment, in 1978–79 and again in 1981, without which my thoughts might never have coalesced.

## SOURCES AND ACKNOWLEDGEMENTS

"Philosophy and the Analysis of Language" originally appeared in *Inquiry*, Vol. **III**, No. 1 (1960), pp. 29–48. It is reprinted here by kind permission of Universitetsforlaget, Oslo, Norway.

"Mathematical Ideals and Metaphysical Concepts" appeared in *The Philosophical Review*, Vol. **LXIX**, No. 3 (1960), pp. 376–385; "The Structure of Scientific Revolutions" also came out in that journal, Vol. **LXXIII**, No. 4 (1964, pp. 383–394. Both are reprinted here by kind permission of *The Philosophical Review*.

"The Paradigm Concept" appeared in *Science*, Vol. **172**, pp. 706–709 (14 May 1971), and is reprinted by permission of *Science*. Copyright 1971 by the American Association for the Advancement of Science.

"Meaning and Scientific Change" originally appeared in *Mind and Cosmos*, Robert G. Colodny, editor, published in 1966 by the University of Pittsburgh Press; used by permission of that press.

"Notes Toward a Post-Positivistic Interpretation of Science" appeared in P. Achinstein and S. Barker (eds.), *The Legacy of Logical Positivism*, Baltimore, Johns Hopkins University Press, 1969, pp. 115–160, and is reprinted here by permission of that press.

"Space, Time, and Language" is from B. Baumrin (ed.), *Philosophy of Science: The Delaware Seminar*, Vol. II, New York, John Wiley, 1963, pp. 139–170. The copyright for that work has been transferred to the University of Delaware Press, which has kindly granted permission to use the article here.

"Unity and Method in Contemporary Science" was originally published as "Unification and Fractionation in Science: Significance and Prospects," in *The Search for Absolute Values: Harmony Among the Sciences*, Proceedings of the Fifth International Conference on the Unity of the Sciences (Washington, D.C., 1976), pp. 867–880, and is reprinted here by permission of the International Cultural Foundation Press, copyright 1977 by the International Cultural Foundation, Inc.

"What Can the Theory of Knowledge Learn from the History of Knowledge?" was originally in *The Monist*, Vol. **LX**, No. 4 (1977), pp. 488–508. It is reprinted by permission of *The Monist*, La Salle, Illinois.

"The Character of Scientific Change" originally appeared in T. Nickles (ed.), *Scientific Discovery, Logic, and Rationality*, Dordrecht, D. Reidel Publishing Co., 1980, pp. 61–116.

"The Scope and Limits of Scientific Change" is to appear in *Logic, Methodology and Philosophy of Science VI*, edited by L. J. Cohen, H. Pfeiffer, K.-P. Podewski, and J. Los, and is reprinted here by permission of the publishers, North-Holland Publishing Company.

"Scientific Theories and Their Domains" appeared in F. Suppe (ed.), *The Structure of Scientific Theories*, Urbana, University of Illinois Press, 1974, pp. 518–565, copyright 1974 by the Board of Trustees of the University of Illinois, and is reprinted here by permission.

"Remarks on the Concepts of Domain and Field" is a revised version of the opening pages of "The Influence of Knowledge on the Description of Facts," *PSA 1976*, Vol. 2, East Lansing, Philosophy of Science Association, 1977, pp. 281–298. "Reason, Reference, and the Quest for Knowledge" appeared in *Philosophy of Science*, Vol. **XLIX** (1982), pp. 1–23. Both these articles are reprinted here by permission of the Philosophy of Science Association.

The other articles contained in this volume have not previously been published. "Interpretations of Science in America" was a Sigma Xi National Bicentennial Lecture in the years 1974–77. "Alteration of Goals and Language in the Development of Science" is a completely rewritten and greatly extended discussion of a case dealt with in "The Influence of Knowledge on the Description of Facts." "The Concept of Observation in Science and Philosophy" is a brief summary of an article appearing in *Philosophy of Science*, Vol. **XLIX**, pp. 485–525 (December, 1982); that article, in turn, is part of a chapter of a book of the same title, to be published by Oxford University Press. "Modern Science and the Philosophical Tradition" has been presented as a public lecture on various occasions.

# INTRODUCTION

Section I of this Introduction summarizes some of my criticisms of certain important movements and doctrines in the philosophy of science of the last three decades or so. It also includes a sketch of the ways I have conceived some of the fundamental problems of the philosophy of science in the light of those criticisms. Section II outlines the view of scientific change which emerges in the essays included in this volume, while Sections III and IV, in addition to some further points to be found in the included essays, outline a broader viewpoint, one which is for the most part, especially with regard to the ideas discussed in Section IV, not represented explicitly in these essays. (To avoid confusion with the "Parts" into which the body of this book is divided, I refer to the divisions of this Introduction as "Sections.") I must emphasize that that larger viewpoint is only sketched here, largely without the detailed development and argument it requires; its full elaboration and defense will be found in other writings, most of them not yet published. But without this overview, many of the ideas found in the present essays might appear fragmentary and unrelated; it is therefore offered here for the sake of completeness, in order to provide the reader with a framework within which the ideas developed in the present papers should be understood. Throughout the Introduction, I refer to places in the included articles where the ideas mentioned are developed, and I also refer, especially in Sections III and IV of the Introduction, to other recently-published and forthcoming work.

## I

Logical Empiricism dominated the philosophical interpretation of science in the United States from the 1930's until well into the 1950's. (Logical Empiricism is often referred to as Logical Positivism; since my work focusses on the connections of that movement with traditional empiricism, I will generally use the former term except when explicitly comparing or contrasting Logical Empiricism with traditional empiricism, or when discussing doctrines characteristic of the former but not of the latter movement.) Despite numerous internal disagreements, often of a quite serious nature, adherents of that movement agreed on certain very general approaches, or at least on a very general program for the interpretation of science. However, beginning in the 1950's and reaching a peak in the early 1960's, those more fundamental approaches themselves began to be criticized, and by about 1970 had been widely rejected. In my own view (partly presented in Part I of "Notes Toward a Post-Positivistic Interpretation of Science" and "Interpretations of Science in America"), the most important weaknesses of the Logical Empiricist program were the following. First, its focus on the formal (logical) structure of theories ignored the developmental aspects of scientific ideas, and positivists even denied that those developmental aspects were of any philosophical interest. On the contrary, it seemed to me (in company with a number of other writers at the time) that there are often reasons for the introduction of new scientific ideas, and that those reasons need not necessarily be the same as those involved in the justification of those ideas. Second, the Logical Empiricist distinction between "scientific" terms – those occurring "within" science – and "metascientific" ones – those used in "talking about" science, – and the positivist insistence that the primary concern of the philosophy of science was the analysis of the latter, seemed to me highly questionable. Could it not be that "metascientific" concepts themselves change, and, further, that those changes come about, in at least some cases, in response to new substantive views about nature? New investigations by historians of science, as well as my own studies, seemed to me to indicate that such changes might be deep and pervasive. And finally, the Logical Empiricist – and traditional empiricist – distinction between 'theory' and 'observation,' at least in the forms in which it had generally been understood, as a mutually exclusive (as well as collectively exhaustive) classification, was indefensible. The problem was not merely that – as many critics were asserting at the time – there was no clear criterion for making the distinction – that many, and perhaps all, of the positivist-empiricist examples of "observational" terms arguably contained a "theoretical" component, and that many (at least) of their examples of "theoretical" terms sometimes functioned, in science, as "observational." Those objections seemed to me to be essentially correct (though their bases or implications seemed to me not to be well understood). But there was a more fundamental objection, which could not be interpreted as a mere failure to make a distinction adequately. Rather, it questioned the very program of making the distinction as an unalterable, mutually exclusive one. For even if there were any such thing as an observation term wholly pure of any theoretical presupposition or component, that term would, by that very purity, be rendered completely irrelevant to the theories with which those observation terms were supposed to be concerned. An important aspect of this argument, of course, was the failure of the positivist movement to specify how observation terms *could* be relevant to theoretical terms. The latter were admittedly not definable in terms of the former; and the "partial interpretation" view proposed by Carnap seemed to suffer from so many difficulties as to be untenable. But on the other hand, to deny the distinction between a pure, uninterpreted observation-language on the one hand and a theoretical language on the other would threaten to betray the original motivation of the entire empiricist movement: namely, to show how the meanings of our theoretical ideas could be objectively based in experience. Similar questions arose with respect to the *acceptability* of theoretical claims (as distinguished from the *meanings* of theoretical terms). Such acceptability or rejectability seemed to me, from my examination of cases in science and its history, to require antecedent "theoretical" beliefs; yet if acceptance or rejection was to be "objective," should not the "observations" on which empiricism rested that acceptance or rejection be *free* of bias by any antecedently-accepted "theory"? There thus seemed to be a conflict between what I later called (in "Interpretations of Science in America") the *Condition of Objectivity*, according to which the bases of test (acceptability) of a scientific idea must be "independent" of the idea to be tested, and the *Condition of Relevance*, according to which the bases of test (the "observational evidence") had to be "relevant" to the idea to be tested. The original positivist and empiricist programs emphasized the former condition at the expense of the latter; and that very fact meant that *that* version of the empiricist-positivist program, at least, was misguided in principle.

During this period of criticism of Logical Empiricism, the 1950's and early 1960's, a number of alternative views were being proposed. The positions advanced by Stephen Toulmin, Paul Feyerabend, and Thomas Kuhn were particularly influential. But while I was sympathetic, at least in part, to their recommendation that philosophy of science must rest on examination of the history of science, the conclusions those writers thought followed from that examination did not seem in fact either to follow or to be acceptable. Their views were often extremely ambiguous, admitting of a variety of interpretations; some central ideas were not explained at all. Far from following from an examination of the history of science as they claimed, their views seemed shaped by the fundamental ideas involved, vague though they were; their conclusions rested ultimately on conceptual unclarities and confusions. For example, Feyerabend and Kuhn claimed that some presupposed unitary "high-level background theory" or "paradigm" functions in every specific scientific situation within a scientific "tradition" or "community" to shape or determine or govern all aspects of science as practiced by members of that tradition or community, including the meanings of terms, what counts in that tradition or community as observational, the problem-field, the methods of approach, the programs of research, and the standards of possible and correct solution of problems. (Because of the generality, unity, and pervasiveness of this supposed governing viewpoint, I have referred to their view as *Global Presuppositionism*.) But they never attempted to make clear the precise way or ways in which such "shaping" or "determination" or "governance" takes place. Is the relation supposed to be one of logical implication, or is it weaker than that, perhaps only one of psychological association? Or does the manner of governance vary from circumstance to circumstance within the tradition or community? This idea of an overarching governing viewpoint had disastrous implications for their interpretations of science, in particular with regard to two central notions. In the first place, since the meanings of all terms ("theoretical" and "observational," if not also logical ones) were supposed to be "determined" by the high-level background theory or paradigm, Feyerabend and Kuhn concluded that the meanings of all those terms differ incommensurably from one high-level theory or tradition or community to another. Yet, though the concept of "meaning" thus played a fundamental role in their views, they made no effort to analyze it. (Or at least no successful effort: *cf.*, "Meaning and Scientific Change.") Any similarities between the use of a term in two different traditions were simply relegated by fiat to being not "part of the meaning." No criterion for distinguishing what is from what is not "part of the meaning" was given; no account was provided of the fact of the similarities of usage. Secondly, although their lack of analysis of the sense or senses in which the governing viewpoint governed left their notion of scientific reasoning vague, it nevertheless seems that, however that claim is interpreted, it must follow that there could be no reason for the replacement of one such governing viewpoint by another. For even what counts as a "reason" in favor of or against a scientific idea (or problem or method, *etc.*) is supposed to be determined by the governing viewpoint; and by the incommensurability thesis, what one tradition counts as a "reason" would be incommensurable with what is so counted by another tradition. Given their lack of analysis of these and other fundamental ideas, the most reasonable interpretations of their positions imply an extreme relativism in which there is no progress, or even any knowledge, in science. Such an outcome could only be expected from positions which deny the existence or possibility of *any* theory-independent observation-language without at the same time showing how the objectivity of scientific test could be preserved. The problem with their views was thus the reverse of that of positivism: they focussed on the Condition of Relevance to the exclusion of the Condition of Objectivity. (These criticisms are detailed in "The Structure of Scientific Revolutions," "The Paradigm Concept," and "Meaning and Scientific Change," and, with increased generality, "Interpretations of Science in America." More restricted but nevertheless similar views of Toulmin are examined in "Mathematical Ideals and Metaphysical Concepts.") Since in the final analysis the views of Feyerabend and Kuhn flowed from the above vague and confused ideas rather than from any studies of cases from the history of science, it seemed that their relativistic implications might be avoidable, and that a better view might involve a more adequate analysis of the concepts and roles of "presupposition," "meaning," and "reason."

The complementarity of both the difficulties and insights of the two warring parties, Logical Empiricism and its "Global Presuppositionist" critics, emerged in "Interpretations of Science in America," and reflected what I had already seen as a program for the philosophy of science. On the one hand, the problem was to develop a view which would preserve the objectivity and rationality of science, and the possibility of its attaining knowledge – goals which the empiricist-positivist tradition had rightly tried to achieve. Yet those aims could not be fulfilled through the paths those movements had taken: the pure, uninterpreted "given," the theory-free "observation-language" by which they had attempted to satisfy the original motivation of empiricism, to account for the objectivity of the knowledge-seeking enterprise, would have to be rejected. Account would have to be taken of the roles of presupposed beliefs, and the extent and character of those roles. The notions of objectivity and reason, and of the possibility of scientific knowledge, would have to be reinterpreted in the light of those roles – but not, however, rejected in favor of the relativism into which the critics of empiricism and its positivist descendants had fallen. The ways in which the presupposed beliefs are chosen would have to be understood (an understanding which the critics of positivism had in effect denied was available). And while some philosophers of science seemed on the verge of abandoning the theory-observation distinction entirely because of the criticisms that had been levelled against it, I thought that it would have to be reinterpreted and reassessed in the light of the roles played by presupposed beliefs in the determination of what was to count as "theoretical" and as "observational."

The problem was of course not merely a matter of philosophical reflection; it was a matter of constructing a view of science which would be adequate to the ways in which science itself proceeds. The positivistic tradition too often ignored real science in its focus on purely logical issues; its critics distorted science and its history through conceptual confusions. Yet science itself has managed to avoid relativism and skepticism while at the same time developing, more and more, especially in the twentieth century, an attitude of complete openness to new ideas, of willingness to abandon any idea in favor of any new one, no matter how strange or contrary to presupposition, whenever such changes were called for. And increasingly in the twentieth century, scientists have seen such changes as governed by observation. Clearly, the construction of a new view of science would have to be based on a close examination of cases in science and its history as well as on attention to philosophical issues.

In the years during and following the writings included here in Part I, I made a number of studies of cases in the history of science and contemporary science. Most of those studies were largely exploratory in nature, trying partly to extract philosophical lessons from the cases and partly to test various philosophical interpretations which seemed to offer some promise of dealing with the issues I have just outlined. However, my view of the role of such case studies gradually evolved beyond using them merely as bases for criticism or generalization. That later conception of the role of case studies diverges in fundamental ways from that of most historians and philosophers, as will be described in Section III, below. (It and its criticisms of usual conceptions are detailed in "What Can the Theory of Knowledge Learn from the History of Knowledge?")

Of the case studies made during those years, only a few have been included in the present volume. Two such papers dealing with case studies, "Scientific Theories and Their Domains" and Part II of "Notes Toward a Post-Positivistic Interpretation of Science," require special comment. The analyses of cases given in those two articles have been important for the development of my views, in ways that will become apparent later. Nevertheless, as they stand, those analyses are limited in certain ways. In "Scientific Theories and Their Domains," "patterns of reasoning" play a prominent role in the interpretation of the cases examined. But those patterns are presented there without any indication of their source or justification, and I would now insist that such be provided. The concept of a "domain" itself has undergone considerable refinement over the succeeding years (as will be seen in "Remarks on the Concepts of Domain and Field" and "Alteration of Goals and Language in the Development of Science.") For some time, indeed, I rejected the notion; I only readopted it around 1980, as I gradually came to see that, properly understood, it provides an important ingredient in the analysis of 'reason' for which I was searching (see below, Section II). Again, Part II of "Notes Toward a Post-Positivistic Interpretation of Science," and, in a somewhat different way, Part IV of "Scientific Theories and Their Domains," present an interpretation of scientific "idealizations" or "simplifications" – a group of functions of scientific ideas to which I now refer more generally as "conceptual devices." At the time I wrote those essays (the last half of the 1960's), this topic was in many ways isolated from the problems with which I was centrally engaged, as I have outlined those problems above. That idea too, however, like that of "domains," has now found its place in the larger context of thought which I will outline in Section III.

As my studies of scientific cases developed, I gradually came to understand more definitively what I had supposed in a general way from the outset: that the issues with which I was concerned were only special cases of far more general ones in the history of philosophy; that the doctrines which I thought must be opposed were only specific versions of ones which had permeated philosophy since the time of Plato. These generalizations of issues and doctrines began to be presented in "The Character of Scientific Change" (written in 1978). They, rather than the more specific and transitory doctrines of Logical Empiricism and its critics, now became the focus of my work. I now saw the empiricist doctrine of a pure, uninterpreted and unassailable given and the positivist doctrine of a theory-free observation-language as being in the same company as other "absolutist" views in the history of philosophy: the Platonic claim that there are unalterable Ideas in terms of which experience must be interpreted; the Kantian view that experience presupposes certain Categories and Forms of Intuition; the view that science, or more generally, the knowledge-seeking (and/or the knowledge-acquiring) enterprise, must necessarily accept certain presuppositions, like the "Principle of the Uniformity of Nature" or the "Principle of Limited Independent Variety"; the idea that there is an unalterable "scientific method"; the view that a certain particular set of concepts must appear in any fundamental scientific theory; that the goals of science are established once and for all, presumably at the beginning of inquiry, or perhaps by virtue of the "very concept" of an inquiry. But also in that company was the Logical Empiricist notion of a set of unalterable "metascientific concepts" which are definitory of science, past, present, or future. All such views have in common the idea that there is something about the scientific (or, more generally, the knowledge-seeking or knowledge-acquiring) enterprise that cannot be rejected or altered in the light of any other beliefs at which we might arrive, but that, on the contrary, must be accepted before we can arrive, or perhaps even seek, such other beliefs. I came to call this idea "the Inviolability Thesis." But what could be the justification of such allegedly inviolable constituents, methods, presuppositions, or whatever of science? My own investigations had convinced me that an understanding of "observation" was at hand which was far more adequate, both to philosophical issues and to the way science has developed in its history, than the "pure, uninterpreted – and inviolable – given" or "theory-independent observation-language" doctrines or any of the alternative views which had been proposed to replace them. (See "The Concept of Observation in Science and Philosophy," of which only a brief summary is included in the present collection. The full article is in *Philosophy of Science*, December, 1982; that article will be part of Chapter II of a book of the same title, *The Concept of Observation in Science and Philosophy*, forthcoming from Oxford University Press.) And as for necessary presuppositions, whatever their claimed nature, history is littered with the ruins of allegedly *a priori* necessities or impossibilities which have been overthrown by scientific developments. One way of attempting to justify such necessities has been through "transcendental" arguments. These have in common with scientific explanations the attempt to show how experience, or inquiry, or whatever, is possible. But unlike explanations in science, a transcendental argument claims that the explanation given is the *only* one that could be given – that it is *necessary* as an explanation of the phenomena in question. And the argument that would establish *that* claim can only be an *a priori* one, subject to all the doubts that such arguments are heir to. Some of the general reasons for suspicion of *a priori* claims are given in "Modern Science and the Philosophical Tradition."

The relativism into which the views of Feyerabend and Kuhn degenerate, on the other hand, has its kindred in the various forms of relativism and skepticism that have arisen so often in the history of philosophy. The aim of variants on the Inviolability Thesis from Plato to the present day has always been at least partly to counter skeptical and relativist arguments; and despite the failures of Inviolability counterarguments, this motivation behind them seemed to me a valid one. Relativist and skeptical arguments have usually been as confused and unconvincing as those of their absolutist opponents; and, like the latter, they take no account of the evident achievements of science. The real point of both skepticism and relativism lies in their exposure of shortcomings in our understanding of the nature of knowledge and of the knowledge-seeking process, shortcomings which proponents of the Inviolability Thesis, in any of its forms, have done nothing to alleviate. Specific arguments against relativism and skepticism are found in many of the papers included here, and new ones, in some cases more powerful than the present ones, will be given in the forthcoming book, *The Concept of Observation in Science and Philosophy*.

The problem with which I have been most centrally concerned can therefore be summarized as follows. Given the failure of all absolutist and relativist-skeptical arguments hitherto presented, can an account of the knowledge-seeking and knowledge-acquiring enterprise be given which, while not relying on any form of the Inviolability Thesis, will also not collapse into relativism or skepticism? More specifically, can an account be given of the scientific enterprise which will not make that enterprise subservient to some unalterable given or presupposition? And the focal issue in the attempt to construct such a view must be this: Is it possible to understand science (or, more generally, the search for knowledge) as able to proceed *rationally* without presupposing criteria of what is to count as "rational," criteria which could not be *arrived at* in the course of seeking knowledge, but which must be assumed in order to engage in that enterprise at all, or at least to engage in it successfully? It has been the supposed impossibility of constructing such a view that has led many philosophers to adopt some version or another of the Inviolability Thesis, as the only possible alternative to relativism or skepticism.

My investigations have thus stemmed not from an assumption that the Inviolability Thesis is unavoidable if we are to escape relativism and skepticism, but rather from the very opposite standpoint: from the attempt to ascertain how much of the knowledge-seeking and knowledge-acquiring enterprise can be understood without accepting anything inviolable. If as a result of these studies it turns out that we must accept some such inviolable principles, that will be a conclusion, not a starting-point, of the inquiry. My starting-point has not, of course, been chosen gratuitously; the reasons for its choice, as will be seen in the essays included in this volume, lie partly in the failure of the alternative views and partly in an examination of science and its development.

With this statement of issues providing a background, then, I will now outline the general view of science and scientific change which has emerged, and which is presented in a more detailed way in the papers included here.

## II

At any stage in human history, beliefs are available which have (thus far) proved successful and free from reasons for doubt. (The roots of such beliefs will be detailed in the book, *The Concept of Observation in Science and Philosophy*; in particular, it will be shown there that no bars to the attainment of knowledge necessarily result from the employment of such beliefs. As to "success" and "reasons for doubt," more will be said in what follows.) Increasingly, such views have become the bases on which we build our further beliefs. The ways in which those views achieve this status are of primary importance for understanding the nature of science and scientific reasoning. For a central feature of the development of science – a feature which indeed has evolved into one of the major distinguishing marks of what we call "science" – consists in the acquisition of further such beliefs, of relevance-relations between them, and of the organization of such beliefs (those which have been found to be successful and free from doubt) into areas for investigation and well-founded beliefs relevant thereto. That organization becomes more and more characteristic of the knowledge-seeking enterprise; and it comes more and more, as science develops, to be made in terms of beliefs which have been found to be relevant to one another. It is the process by which *areas or fields of scientific investigation are formed*. Increasingly, as science develops, such areas come to consist of two distinguishable parts: (1) a body of information to be investigated (this I call the "domain" of the area or field – see "Scientific Theories and Their Domains," "Remarks on the Concepts of Domain and Field," and Part II of "Alteration of Goals and Language in the Development of Science"); and (2) a body of "background information," that is, a body of successful and doubt-free beliefs which have been found to be relevant to the domain. (This idea plays a role in most of the essays included in Part III, but is discussed most systematically in the article, "The Concept of Observation in Science and Philosophy"; its role can be gathered to a considerable extent from the summary of that article which is included in the present volume.)

But it is precisely these developments – the formation of domains and of background information relevant thereto – that *constitute* the development of the rationality of science. For one of the most fundamental aspects of the idea of a "reason" in general is the following: to count as a reason, a claim must be *relevant to the subject-matter under consideration or debate. And thus the clear delineation of a subject-matter, and of the body of other claims relevant to that subject-matter, itself constitutes the development of a science based on reasons.* ("The Scope and Limits of Scientific Change"; "Modern Science and the Philosophical Tradition.")

The development of science thus consists in a gradual discovery, sharpening, and organization of relevance-relations, and thus in a gradual separation of the objects of its investigations and what is directly relevant thereto from what is irrelevant to those investigations: a gradual demarcation, that is, of the scientific from the non-scientific. Indeed, to the extent that an area of human activity manifests the sorts of developments I have been describing, to that extent the area is considered paradigmatically "scientific." In other words, *this* is what we have come to *call* "scientific." In that development, science aims at becoming, as far as possible, autonomous, self-sufficient, in its organization, description, and treatment of its subject-matter – at becoming able to delineate its domains of investigation and the background information relevant thereto, to formulate its problems, to lay out methods of approaching those problems, to determine a range of possible solutions, and to establish criteria of what to count as an acceptable solution, *all in terms solely of the domain under consideration and the other successful and doubt-free beliefs which have been found to be relevant to that domain; that is, to make its reasoning in all respects wholly self-sufficient.* (It must be understood, however, that this is not the *only* distinguishing feature of science; other characteristics have also been found to be successful and doubt-free, and therefore worthy of being made standards for further development – *e.g.*, when properly understood, coherence and precision, particularly mathematical precision.) This process employed by science in approaching the formulation and treatment of its subject-matter and problems I call, for reasons which will be given later, the *internalization of considerations*. It is in essence the development of the rationality of science, of what it is for an argument to constitute scientific reasoning.

In the light of this process of internalization, it is no wonder that the development of science involves changes not only in substantive views about nature, but also in the descriptive language of science; in the body of problems regarding nature so described and classified, and in the criteria of genuineness and importance of those problems; in the methods by which those problems are approached; in the standards or criteria of what can count as a possible solution (explanation) of those problems, and also as to what can count as an acceptable solution of them, including what can count as evidence for or against a proposed solution; and even in the conception of the goals of science. For such changes, insofar as they are unequivocally scientific, are aimed at increasing the autonomy, the "internalization," the self-sufficiency of scientific argument.

Many views of science which have been proposed in the 1970's and early 1980's have ignored these important features of science. They have, for example, continued the positivist tactic (also shared by Feyerabend and Kuhn, in spite of the latter's emphasis on the unity of science within a tradition) of explaining science in terms of distinct "levels" of scientific thought and activity, each level having its own distinctive methods of procedure and justification. Usually the "higher level" is seen as in some way governing the "lower" one, and in many views the "higher" level is immune to revision in terms of what happens on the "lower." Such interpretations of science are fundamentally misguided. For a central aim of science is – has become – wherever possible to *remove* such distinct "levels," if they exist, in favor of an integrated approach, "internalizing" the separate levels to achieve an *interaction* of ideas (methods, standards, *etc.*) in which all of them are subject to revision or rejection in the light of what we learn about nature.

In this process, problems and their solutions lead, on occasion, to the rejection or modification of background beliefs, or to the addition of new background information relevant to a given domain, and these changes lead in turn, sometimes, to profound alterations of the fabric of science. Areas for scientific investigation are reformulated, split, or unified, fragments being reattached to other areas in the light of newly-found or newly-understood relevance-relations. Even the descriptive language of domains may in some cases be radically revised to incorporate new beliefs which, wherever possible, have shown themselves to be well-founded in the ways I have outlined. ("Alteration of Goals and Language in the Development of Science.") The problems associated with particular domains become altered, as do the lines between recognized "scientific" problems and questions that are classed as "non-scientific." What counts as "observation" of a subject-matter, too, may be altered, as we shall see shortly. New background information, or old background information formerly considered irrelevant, is found to be relevant to a particular domain. Old methods are rejected or reinterpreted, new ones introduced; new standards of possibility and acceptability arise.

Thus, the more science has attained in the way of well-formulated domains and clearly relevant background information, the more it has available as a basis for pushing forward, for building new ideas on the basis of the best beliefs it has available: those which have shown themselves to be successful and doubt-free (and which satisfy other constraints which have been developed). The more science learns, that is, the more it becomes able to learn. And it is thus that science *can attain* (and to a considerable degree has attained) what is most appropriately described as a *rationale of development*. For on the basis of those well-founded (successful, doubt-free, *etc.*) beliefs, including its conceptions of its subject-matter as well as information relevant thereto, science is able to develop new hypotheses, new problems, new methods, new standards, and even new goals for itself, or at least to modify or refine its earlier ones where such changes are called for. These procedures do not constitute a "logic of discovery," for they depend wholly on the content of belief rather than on purely "formal" considerations, and develop as that content develops; and they do not *necessarily* produce unique and unambiguous solutions of problems (for instance), though they sometimes may. Nevertheless, they show that the worries of positivism about reasoning in the "context of discovery" are groundless. There is no need for the "pure observation language" which (as we shall see in Section III, below) produced those worries, and no such language exists; and the fact that our conceptions of scientific subject-matter, and of observations of it, have the richness of interpretation provided by background information makes it possible for science to reason in advances to new ideas. But worries stemming from the "loading" of background beliefs into what counts as subject-matter and observation are also dispelled. For we see that it is not just *any* background beliefs that can be used in shaping the course of science; there are conditions governing what can be so used, conditions which have been arrived at in the course of inquiry and which have become ever more stringent with the development of science. The scientific enterprise is thus a process of building, or rather of coming more and more to be able to build, on the best beliefs it has available. Insofar as science is able to proceed in the light of its best beliefs, its arguments and alterations are rational (though *that* it is able to do so is a contingent fact about nature as we have found it). The relativism into which Kuhn's views collapsed is thus escaped, even while all aspects of science are left open, in principle, to revision or rejection, so that the Inviolability Thesis is, at least to the extent of my discussion here, also avoided.

For the processes I have described neither rest on nor produce any guarantees. We *may*, as far as we know, always find reason for doubt concerning any aspect of scientific belief, including our conceptions of the subject-matter we are examining. Even what counts as a "reason for doubt," and criteria for deciding the degree of seriousness of a doubt, are themselves subject to the possibility of alteration or rejection in the light of new beliefs arrived at through application of those very criteria. For example, it is a central aspect of the view expressed in these papers that, in the development of science, we have learned *not* to take as "reasons for doubt" any alleged "reasons" which apply indiscriminately to any proposition whatever – to both a proposition and its negation. Thus "philosophical" doubts, like the possibility that a demon might be deceiving me, or that I may be dreaming, play no role in science. Such skeptical doubts, we have found in the actual practice of seeking knowledge, are only misleading ways of reminding us that, as a matter of learned fact, doubt *may* arise. But the possibility of doubt arising is not itself a reason for doubt of any particular proposition; more exactly, it is no reason to abstain from using the best beliefs we have – those which have been found to be successful and free from *specific* doubt – to build on.

What is to count as success, too – what is to count as a reason *in favor of* a belief, – is something that can change with the accumulation of new well-founded beliefs. In competition, for example, one criterion of success may prove unsatisfiable or only poorly satisfiable (perhaps, but not necessarily, in the light of some further criterion of successful satisfaction), while another, even though initially considered less important, is, so that the first is abandoned or demoted in importance while the second becomes primary.

Finally, as is seen most clearly in "The Alteration of Goals and Language in the Development of Science," even the aims of science, its goals, can undergo change in response to the development of successful and doubt-free beliefs. Not only can aims be surrendered with the abandonment of unsuccessful criteria of success, but new ones can be introduced as new criteria of success develop. More generally, out of successful ideas and approaches spring normative principles, not only on such abstract levels as "Attend only to doubts that are directed at specific beliefs," or "Employ as background beliefs only those which have proved successful and free from specific doubt," or "Aim at the internalization of scientific reasoning," but also more specific principles such as "Try to explain all phenomena in terms of matter in motion," or "Try to construct explanations of elementary particles and forces in the form of renormalizable locally gauge-invariant field theories." Successful approaches become normative guiding principles as well as beliefs about the world (it is a mistake to separate sharply the descriptive and the normative in science). And, as the last two examples show, such principles too can change.

Science has more and more found it possible to achieve self-reliance in its reasoning, and, with the successes which have accompanied that possibility, to seek such self-reliance as a goal. This tendency can be seen in much of the history of science. In early phases of the development of an area of science, it is often unclear what considerations are or might be relevant to debate and its resolution; domains of investigation are not sharply delineated, but tend often to be broadly and vaguely inclusive. There are, in such phases, therefore, few clear limits on the sorts of considerations that can be brought to bear on an issue, or even on whether an alleged issue is an issue. Thus, anthropomorphic, religious, political, analogical, or logical considerations may be brought to bear on a question by one or another thinker or group, and specific considerations within such categories may differ from thinker to thinker or from group to group. Even when arguments are presented in favor of the relevance of such considerations, they themselves tend to be based on weakly-argued further considerations, the clarity and relevance of which are themselves questionable and often questioned. In such situations, it is not at all clear how to choose between competing viewpoints, or even to see how they may be competing. And though the presence of such difficulties may not always be apparent to those working at the stage in question – each side in a dispute, for example, may see *its* arguments for relevance (where they exist) as completely compelling, or there may be, at the time, only one side presented, the existence of assumptions which may easily be questioned and denied not being seen – those difficulties may become apparent in time.

The sort of situation I have in mind is apparent in the case of early Greek natural philosophy, where the possibility of agreement about such questions as the origin of the world (or worlds), the divisibility or indivisibility of matter, the nature and pervasiveness of change, and the interpretation of specific natural phenomena seemed so remote that reaction against the very undertaking of such inquiries finally set in. Furthermore, in the treatment of such issues, the object of investigation tended to be, vaguely, nature as a whole rather than any specific, well-defined body of beliefs. The very formulations of the issues, where clear, tended to vary from one thinker to another, and the appropriate tools for their resolution was itself a matter of disagreement.

But the characteristics I have attributed to "early" phases of the knowledge-seeking enterprise are by no means absent even from the most modern period. For the ideal of complete autonomy of scientific investigation has not, even yet, been realized. Since the degree of such autonomy is a function of the available background information – the body of successful, doubt-free beliefs which have been found to be relevant to the domain under consideration – it is evident that that information will, in at least some cases, to some extent, be insufficient to permit the definitive formulation of domains, problems, methods, and standards of possible and acceptable solutions. And where our knowledge is incomplete, we must look elsewhere for guidance to supplement that knowledge: to less well-founded beliefs, or to beliefs which, however well-founded, have not been shown to be clearly relevant to the domain under consideration.

A detailed example is studied in Chapter III of *The Concept of Observation in Science and Philosophy*. In that case, even though background information in the 1920's had imposed many clear restraints on the formulation and range of possible solutions of the problem of the source or sources of stellar energy, that background information could not provide adequate guidance for the selection of any one of those possibilities. Again, theories of the strong and weak interactions were constructed in the 1930's and following decades on "analogy" with the highly successful quantum electrodynamics, even though the field-theoretic exchange-particle approach of the latter could not be expected necessarily to be applicable to those other domains. On other occasions, appeal may be to ideas still less well-established or still less clearly relevant, even to ideas which are "external" to what has been shown to be "scientific."

It is important to emphasize that the analysis of "reason" here is independent of the specific character of those reasons. What science has *found* is that the development of successful beliefs about nature comes through contact with nature: through observation of it. "Observation" here, as is argued in "The Concept of Observation in Science and Philosophy," does not mean the passive collection of uninterpreted sensory "givens"; in that respect, the doctrine presented there is a "rational descendant" of traditional empiricism. What is to count as observational, how to make observations, and how observations are to function in the acceptance, rejection, and modification of beliefs, are things we have learned through contact with nature; and we have learned that those things must depend on what we have previously learned. (The classical concept of "objectivity" depended on the existence of a pure, "uninterpreted" given. This new account of observation, and more generally the account of scientific reasoning given here, provides the basis for a new view of what it is to be objective: background beliefs may be employed to "interpret" our experience and our problems – and indeed are necessary in order to do so; but objectivity is achieved only if the background beliefs so utilized satisfy strict conditions.) Correspondingly, the inadequacy of *a priori* and absolutist approaches to the study of nature is also something that has been learned through repeated rejections of claimants to inviolability, as well as through the repeated introduction, especially in the science of the past one hundred years or so, of ideas which have violated our fundamental predispositions. It has, in short, become a normative guiding principle of science that we should be open to the possibility of altering or rejecting *any* of the ideas with which we approach nature, that we should suspect *any* claims that a particular proposition, theory, method, standard, or goal is an inviolable feature of our view of nature. Even logic and mathematics must be subject to this openness, and must be interpreted in such a way that they are, in principle, subject to the possibility of change. These remarks do not imply, however, that we may not ultimately *find* ways of learning about nature by means other than observation of it; nor do they imply that we may not *find* that there is necessary and inviolable truth after all, and that we can attain it. We might discover, for example, that by thinking under certain conditions (for instance, by sitting in a certain stove-heated room), without any known physical interaction with the objects of our thought, we might come to certain conclusions which are invariably borne out. (We might even find that, given the known laws of nature, there *could* not be any interaction, under the given circumstances, between our thoughts and the objects in question.) It might then become a normative principle to accept all further conclusions arrived at under those conditions. Or, again, we might ultimately find that the view of nature at which we arrive, and the logic and mathematics associated with that view, are so tightly bound together, and so constraining on alternatives, that we see the impossibility of things having been otherwise; and we might even come to see that that ultimate view could have been anticipated in some way by careful enough thought. But if such possibilities do come about, they will have to be arrived at through observation. As matters stand now, we must rely, in our search for knowledge about nature, on observational contact with nature; and that contact and the conclusions we draw therefrom are, so far as we have learned, open to the possibility of doubt and rejection.

Insofar as the present view is "empiricist," then (or rather, a rational descendant of empiricism), it is so on empirical grounds, and is so on the basis of considerations subsidiary to the analysis of what it is to be a "reason." In other words, that all our knowledge of nature comes through observation of nature is not an ultimate doctrine, but a derivative one. Furthermore, that doctrine is itself not necessary and inviolable, but expresses the approach to inquiry about nature which we have learned satisfies successfully what it is to be a "reason" for or against a belief about nature. (The empiricism which is being advocated may thus be described as "contingent empiricism.") But it must be emphasized that the concept of what it is to be a "reason," too, is a product of experience: we have learned that things in nature have relevance-relations to one another, and that, along with such ideas as "success" and "understanding" which have codeveloped with our view of nature and of "reason," it is through employment and systematization of discovered relevance-relations ("internalization") that understanding and successful dealing with nature are achieved.

In summary, then, science is open to change in all the respects I have mentioned – in subject-matter, problems, methods, standards, goals, *etc*.; and such changes come about, in the most characteristically scientific cases, *for reasons*. More explicitly, the considerations which lead to such changes are specified in the light of background beliefs which have been found to be successful, free from specific (and compelling) doubt, and relevant to the subject-matter at hand, that subject-matter itself having been developed in terms of such background beliefs. Scientific change is thus not restricted to the collection of new observations or development of new substantive views of the world: we not only learn about nature, we also learn how to learn about it, and how to think and talk about it, in the light of what we have already learned. And given our circumstances as we have found them (see Section III, below), what provide the immediate, though not necessarily sufficient, reasons for change are observations of and relevant to that subject-matter, where what counts as an observation is also determined by background beliefs satisfying the conditions I have stated. (These points about observation are detailed in "The Concept of Observation in Science and Philosophy.") In Section III, we shall find still deeper arguments leading us to refer to the above-described developmental procedures as "rational"; for the ideas of "success," "freedom from doubt," and "relevance" – and therefore of "reason" itself – will be seen to be intimately intertwined with our most pervasive notions of "knowledge" and "truth."

## III

The position I have outlined thus far agrees with that of such writers as Feyerabend and Kuhn in that, in all its aspects, the knowledge-seeking and knowledge-acquiring enterprise involves the employment of antecedent beliefs. It departs from their views, however, on two major counts.

(1) That all aspects of science – its subject-matter, problems, *etc*. – are interrelated was a fundamental insight of Kuhn's. But he tried to ground that

interrelatedness in some single all-determining "paradigm," a view which, as we have seen, distorts the development of actual science and leads to relativism. Rather, the interrelatedness lies in the fact that all those aspects, including what counts as a reason, interact in their functioning and are open to revision in the light of what we learn about nature. Background beliefs are employed in specific ways in specific problem-situations, and a particular belief may function in different ways in different situations. And in any given problem-situation, many such beliefs function in overlapping and varying ways, even though some play, especially in the light of past successes but also on other grounds, more important roles than others. But although some beliefs function very widely – that is, in many situations – and deeply – that is, in shaping the character of thought and activity in particular situations – as compared with others, there is no one belief or set of beliefs ("high-level background theory" or "paradigm") which functions in one single way to "determine" ("shape," or whatever) every scientific activity in every situation in a "tradition." The "presuppositionism" involved in the present view is therefore "local" rather than, as in the case of Feyerabend and Kuhn, "global." (However, the term "local" must not be understood as precluding the use of some background beliefs more widely or in more fundamental ways than others.) Also, there is, *as a matter of historical fact*, no point in the history of science after which all of the background beliefs which had previously been employed are rejected; there are no "revolutions" in the Kuhnian sense of replacement of an old "paradigm" by a new one which is incommensurable with the old. (Although some aspects of this first point of divergence from the views of Feyerabend and Kuhn are presented in the essays included here, a fuller and more systematic detailing will be found in Chapter IV of the forthcoming book, *The Concept of Observation in Science and Philosophy*.)

(2) In contrast to their role in Kuhn, and presumably also Feyerabend, even the highest-level (in the sense, now, of most widely-adopted and pervasively functioning) background beliefs are, in *mature* science, subject to ever-stricter constraining conditions: as we have seen in Section II, above, to the extent to which science has proceeded in "internalizing" the considerations on which it is able to rely, only those beliefs may be employed which have been found to be *good reasons*; that is, which satisfy conditions which may be schematically referred to as "success" and "freedom from specific and compelling doubt," and which have been found to be, in specific and known ways, relevant to the precisely-formulated subject-matter at hand. (It must be remembered also that additional conditions have been added as

having proved successful, most notably, "precision," though that notion requires careful analysis due to the fact that science often remains satisfied with "order-of-magnitude" estimates. This point is related to the role of "conceptual devices" in science.) Of course, as I have already remarked, it is not always possible for science to deal with (or to raise) its problems solely on the basis of such stringent constraints; sometimes it must rely on less well-established beliefs (appropriately called "hypotheses") or even on "speculation" or "analogies." But with the increasing successes achieved through seeking and respecting such constraints, continued respect of them wherever possible has become a normative guiding principle of science; and it has become a goal of science to try to achieve a state where its reasoning can be fully autonomous and integrated.

On this second point, that science of all human activities is the paradigmatic case of an enterprise proceeding according to reason, my view is in essential agreement with the most widespread opinion of scientists and non-scientists. But although Logical Empiricism agreed that science proceeds by reasoning, it characteristically tended to limit that reasoning to the justification or rejection of a hypothesis *after* it has been proposed; there is, according to most adherents of that movement, no "logic of discovery." In the view presented here, however in cases which are of a *characteristically scientific sort* (*i.e.,* ones in which the "internalized" considerations are adequate to the task), science is able to, and often does, introduce new hypotheses (and problems, methods, goals, *etc.*) for good reasons; reasoning does not occur solely in the case of justification or rejection after a hypothesis has been advanced.

The bases of the positivist rejection of a logic of discovery – or, more accurately, a rationale of development) are instructive. Traditional empiricists, in company with their rationalist opponents, tended to maintain the existence of a rationale of discovery, but attempted to construct that rationale on the basis of a supposed pure, uninterpreted "given" in experience. Positivist doctrines in the 1930's through 1950's rested heavily on criticisms of those attempts to ground a logic of discovery on a pure "given"; but the conclusion drawn from those criticisms was not that science must be understood otherwise than as resting on such a foundation, but rather that there is no logic of discovery. A review of their criticisms of those traditional empiricist methods will be given in Chapter IV of *The Concept of Observation in Science and Philosophy*. It will be found that, when reinterpreted in the light of the views outlined here and developed further in that work, such widely-rejected traditional ideas as Mill's methods of experimental inquiry, operationalism, and even the positivists' own early but abandoned verification principle, have achieved an important place in the knowledge-seeking enterprise. That is, their sources, value, and limitations, as well as the roots and misguidedness of the criticisms made of them, can all be clarified.

Science thus develops through a give-and-take interaction between the methods with which it approaches nature and what it learns about nature, or at least claims to know on the basis of the best reasons it has available. Included in that interactive development are, as we have seen, the subject-matter, the problem-structure, the standards, and the goals of science: in all these aspects, science is subject to change. And it has learned to make those changes, wherever possible, in the light of reasons – reasons which, for us, consist of observations of nature.

As is seen most clearly in "Alteration of Goals and Language in the Development of Science," "The Concept of Observation in Science and Philosophy," and "Reason, Reference, and the Quest for Knowledge," even the language in terms of which we describe and otherwise talk about nature is often altered by what we learn about it in the course of scientific investigation. In the process of learning about nature, science continually revises the language in terms of which it talks about nature, to make that language conform more closely to what we have the best reason to believe nature is like. The traditional doctrine of conceptualism was thus partly right: we do extract our concepts of nature through observation of nature. But we do not simply "abstract" those concepts by perception and reflection thereon. Rather, learning through observation, we gradually forge concepts which reflect what we have found to be the case, or at least believe to be the case on the basis of the best reasons we have available; and then we seek to learn further about nature, and perhaps will be led to further revision of the concepts we have forged, and so on. Thus the present view may be termed "Bootstrap Conceptualism." (However, as will become clear later, there is no guarantee that our concepts will converge to a set which will no longer, as a matter of fact, be modified further.)

Because of linguistic conservatism and other factors, changes of vocabulary do not always come about even when they would otherwise be advisable in the light of what we have learned. (A discussion of conditions under which they generally do and do not come about will be given in Chapter IV of *The Concept of Observation in Science and Philosophy*.) But when they do take place – as they did, on a very broad scale, in the reform of chemical nomenclature in the late eighteenth century, – their impact on the knowledge-seeking enterprise can be profound.

Even when the same term continues to be employed, changes in knowledge will often be reflected in changes in the use of that term. Thus "Reason, Reference, and the Quest for Knowledge" contains a brief discussion of changes in the use of the term 'electron' from Stoney to Feynman – changes in what was said at successive stages about electrons, including changes in criteria for identifying what is to count as an electron. Continued use of the term 'electron,' despite those radical alterations, was justified by a "chain-of-reasoning connection" between the successive uses: for there were *reasons*, in the sense outlined in this Introduction, for the changes being made. There was a continuity of development; and it was that continuity, or, more importantly, the chain of reasons which produced that continuity, that alone justifies our speaking of "the concept" or "the meaning" of the term, and our speaking of the term having "the same reference."

The idea of chain-of-reasoning connections has a number of important philosophical implications. In the first place, it provides the basis for a better understanding of "meaning" and "reference" than philosophers have been able to provide. The "concept" of "electron" – the "meaning" of the term 'electron' – is *not* something held constant throughout the history of the use of that term (or even just from Stoney to Feynman); there *need* not be any constant set of necessary and sufficient conditions for something's being an electron. (Of course there *may* be common elements, but the important point is that it is not necessary that this be the case, and that if there are, those common elements have no special status.) What counts as an electron is determined by the best beliefs available at a given time; rather than there having (necessarily) been a constant "meaning" throughout the "tradition" from Stoney to Feynman, it is more accurate to say that the "concept" of "electron" is just the family of uses of that term which are related in an ancestry-and-descent (or cousinhood) relation by a chain-of-reasoning connection. I take this to be an application to scientific change of Wittgenstein's notion of a "family resemblance," except that here the basis of that "resemblance" is not left as a vague metaphor: the relationship is governed by reasons, and is clear to the extent that it is clear, at a given stage of scientific development, what is to count as a reason in some specific case. Similarly, what we are *talking about* (referring to) throughout a "tradition" in science need not be one specific something; what we are talking about at any given stage is determined by the best information on the subject (*e.g.*, of electrons) at that stage, and is related, if it is at all, to what people previously talked about in talking about that subject by a chain-of-reasoning connection. (The present view is one of "Bootstrap Semantics" as well as "Bootstrap

Conceptualism.") Nothing else is available to us to determine what we "mean" or are "talking about," and nothing else is needed to explain our activities; in particular, we need not suppose a something-we-know-not-what (an "essence") whose nature we seek. ("Reason, Reference, and the Quest for Knowledge." The remarks in the present paragraph would be supplemented, in a fuller account, by an extension to the "meanings" and "references" of terms in theories – *e.g.*, of electrons – that are not accepted fully or at all.)

It is important to note that this viewpoint makes the notion of a "reason" more fundamental, for the understanding of science, than the notions of "meaning" and "reference." What in such papers as "The Structure of Scientific Revolutions" are objections to *mis*uses of the concept of "meaning" in philosophy of science, broaden to attacks on the concept of "meaning" itself in "Meaning and Scientific Change." And those objections become coupled, in "Reason, Reference, and the Quest for Knowledge," with parallel criticisms of the role assigned to the concept of "reference" as a fundamental tool for interpreting science, its language, and its reasoning. The situation is rather the reverse: it is the concept of "reasoning" that is fundamental to the interpretation of science, and the interpretation of scientific language is derivative from that. The present view must be seen as opposed to the tradition, so dominant in twentieth century Anglo-American philosophy, which holds that the philosophy of language is central to philosophical thought. As philosophy of science is not a branch of logic, as it became under logical empiricism, so also it is not subservient to the philosophy of language, to be approached only with the tools ("meaning" and "reference") provided by the latter. On the contrary: far from being useful tools for the philosopher of science, those notions have proved a bane, obscuring the dynamics of the knowledge-seeking and knowledge-acquiring enterprise as it actually takes place. Indeed, as I have tried to suggest in this brief discussion, an understanding of our views of nature, and of the ways we have developed those views, can illuminate our understanding of the language we use in talking and thinking about nature.

But it must be understood that the "concept" of "reason" is, like any other concept, specifiable no further than as a family of reason-related criteria. (No vicious circle is present here, if one sees that the development of criteria of rationality is itself a bootstrap process.) The same is true, of course, of such "concepts" as "success," "reason for doubt," and "relevance," and of the supposedly "metascientific" concepts of the logical empiricists. Such "concepts" can be exhibited by examples (from their families, and through the connections between "members" of those families) and discussed in a general schematic way; but "analysis" in the sense of provision of a set of necessary and sufficient conditions, covering all possible cases of application of the term, is not in general possible. ("Not in general," because on the present doctrine it might under certain circumstances be possible to specify or even to discover such conditions.) That view of "analysis" is a philosopher's bad dream, produced by an indigestible view of "meaning." The present view does not replace such analyses with a different such "analysis" in the same spirit. Concepts develop just as do all other aspects of our views of nature; and they are prone to the same sorts of deficiencies and openness that are characteristic of other aspects of our views of nature. These points are recognized and explained in the present account. (The remarks made in this paragraph, together with the concept of a "concept" as a family of reason-related criteria, makes it possible to see how, and more importantly why, there is a point to the analytic-synthetic distinction, as well as how and why it is mistaken.)

During the 1950's, views asserting the philosophical centrality of the analysis of language, particularly of ordinary language, had come to dominate the philosophical scene outside the more technical areas of philosophy of science. "Philosophy and the Analysis of Language" was a critical examination of certain fundamental aspects of the relations between language and the world as those relations were seen by Russell, Wittgenstein (both early and late), and Ryle; it attempted to expose the underlying assumptions of those linguistic philosophies. The question of the relation between language and the world arises again in "Space, Time, and Language," this time in connection with the possible ways in which everyday language might be related to the language and theories of science – a problem which, as is evident from what has already been said above, has remained central to my thinking ever since. Although I rejected the contention that "ordinary language" cannot be violated, the writings of the later Wittgenstein have influenced the views I have sketched on at least three points, though on each of these my thinking diverges significantly from his. I have already spoken of one of those points, the application of his idea of "family resemblances" to the interpretation of science and scientific change – even though this application, unlike Wittgenstein's, rests on a prior notion of what it is to be a "reason," and thus constitutes a clarification and not a mere application of his idea. The second way in which Wittgensteinian ideas are reflected (to at least some extent) here is the following. According to a widely cited slogan, the philosopher of science must pay attention to what scientists "do" rather than to what they say. I believe, however, that we must attend to both, though of course with a great deal of critical awareness. It is true that when scientists discuss general issues about methodology, their remarks are often naive; nevertheless, their remarks must not be dismissed out of hand. And further, when scientists use terms like "explanation," "observation," "theory," "evidence," and "confirmation" in working contexts, those uses must be taken very seriously indeed. The importance of this point is developed most fully in "The Concept of Observation in Science and Philosophy" (and in Chapter II of the book of that same title). There it is shown that, despite the initial oddity of the use of the term 'observation' in a certain sophisticated scientific context, a coherent interpretation of that usage can be given according to which it constitutes a *reasoned* extension of and departure from certain ordinary uses, and, furthermore, that an understanding of that reasoned extension and departure exposes fundamental mistakes in usual philosophical conceptions of the problem of observation and its role in science. (Here I depart from the general trend of "ordinary language" philosophy, according to which ordinary language cannot be violated.) In general, of course, whether such a coherent interpretation of usage can be found can only be determined by detailed investigation; the mere fact that a certain individual or group *uses* an expression in a certain way is no guarantee that that usage is coherent, much less that it plays a role in our interactions with the world around us. (Wittgenstein, I assume, would have held that if a group uses a term, that use must be coherent.) I will mention the third influence of (and departure from) Wittgensteinian thought later, in discussing the functions of philosophy.

The idea of "chain-of-reasoning connections" disposes of the problems of "incommensurability" which have produced so much difficulty and confusion in philosophy of science, and which have been the source of so many relativistic and skeptical views of science during the last two decades. Complete incommensurability is of course a myth: no comparison whatever would be possible between two absolutely incommensurable (*i.e.*, incomparable) ideas. We could not even say they disagreed; we could not even say that the two ideas were (for example) "theories," or that they disagreed about what an "explanation" is, since we would presumably have no trans-theoretical criterion for identifying what each took to be an "explanation." We could not, indeed, even say that the two were "ideas." As I have argued in numerous papers, there has always been at least some (and in most cases a great deal of) common thought spanning even the most radical "revolutions." It is true that if we isolate a particular idea (concept, proposition, or set of propositions) at two widely different periods, we might find nothing in common. For, according to the views I have been presenting, it is in principle possible that *every* aspect of an idea (*e.g.*, *every* property attributed to electrons) might be rejected and replaced, for good reasons. But as long as there are such reasons, an understandable relationship holds between the two uses – a "chain-of-reasoning connection." If there were no such connection, we would not even be able to ask intelligibly that someone *consider a particular idea in two occurrences, at two different periods, between which two uses there is nothing in common*. Our ability to talk about "the same concept" in two such occurrences rests, if the development of science is to be intelligible, on the existence of such chain-of-reasoning connections. If there are any *such* incommensurabilities in science, they do not preclude talk of "the same concept" (or "the same reference"), though the price is an abandonment of prevailing views of "meaning" ("concept") and "reference." But in view of the failures of those views, it is a small price to pay, especially when we have an alternative view which makes sense of the development of science, recognizing both change and continuity therein. Later ideas in science are often *rational descendants* of earlier ones, even if they abandon a great deal, or even all, of what was in those earlier ideas.

The depth and extent of scientific change imply that we cannot use historical cases uncritically as bases for generating an interpretation of science. The mere fact that a certain scientist or group of scientists in the past thought and worked in certain ways cannot be used as a basis for generalizing to an interpretation of science in general; for those ways may well have been altered drastically later, or even abandoned or relegated to being "external" to science. Nor can past cases be used uncritically to refute an interpretation of science, if that interpretation is supposed to be limited to the science of today. (These and related issues are discussed in detail in "What Can the Theory of Knowledge Learn from the History of Knowledge?") Note that the idea of "internalization of considerations" is explicitly directed, in the very term I have chosen for that process, at a resolution of the debate among historians of science as to whether scientific change is governed primarily by "internal" or "external" considerations. On the view presented here, science attempts to *become* autonomous in its reasoning, to rest its arguments solely on "internal" considerations; and this *goal* has been adopted because science has been *able* to achieve such autonomy to a considerable extent, and has found it possible to achieve great *success* by making its reasoning rest autonomously on considerations having the characteristics I have described.

## IV

What, then, about science and the "search for truth"? In the history of philosophy, three general types of theories of truth have been proposed: *correspondence theories*, according to which a belief or proposition is true if and only if it "corresponds to reality"; *coherence theories*, according to which a belief or proposition is true if and only if it "hangs together" with other beliefs or propositions (presumably with the largest number of them); and *pragmatic theories*, according to which a belief or proposition is true if and only if it "works." Correspondence theories focussed on the intuitive idea that whatever makes a belief or proposition true or false must be *distinct from* the belief or proposition itself, and in particular from the reasons for or against the belief or proposition. (We might have excellent reasons for a belief, and yet the belief still be false.) But in thus breaking the connection between reason and truth, correspondence theories tended to leave it impossible ever to know, or even have reason to believe, that we had attained truth, no matter how excellent our reasons in favor of a belief. They thus became easy prey to various forms of skepticism. Coherence and pragmatic theories, realizing that our concept of truth, like all our concepts, must be structured in terms of the circumstances in which we find ourselves, and in particular in terms of the reasons we have for believing, emphasized that there must be a relation between our concepts of "reason" and "truth." But the relation they postulated was one of identity, of definition. 'Truth' was *defined* as that which "hangs together" or "works." Those theories thus neglected the intuition on which correspondence theories had concentrated, the need for the truth-making factors to be distinct from, independent of, reasons for belief; and thus they verged on the brink of relativism. (Suppose we had two conflicting theories which worked equally well, or which cohered equally within themselves: would they then not be, by definition, equally "true"?) It is thus clear that both the appeal and the weaknesses of the two sets of theories of truth – correspondence theories on the one hand, and coherence and pragmatic theories on the other – are complementary. They focus on complementary intuitions about the relation between 'truth' and 'reason,' and they have corresponding and complementary weaknesses. But the three major requirements that science has developed for acceptability of a belief – those to which I have referred schematically as "success," "freedom from doubt," and "relevance" – together capture the insights of the three traditional types of theories of truth while collectively avoiding the weaknesses to which the latter were subject. "Success" and "relevance" capture the heart of the appeal of, respectively, the pragmatic and coherence theories; they encapsulate the idea that there must be *positive reasons* for acceptability of a belief or proposition as true. But "success" and "relevance" do not *define* 'truth'; for no matter how successful an idea might be, no matter how definitively its relevance to other successful ideas has been established, *it is always possible in principle that specific reasons for doubt may arise*. And it is that possibility that *breaks* the connection between 'reason' and 'truth,' and thus captures the essential intuition of the correspondence theories. But at the same time that it breaks the connection, it does not destroy it entirely; for we must nevertheless accept as true what we have best reason to believe and no (specific) reason to doubt. Hence both the connection and the disconnection between 'reason' and 'truth' are maintained. The insights of the three traditional theories of truth are preserved, while their individual shortcomings are escaped. (And so the present view cannot be identified exclusively with *any one* of the traditional views, for example with pragmatism.) Indeed, the following relationships now become evident:

| *theory of truth* | *condition of acceptability (theory of reason)* | *epistemic function* |
|---|---|---|
| correspondence | freedom from doubt | description |
| pragmatic | success | prediction |
| coherence | relevance | explanation (order) |

Chapter V of *The Concept of Observation in Science and Philosophy* will detail how these nine concepts and associated ideas are illuminated through an understanding of the relationships I have indicated here. Those relationships themselves will both illuminate and be illuminated by the larger view of the knowledge-seeking and knowledge-acquiring enterprise that has been sketched in this Introduction. In particular, in keeping with the spirit of the approach presented here and in that book, it will also be shown *why* we have developed this particular network of concepts, linked in these particular ways.

This sketch of the relations between "reason" and "truth" does not involve any assumptions as to the nature of the "truth" being sought. The universe with which science is concerned, supposing it to be concerned with "a universe" at all, must not be supposed *as a matter of a priori necessity* to have any specific character. It must not, for example, be assumed, *independently of what we learn in our studies of that universe*, that what we are studying are to be understood as "substances," "essences," "individuals," "objects," or anything else that might in any way prejudge the character of what science might learn. (*Cf.*, for example, "Reason, Reference, and the Quest for Knowledge.") We must not even suppose, as a matter of *a priori* necessity, that there is a single entity which is "the universe." ("Unity and Method in Contemporary Science.") The views at which modern science has arrived are too bizarre, in contrast to *anything* we might have anticipated, to make it advisable or even possible to build such assumptions into an account of what science *must* be and seek. ("Unity and Method in Contemporary Science"; "Modern Science and the Philosophical Tradition.") The things science studies (if indeed that language itself is appropriate) must be determined by scientific investigation, not by *a priori* reasoning.

Furthermore, my sketch of the relations between "reason" and "truth" does not assume in any way that we will *discover* truth about nature. On the basis of the most successful and doubt-free beliefs available to us, we may ultimately come to see that we cannot understand nature in any fundamental way, or at least that there are severe limits to the possibility of our knowing nature as something independent of ourselves, of our thought. Indeed, some of the most widely-accepted interpretations of quantum mechanics do suggest that there are just such limits. Or perhaps we may find (specific) good reasons to believe that there is no "truth" to be found, in the Parmenidean sense of there being some permanence independent of our thought. If any such view were to be acceptable, however, it would have to be arrived at as a product of inquiry, as quantum theory, for example, has been, and not as a product of *a priori* reasoning. The *knowledge* at which science arrives, if it arrives at knowledge at all, may not be a knowledge of "truth" in the sense of knowledge of the way an independently-existing universe exists; it may preclude knowledge of *such* truth. The terms 'knowledge' and 'truth' must not be so defined as to exclude the possibility of our coming to "know" that we cannot learn the "truth" about how things are. In this respect, science is the search for knowledge, which may not be "knowledge of the way things are."

But though the views I have presented thus leave open the possibility that we may learn that we cannot learn truth about nature, they also leave open the possibility that we can arrive at knowledge and even truth, in the sense of knowledge about "the way things are." There are many ways in which we might do so, the most important of which, from the standpoint of traditional discussions in the philosophy of science, are the following. We may arrive at a theory which successfully accounts for all aspects of a given domain, and concerning which theory, as a matter of fact, no doubts ever arise which are not resolved. (The domain in question would, of course, itself have been developed in the light of what we have learned. It might indeed be very comprehensive, and the theory of it might provide a unified account of *all* the objects of our inquiry as they are conceived then. The theory in question would, of course, in conjunction with background information, specify what is to count as an "explanation." The distinction between the "theory of the domain" and "background information" relevant thereto might well have disappeared by then through the process of scientific internalization of considerations.) Furthermore, it might eventuate that, when and if such a stage were reached, no alternative, equally successful and doubt-free theory would (ever) as a matter of fact be available. In such a case, since we would have *no reason* to doubt that theory, then, if its claims were of a certain sort, we would have *no reason* to doubt that it was *true of*, a true account of, "the way things are." The exact characteristics of those "claims of a certain sort" would depend on the exact nature of the conditions established by the domain, the theory, and any relevant background information, as those have developed at that stage in the history of science. Nevertheless, as will be shown in *The Concept of Observation in Science and Philosophy*, it can be reasonably anticipated that such characteristics will be rational descendants of certain characteristics traditionally required of a theory if it is to count as "realistic." Full discussion of this point will involve incorporation of the discussion of existence-claims, in Part II, Section 9 of "Notes Toward a Post-Positivistic Interpretation of Science," into the larger framework of the position outlined in this Introduction.

These points can be clarified by considering the role in modern science of "conceptual devices" (including what are ordinarily referred to, rather vaguely, as "idealizations," "simplifications," or "abstractions," and, in some uses, as "fictions," "models," or "approximations"). Nowhere is the "internalization of considerations" more clearly exhibited than here. For science has come to insist that a concept, proposition, or theory be considered to be a conceptual device (*e.g.*, an "idealization") if and only if there are *specific relevant scientific reasons for supposing that what that concept, proposition, or theory is about cannot really be the way it is treated by that concept, proposition, or theory*. Thus, whether an idea is an "idealization" (for example) is determined on the basis of internal scientific considerations, not by any science-transcending philosophical ones. At earlier stages in the history of science, when the process of internalization was still relatively primitive, such questionably relevant considerations (*e.g.*, "metaphysical" suppositions like "All bodies must be extended") provided the bases for considering a treatment of a subject-matter as "idealized" (*e.g.*, considering the treatment of a certain object as a point-mass as being an "idealization"). But physical science, at least, has by now reached the point where it is able to insist that, if we are to consider a treatment as "idealized," there be *unambiguously scientific reasons for doubt* that things are the way they are being treated. (This shows that there can be reasons for doubt concerning an idea, and yet that idea still be useful, if certain *specific* conditions are satisfied.) It has arrived at this point because it has *learned* that that demand *can* be satisfied in such a way as to lead to success in the search for knowledge. Fuller discussion of the conditions at which science has arrived for considering an idea to be a conceptual device (including discussion of the utility of such devices) are found in Part IV of "Scientific Theories and Their Domains" and, especially, in Part II of "Notes Toward a Post-Positivistic Interpretation of Science." However, neither of those discussions is fully general (neither speaks in terms of "conceptual devices" or makes certain appropriate distinctions between types or functions of conceptual devices), and neither incorporates the idea of conceptual devices into the larger framework of the ideas outlined here by relating this idea to that of the internalization of considerations.

The above analysis shows that the notion of a conceptual device is dependent on the contrast between ideas for which we have (specific, *etc.*) reason for doubt and those for which we do not. In the former case, we have *reason* for calling the ideas "idealizations" (or for speaking of them in some such terms); in the latter, we do not. By referring to both types of ideas indiscriminately as "idealizations" or whatever, instrumentalist interpretations of science (just like skeptical and relativistic ones) fail to note the crucially-important distinction between having specific reasons for doubt and the (mere) possibility that such reasons might arise. As I have said, the mere possibility that reasons for doubt might arise is not itself a reason for doubt. In particular, it is not, by itself, a reason for doubting the truth of an idea, or, more generally, that the idea constitutes knowledge. Traditional controversies between "realism" and "instrumentalism" have ignored the fact that "realistic" and "instrumentalistic" concepts, propositions, and theories, are thus *related*, and that indeed the latter can only be understood in terms of the former, or at least in terms of the possibility that we might have "realistic" ways of treating nature. They have also tended to ignore the fact that realistic ideas and conceptual devices can coexist in science in application to the same domain, and can interact in the ways they function; and that what is today considered, for the best of available reasons, to be realistic (or a conceptual device) may, at some later period, be considered, for what are then the best of available reasons, to be a conceptual device (or realistic). In fact, in those respects in which certain ideas in science at present clearly satisfy conditions of not being conceptual devices (as being existence-claims free from reasons for doubt), in those respects and to that extent the science of today is realistic with regard to those ideas. No other formulation of "realism" is defensible, and no other is needed.

Thus the present viewpoint opens the *possibility* of science being or becoming "realistic," in a sense which is a rational descendant of traditional realistic interpretations. It does not guarantee, on *a priori* grounds, either that it must be or that it cannot be, that it is or that it is not. It does not guarantee, for example, that our theories will converge to a unified account of nature. Nor does it guarantee that our ideas, or any particular one, must change; it only shows that they may. Indeed, the present account does not provide guarantees for anything. On the contrary, *its entire direction is to show that the possibilities of what we might find (and even of how we might go about finding it) are completely open in principle.* The present view does not rule out, on *a priori* principle, even the possibility that science might find means of ruling out certain alternatives, or ruling out the possibility of doubt ever arising with regard to others: the fallibilism of the present view, like its empiricism, is contingent, fallible. But if science does find such means, it will have to do so, given what we have learned so far about how to go about inquiry, through the fallible methods which it has learned. (Naturally, none of this implies that "science might eventually become football." For although reasons for changes in science *may* arise, nevertheless, on the level of specific possibilities, if we have no *specific* reason for anticipating a particular change, we have *no reason* for concern about it; the mere possibility of a particular change is by itself no reason for anticipating that change or even, in any sense – except, of course, and importantly, of reminding us that we may be wrong – that can affect our uses of the specific beliefs we have and the specific ways in which we engage in the knowledge-seeking enterprise – except, again, in making us open to the possibility that those ways may prove mistaken. This sort of worry is of the same kind as skeptical and relativistic worries, and must be given the same general response: there is *no reason* why we should need an ironclad way of ruling them out, even if such a one could be found; and they are, by themselves, *no reasons* to doubt any *specific* belief or to expect any *specific* possibility.)

In its refusal to offer guarantees, the present view is in opposition to the major trend of our philosophical tradition. For the dominant theme of that tradition has been to try to show what it is *necessary* for human beings to accept if they are to find knowledge, or to engage in inquiry, or to have experience. Or else it has tried to show what it is *impossible* for human beings to think, or how it is impossible for them to inquire. The only alternatives, it has seemed to that tradition, have been skepticism or relativism. In the light partly of the failure of all such arguments, and partly of a study of the radical innovations of scientific thought in its history (*cf.*, "Unity and Method in Contemporary Science"; "Modern Science and the Philosophical Tradition"), the present approach adopts precisely the opposite standpoint. *It attempts to remove barriers to the possibilities of human thought*, while at the same time showing that the removal of such barriers, the refusal to countenance the traditional search for *a priori* absolutes, need not force us into relativism or skepticism. As here conceived, the philosophy of science – and more generally, the theory of knowledge – attempts, first of all, to show that arguments about how or what it is necessary or impossible for human beings to experience, think, or know, are mistaken. (This is my third point of agreement with the later Wittgenstein: to suspect "the hardness of the logical must," and to view the removal of that "hardness" by intellectual therapy as a prime function of philosophy. But neither the diseases nor the appropriate therapies are, on my view, necessarily purely linguistic.) But second, and more positively, the philosophy of science, and more generally the theory of knowledge, also attempts to provide a coherent account of the knowledge-seeking and knowledge-acquiring enterprise which shows how knowledge without *a priori* absolutes, necessities or their opposites, is possible. (Here again I depart from a general trend of Wittgenstein's views: philosophy has a positive, constructive function as well as a negative, therapeutic one.) Perhaps this brief Introduction, supplemented in part by the essays following, will suggest at least the outline of such an account.

# PART I

# CRITICAL PAPERS

CHAPTER 1

# PHILOSOPHY AND THE ANALYSIS OF LANGUAGE

Both Wittgenstein, in the *Tractatus Logico-Philosophicus,* and Russell, advancing the philosophy of Logical Atomism, maintained that statements are, or purport to be, records of facts. Wittgenstein held that philosophers, by improperly interpreting language, create for themselves pseudoproblems, and that, to avoid confusion, we should throw statements into a form in which their true function, that of picturing facts, would be revealed more clearly and readily than it is in ordinary language.

Russell agreed that the statements of ordinary language should be translated into another form. But for him the reason for such translation was not just that ordinary language, while it functions perfectly well in ordinary life, misleads philosophers, but also, and more important, that ordinary language really gives an incorrect portrayal of facts. And only by translating the statements of ordinary language into a form which *does* reflect facts accurately can philosophical progress be made.* For Russell, such progress was not merely (as it was for Wittgenstein) of the negative sort that consists of the elimination of confusion, but of the positive sort that consists of the discovery of new information about facts.[1]

In this paper, I wish, first, to consider these two views, showing some of the reasons why they are open to severe criticisms, not all of which have yet been made fully clear; and second, to show how, by dropping or modifying some of the fundamental theses of these two views, certain positions highly influential in philosophy today have arisen.

I will begin my discussion with a study of one of the most famous and influential articles of what might be called the "Transition Period" of Twentieth Century philosophical analysis — the period, that is, between the *Tractatus* and Logical Atomism on the one hand, and the later views of Wittgenstein and the present views of Austin and others on the other. This article, which even today holds the place of honor in one of our most-used anthologies, is Ryle's "Systematically Misleading Expressions."[2] My reasons for centering attention on this article are as follows. First, it advances, simultaneously and inconsistently, both of the views outlined above; but in spite of this inconsistency, it advances each of its incompatible theses in a clear and powerful way, making plain the changes which it had to make in the original formulations of them. Second, because of its clearness, it wears its difficulties (and hence those of a whole tradition) on its sleeve, and thus points the way to later developments. Thus, through a close examination of this article, we will be able to survey the whole development of at least one side of Twentieth Century philosophy deriving from the above-mentioned views of Russell and Wittgenstein, and to evaluate some of the strengths and weaknesses of the earlier phases of that tradition.

## I

The argument of "Systematically Misleading Expressions" departs from the following thesis:

> There are many expressions which occur in non-philosophical discourse which, though they are perfectly clearly understood by those who use them and those who hear or read them, are nevertheless couched in grammatical or syntactical forms which are in a demonstrable way *improper* to the states of affairs which they record (or the alleged states of affairs which they profess to record). (pp. 13-14)

Although such grammatical forms do not obscure from the ordinary man in his everyday business the true meaning of the expressions, to anyone who tries to analyze them closely they present a vicious trap.

> ... those who, like philosophers, must generalize about the *sorts* of statements that have to be made of *sorts* of facts about *sorts* of topics, cannot help treating as clues to the logical structures for which they are looking the grammatical forms of the common types of expressions in which these structures are recorded. And these clues are often misleading. (p. 22)

Such expressions, Ryle finds, fall into fairly definite groups or classes, each class being misleading in a certain way. Hence they are not simply misleading; they are "systematically misleading," in that they can be classified according to the type of presupposition which, in our attempt to analyze them, they tempt us to make. Thus in speaking of such expressions he says:

> . . . all alike are misleading in a certain direction. They all suggest the existence of new sorts of objects, or, to put it in another way, they are all temptations to us to 'multiply entities.' In each of them . . . an expression is misconstrued as a denoting expression which in fact does not denote, but only looks grammatically like expressions which are used to denote. (p. 32)

Ryle lists several types of systematically misleading expressions: quasi-ontological, quasi-platonic, quasi-descriptive, and quasi-referential. Each type is misleading in its own way: what they all have in common is that they mislead philosophers to add entities beyond what the facts really warrant. We might put Ryle's point metaphorically by saying that philosophers are led to add entities "behind" (quasi-ontological), "above" (quasi-platonic), and "along side of" (quasi-descriptive and quasi-referential) entities which are elements of states of affairs. (Ryle incidentally lists two other types of systematically misleading expressions (pp. 32–33), but these do not seem to differ radically from the above, for they too suggest to the analyst a multiplication of entities.)

This trap must and can be avoided; for

> what is expressed in one expression can often be expressed in expressions of quite different grammatical forms, and . . . of two expressions, each meaning what the other means, which are of different grammatical forms, one is often more systematically misleading than the other.
>
> And this means that while a fact or state of affairs *can* be recorded in an indefinite number of statements of widely differing grammatical forms, it is stated better in some than in others. (p. 33)

From these considerations Ryle draws a conclusion pertinent to philosophy.

> Such expressions can be reformulated and for philosophy but *not* for non-philosophical discourse must be reformulated into expressions of which the syntactical form is proper to the facts recorded (or the alleged facts alleged to be recorded). (p. 14)

## II

Underlying Ryle's argument is a theory of the relationship between language and the world which statements are about. This theory is a development, often in the same terminology, of certain views which were for the most part due to Wittgenstein, and which were presented in the *Tractatus* and Russell's papers on Logical Atomism. Ryle, however, often gives detail where Wittgenstein gave only bare sketches; he also tries to avoid some of the objections which had been or could be raised against the views of Russell and the *Tractatus*.

For Ryle, as for Wittgenstein and Russell, every significant statement is, or rather purports to be, a record. What a statement records they called a "fact," a "state of affairs," or "(what is) the case," using these expressions interchangeably. (But *cf.*, below, Note 3.)

Developing this Wittgensteinian theory of the relation of language to reality in detail, Ryle states that in order to qualify as purporting to be a record, a sequence of words must fulfill two conditions: (1) it must have certain constituents, and (2) these constituents must be arranged in a certain order or structure. (p. 14) Presumably corresponding to these two conditions of a significant statement, there are two aspects or components of a state of affairs: (1) a "subject of attributes" or several such subjects, together with their attributes, and (2) the "logical form" or "logical structure" of the fact. Sometimes Ryle also speaks as though expressions in general — single words and phrases as well as whole statements — record facts (though he also speaks of "events" in this connection[3]); but this usage does not seem consonant with his main trend of thought.

Ryle fails to analyze for our benefit the "certain structure" which a significant statement must have in order to be significant; but the constituents of a true statement seem for him to record either the subjects or the attributes, depending on unspecified syntactical factors. To parallel this, we would expect to find him saying that the grammatical form of the statement bears (or ought to bear) some kind of reference to the logical structure of a fact. On this point, however, there is a great deal of obscurity in the article. Certainly Wittgenstein supposed that the logical syntax of language parallels the structure of facts[4], and held that the structure of facts would be most clearly reflected by some syntax such as that provided by the system of *Principia Mathematica* (*cf.*, *Tractatus*, 3.325). But much had happened to logic and philosophy in the decade since the publication of the *Tractatus Logico-Philosophicus*. For one thing, Gödel's Theorem had done much to shake the confidence of those who idolized logical systems. Furthermore, difficulties had already been revealed in the supposition that everything in ordinary language can be cast into an extensionalist syntax without gain or loss of meaning; and finally, and most important, the possibility of syntactical systems other than that of *Principia* was beginning to be understood. Why, then, should the grammatical structure provided by *Principia* have any special priority, in that the structure of facts should be represented better or more naturally by it than by any alternative system? Where, indeed, had Russell and Wittgenstein gotten the incredible assumption that the world must conform to the specifications of logic? (No wonder Russell said that "My philosophy is the philosophy of Leibniz"!) Only one step more needed to be taken to reach an even more fundamental question: Why should it not be that the choice of syntax is simply a matter of convention, rather than of the nature of things?

That Ryle was aware of these difficulties is shown in the "consequential" question he raises toward the end of the article: ". . . is this relation of propriety of grammatical to logical form *natural* or *conventional?*" (p. 34) On the one hand, he does "not see how, save in a small class of specially-chosen cases, a fact or state of affairs can be deemed like or even unlike in structure a sentence, gesture or diagram." (p. 34) For in a fact, he "can see no concatenation of bits such that a concatenation of parts of speech could be held to be of the same general architectural plan as it." (p. 34) Certainly here is a denial of the Wittgensteinian view that "In order to understand the essence of the proposition, consider hieroglyphic writing, which pictures the facts it describes. And from it came the alphabet without the essence of the representation being lost." (*Tractatus*, 4.016) And it makes us wonder about Ryle's own statement, in this very same article, that systematically misleading expressions

must be reformulated into expressions of which the syntactical

> form is proper to the facts recorded (or the alleged facts alleged to be recorded). (p. 14)

Yet on the other hand, Ryle finds that "it is not easy to accept what seems to be the alternative that it is just by convention that a given grammatical form is specially dedicated to facts of a given logical form." (p. 34) Thus he feels himself unable to reject entirely the idea that there is at least some kind of parallel between grammatical form and the form of facts, though he is unable to describe the exact locus of that parallel.

But not only is the relationship between statements and the facts they record obscure; Ryle's account of "facts" themselves is also unclear — again, no doubt, an effect of the critical analysis to which that notion had been subjected (it is to be noted particularly that Ryle speaks only of "facts," never — as did Russell and Wittgenstein — of "atomic facts"). For we are told that "a fact is not a collection — even an arranged collection — of bits in the way in which a sentence is an arranged collection of noises or a map an arranged collection of scratches. A fact is not a thing and so is not even an arranged thing." (p. 34) Contrast this with the simple (perhaps naïve) view expressed by Wittgenstein in the *Tractatus:* "The configuration of the objects forms the atomic fact" (2.0272); "The way in which objects hang together in the atomic fact is the structure of the atomic fact" (2.032). Ryle has told us that we are not to understand the term "fact" as Wittgenstein did; but he has not told us how we are to understand it.

Sometimes, too, he speaks as though there were only one kind of fact — the kind which we should shape statements to meet — and sometimes as though there were *sorts* of facts, so that even systematically misleading expressions indicate or seem to indicate "sorts" of facts. Indeed, according to some passages, such expressions really only record *real* facts and only seem to record unreal ones. (*Cf.* the following quotations: ". . . they are couched in a syntactical form improper to the facts recorded and proper to facts of quite another logical form than the facts recorded" (p. 14); "'Satan is not a reality' from its grammatical form looks as if it recorded the same sort of fact as 'Capone is not a philosopher'" (p. 19).)

## III

The difficulties of Ryle's account of the relationship between language and the world, and his failure to define the "form" and "content" of each, leave much to be desired; and these problems were not peculiar to his view, either, but were the common heritage of his tradition. But closer inspection of the article leads us beyond these questions, exposing still deeper ambiguities and more profound problems: problems which are, in fact, among the most profound in Twentieth Century philosophy; for, in essence, they are among the questions over which Wittgenstein, reevaluating the ideas of the *Tractatus,* brooded for a decade.

These questions which we must now ask in reference to Ryle's views are: Why should he have supposed it necessary, for the theories which he advances, to introduce the notion of facts? Was he correct in assuming that his view of translation — of analysis — requires some such notion?

And finally, if his notion of facts is superfluous, what remains of his theory?

It would indeed be a mistake to suppose that, if a criterion for distinguishing between two classes cannot be formulated explicitly and precisely, the distinction is useless or non-existent. But particularly when the distinction is a fundamental one, introduced partly for the technical purpose of refuting opponents, philosophers cannot afford to let it rest on an uncritical and intuitive basis; for it is the very distinction that must be defended in order to justify the criticisms which it makes possible. And Ryle must often have felt that, for his special kind of translation, which consists of going from a "more misleading" to a "less misleading" form of expression, he would have to formulate some criterion of misleadingness, or, conversely, of propriety of grammatical form: some clear procedure or technique by which we can decide, at least in many cases, whether or not a statement is misleading, and, further, by which we can go about remedying this misleadingness.

One possible view would be that extralinguistic facts serve as the grounds or standards according to which we, as philosophers, strive to model our ways of saying things. In order to serve as such standards, these facts would have to be known clearly, at least to a careful observer — whatever might be meant by "careful." This view naturally assumes the explication of the exact relationship between statements and the facts which they record; only then could we understand what is meant by "modelling" statements after facts.

A great many of Ryle's statements suggest that this is actually his view: "... there is, after all, a sense in which we can properly inquire and even say 'what it really means to say so and so'. For we can ask what is the real form of the fact recorded when this is concealed or disguised and not duly exhibited by the expression in question." (p. 36) But if in some moods he did hold this, at other moments he must have been deeply troubled by the difficulties – discussed at such length in this century – of defining explicitly the "given" in experience. For a large number of other statements suggest that Ryle believes, not that facts are clearly known to an observer and can therefore be used as the criteria of the form into which philosophers ought to throw expressions, but the entirely different view that expressions properly formulated can give us a clue to the form of facts otherwise not clearly known.

> ... as the way in which a fact *ought* to be recorded in expressions *would* be a clue to the form of that fact, we jump to the assumption that the way in which a fact *is* recorded *is* such a clue. And very often the clue is misleading and suggests that the fact is of a different form from what really is its form. (p. 19)
>
> ... those who, like philosophers, must generalize about the *sorts* of statements that have to be made of *sorts* of facts about *sorts* of topics, cannot help treating as clues to the logical structures for which they are looking the grammatical forms of the common types of expressions in which these structures are recorded. And these clues are often misleading. (p. 22)

Philosophy "must then involve the exercise of systematic restatement. ... Its restatements are transmutations of syntax ... controlled ... by desire to exhibit the forms of the facts *into which philosophy is the inquiry.*" (p. 36; italics mine.)

Thus the reader may from some remarks be led to think that Ryle holds that the common man is able to discern the true form of facts, and that only a few overly analytical philosophers are misled by the form in which those facts are expressed – so that with just the proper amount of attention to the facts, the true form of those facts can be made clearly discernible in the grammatical form, and the philosophers' errors avoided; and this is a view quite reminiscent of the *Tractatus*.[5] But now it appears that Ryle believes, with Russell in his Logical Atomism period, that the true form of facts is to be *discovered,* at least by philosophers, through an analysis of the proper form of expressions; philosophy is again, as it was not for Wittgenstein (*cf.*, *Tractatus,* 4.111), a science.

The philosopher errs, it appears now, not in trying to discover the form of the facts, but in considering the grammatical form of an expression as a clue to the discovery of that form; whereas what would really furnish such a clue (and what should therefore be the object of the philosopher's search) would be the true form of the expression. How, then, are we to find this true form? It is perhaps in line with this new suggestion, that a well-expressed statement would give a clue to the form of facts, that Ryle introduces, at the very end of the article, a new procedure for deciding about the propriety or impropriety of a statement.

> How are we to discover in particular cases whether an expression is systematically misleading or not? I suspect that the answer to this will be of this sort. We meet with and understand and even believe a certain expression such as 'Mr. Pickwick is a fictitious person' and 'the Equator encircles the globe'. And we know that if these expressions are saying what they seem to be saying, certain other propositions will follow. But it turns out that the naturally consequential propositions 'Mr. Pickwick was born in such and such a year' and 'the Equator is of such and such a thickness' are not merely false but, on analysis, in contradiction with something in that from which they seemed to be logical consequences. The only solution is to see that being a fictitious person is not to be a person of a certain sort, and that the sense in which the Equator girdles the earth is not that of being any sort of ring or ribbon enveloping the earth. And this is to see that the original propositions were not saying what they seemed on first analysis to be saying. Paralogisms and antinomies are the evidence that an expression is systematically misleading. (p. 35)

Whether or not a statement is misleading is to be discovered, then, not by an appeal to extralinguistic facts which the statement records more or less properly, but by a direct examination of the statement itself, and the statements which it implies. For example, a statement like, "The wall encircles the city," would, in ordinary contexts, in some sense "imply" the statement, "The wall is of a certain thickness." Now the statement, "The equator encircles the globe," might "seem to be saying" something very like what "The wall encircles the city" says — so that the former would imply "The equator is of a certain thickness" in the same way that the latter implies "The wall is of a certain thickness." But in the equator case, such an implication would contradict (in some sense correlative with the sense in which "imply" is used in these contexts) "something" in the original statement. The only way to escape this contradiction, Ryle tells us, is to see that the original statement was not saying what it seemed "on first analysis" to be saying: that statements of this sort "disguise instead of exhibiting the forms of the facts recorded." (p. 35)

Ryle's new procedure for discovering when an expression is misleading, or when an interpretation is incorrect, offers a more modest prize than we might at first have expected; for we find, now, not what the true form of the expression is, but only what it is not (and hence, not what the true form of the fact is, but only what it is not). But can this new criterion show us, "in a demonstrable way" (p. 14), even this?

According to Ryle, the contradiction between "The equator is of a certain thickness" and "The equator encircles the globe" can be removed only if we realize that the latter does not have the "form" the philosopher thinks it has. Yet why is this the only solution? Could an opponent of a Rylean analysis not say, with perfect consistency, that his own analysis of the proposition is the correct one — that the form he attributes to the proposition is the one which truly reveals the form of the facts? Of course he would not take such a line in regard to cases like those Ryle gives as examples. But he might well do so in regard to cases which he claims to be more difficult and important; and in such cases, we might well, on the present view of analysis, have a problem as to which interpretation is the one that really gives a clue to the form of the facts.

Certainly some such answer would be what a Plato or a Meinong or a Russell of *The Problems of Philosophy* would want to give in reply to a Rylean critique. If the philosopher does take such a line, he will be maintaining that he really understands the statement (the form he attributes to it really revealing the facts), while Ryle does not; and Ryle will be maintaining that he really understands the statement, while the philosopher does not. Ryle cannot then prove the philosopher's interpretation to be wrong unless he assumes his own to be right. But simply to assert the denial of the philosopher's contention as a premise, and so "prove" the latter's contention to be wrong, will not serve to prove anything. The philosopher may well be wrong; but Ryle's argument, at least, will not show him to be so.

Ryle is at liberty, of course, to begin by assuming that he understands correctly the expressions under consideration. Only then he must abandon a number of claims he makes in the paper. He must, as we have just seen, abandon his claim to be demonstrating something about the true form of an expression; and this entails abandonment of the claim to be demonstrating something about the form of facts. Further, he must drop the claim that "there is, after all, a sense in which we can properly *inquire* and even say 'what it really means to say so and so'" (p. 36; italics mine), and that the propositions supposedly deducible, under the philosopher's interpretation, from the expression in question prove "*on analysis,* in contradiction with something in that from which they seemed to be logical consequences" (p. 35; italics mine). For if we understand the expression to begin with — as we must in order to employ the paralogisms and antinomies method — then we need not "inquire" about the meaning, or go through a process of "analysis" to see that it implies or contradicts some other expression which we also understand. And finally, such an approach would make a sham of the paralogisms and antinomies method: for the involved process of finding a contradiction in the philosopher's interpretation of a statement is utterly useless if we know to begin with that that interpretation is incorrect.

Worst of all, still, is the fact that this approach loses the advantage of proving the philosopher in error; and the desire to keep this advantage lies at the heart of the search for a criterion. Some criterion of proper form of an expression seems needed in order to make Ryle's position really an argument. For we would not be able to tell, without first knowing the real intention of a statement, whether or not considering it to have a certain other intention would lead to contradiction; but the real intention is itself what is at issue. Without some criterion, the paralogisms and antinomies procedure amounts either to mere assertion, devoid of any kind of proof or even disproof, or else to an elaborately disguised case of circular reasoning: assuming the denial of a position in order to prove that that position is in error.

The remedies that suggest themselves, however, seem worse than the disease. To reintroduce "facts" as the criteria of misleadingness would again beg the question if the discovery of the facts (or their form) is the goal. And "meanings" or "intentions," even if there were such things, and we could have access to them, and be sure we had not made any error about them, would not settle the present problem; for they would still have to be "correlated" somehow with facts, and this correlation could itself always be questioned consistently by a determined philosopher.[6]

One might hope to salvage the view by a compromise along Logical Atomist lines, by maintaining that we *do* know the structures of *enough* facts to tell us what the structure of language ought to be; and that, once we throw our statements into this form, we will, by inspection of that form, be able to discover new forms of facts (or, perhaps, the forms of new facts).[7] But this line is open to at least two objections: (1) it still hinges on successful analysis of the obscure technical notion of "fact," and its companion notions of "constituents" and "logical form" and their linguistic correlates and the nature of that correlation; and (2) there is no guarantee that all or even any undiscovered facts will necessarily comply with the forms of already known facts.

## IV

What lies at the root of the difficulties we have encountered in Ryle's position? Are they peculiar to him — mere products of his own confusion of two distinct theories? Or is there something more fundamentally wrong with the theories themselves — something which encourages or even perhaps forces such confusion? And will the difficulties be removed if some overhauling of the whole approach is made — say, for example, the abandonment of the view of language as a record of "facts"?

In order to answer these questions, let us briefly summarize the doctrines we have examined. Beneath the views of the *Tractatus* and Logical Atomism and the traditions they engendered are three fundamental theses, the first two of which were held in common by Russell and Wittgenstein, but the third of which is different for each.

1. *Ordinary language, if it is not flatly self-contradictory, is at least vague, ambiguous, and misleading, and generally fails to permit clear and accurate expression of what we want to say.* This doctrine is fundamental to the *Tractatus,* despite Wittgenstein's remark (5.5563) that "All propositions of our colloquial language are actually, just as they are, logically completely in order": the whole trend of that work, and certainly its influence on others, was along the lines of saying that reconstruction of ordinary language is needed.[8] It is not too much, indeed, to attribute this doctrine to almost the entire philosophical tradition from Thales on; for the program of traditional philosophy has nearly always been conceived, at least tacitly, as the replacement of ordinary ways of looking at the world — and of talking about it — with a new and more precise one.[9]

2. *For a proposition to be expressed clearly and accurately is for the sentence in which it is expressed to "picture" or "record" facts, or, more precisely, for it to "purport to picture or record" facts* (Wittgenstein: to represent a possible combination of "objects"). *This picturing relation between facts and sentences has two aspects:*

i. *A "form" of the sentence which corresponds to or represents the "form" or "structure" of the fact.* The "form" of the sentence (which is the true form of the "proposition") is, of course, not necessarily — not even usually, in ordinary language — the grammatical form; it is rather a "logical" one.

ii. *A "matter" or "content" of the sentence which corresponds to or represents the "constituents" of the fact.* Again, the "content" of a fact or proposition was not necessarily ordinary nouns, pronouns, adjectives, or their factual correlates as ordinarily conceived, but, especially for Wittgenstein, something more fundamental.

This doctrine is a version of the old Correspondence Theory of Truth; and the distinction between "form" and "content" of a proposition did not, either, spring full-grown from the brow of Wittgenstein, but is fundamental to Aristotle's whole philosophy, and reappears clearly in the logical writings of Leibniz; it probably underlies the old distinction between "syncategorematic" and "categorematic" expressions, and certainly underlies the Twentieth Century distinction between "Syntax" and "Semantics." The weighty *a priori* considerations (with which I have not dealt explicitly in this paper) which led Wittgenstein to adopt it and its trailers seem, however, to have been brought into the open first by him.

With regard to the third thesis, Russell and Wittgenstein disagreed. (Whether Russell was fully aware of the disagreement is highly questionable.)

3a. *The function of philosophy is to remove misunderstandings which are the products of linguistic confusions.* (Wittgenstein)

3b. *The function of philosophy is not merely to remove confusions, but, more important, to discover the true form (or forms) of facts.* (Logical Atomism)[10]

These, then, are the fundamental doctrines of Twentieth Century analytic philosophy up to the later thought of Wittgenstein.[11] In this essay we have examined some aspects of these doctrines, and have seen how philosophers of that period struggled to make them precise: how they tried to specify in just what ways ordinary language is vague, ambiguous, and misleading; and to pin down the relations which a perfect language would have to the world — to literalize the picturing metaphor which they believed must indicate the character of that relationship; and to state clearly what the goal of philosophical inquiry is, and to lay down the methods of achieving that goal. We are now in a position to understand the basis of the difficulties which we have encountered.

If one begins — as Wittgenstein began, and as Ryle apparently began — with Thesis 3a, he will be driven to propose some criterion of misleadingness; merely to assume a certain interpretation of an expression will not suffice to refute a contrary interpretation. To these philosophers, with their picture theory of language, the most natural thing to say here seemed to be that a statement is misleading if it does not reflect the facts clearly and accurately: that is, the facts (Thesis 2) provide the starting-point of philosophy, the model according to which the ideal language (Thesis 2) is to be shaped. But the notion of "facts" proved so intractable that ultimately it could achieve nothing; and these philosophers were forced more and more — like Ryle — to fall back on other criteria (which, however, they still thought of as somehow correlated with "facts"): such were "propositions," "(real) meanings," "(real) intentions." And this policy, of course, begs the question; for, stripped bare, it tells us that we can find the real meaning by looking at the real meaning. By this procedure, as we have seen, no *proof* is offered that traditional philosophers were mistaken or misled; the assertions made on the basis of circular reasoning remain just that: mere assertions.

If, on the other hand, one begins with Thesis 3b, supposing to begin with that the logical form of the proposition (the real meaning of the sentence) is known, and uses this to determine what the (forms of the) facts are, he falls into similar difficulties. For how do we tell that *this* is the "proper" form of language, the real meaning of the sentence? Surely not by noting that it conforms to the facts! For that would be saying that we discover the facts by examining sentences which are shaped according to the facts. Yet this is exactly the position in which we found Ryle at one juncture.

In short, it appears that whichever aspect of Thesis 2 we take as primary — whether, in line with Thesis 3a, we take facts as our starting-point and the ideal language as our goal, or, in line with 3b, we take the ideal language as known and the discovery of the facts as the goal — we are driven into circularity: we seem forced to assume *both* that we know the facts, and so can discover the true meaning or real intention or true form of the proposition, *and* that we know the true meaning or real intention or true form of the proposition, and so can discover the facts.[12] It is this tendency, arising necessarily from the weaknesses of the two positions, and not a mere mental lapse, that accounts for the inconsistencies which we have found in Ryle's paper, and which, I believe, close attention will reveal in so many of the basic papers of the early and middle periods of the analytic movement.

## V

It is impossible not to ask, at this stage, whether the later developments of analytic philosophy have succeeded in overcoming these weaknesses. To deal with such a question is obviously far beyond the scope of any single essay; but a brief account of the course of development, with regard to the above theses, of later philosophical analysis, can serve as a background for further and more detailed investigations.

Some of the most important developments have emerged from a repudiation of Thesis 1[13]: many philosophers now hold that when we properly understand the roles, the functions, the jobs, the uses (let us avoid the difficult and problem-generating term "meanings") of ordinary expressions, we will see truly, as Wittgenstein had seen earlier, that "All propositions of our colloquial language are actually, just as they are, logically completely in order."

It must not be supposed, however, that the abandonment of Thesis 1 by many philosophers has resulted in agreement among them as to what the function of philosophy is. But behind the many minor differences as to what that function is, two divergent views can be discerned, one claiming kinship with the later thought of Wittgenstein, the other (agreeing with the first group in abandoning Thesis 1) being led by Professor Austin.

(1) This latter group, which really seems to be in the majority at present, will be discussed first, as it is not as much a departure from the views of the early Russell as is sometimes supposed. For, with these philosophers, the "ideal language" of Russell — an artificial construction by philosophers — has been dropped; but in its place, playing a role analogous to that of Russell's ideal language, is ordinary language.

Such a view could, in the light of the criticisms noted in this essay, only be expected to evolve from Logical Atomism; after all,

> . . . our common stock of words embodies all the distinctions men have found worth drawing, and the connexions they have found worth marking, in the lifetimes of many generations: these surely are likely to be more numerous, more sound, since they have stood up to the long test of the survival of the fittest, and more subtle, at least in all ordinary and reasonably practical matters, than any that you or I are likely to think up in our armchairs of an afternoon — the most favored alternative method.[14]

Thus, "If a distinction works well for practical purposes in everyday life (no mean feat, for even ordinary life is full of hard cases), then there is sure to be something in it, it will not mark nothing." (Austin, p. 11) It follows that

> When we examine what we should say when, what words we should use in what situations, we are looking again not *merely* at words (or "meanings", whatever they may be) but also at the realities we use the words to talk about: we are using a sharpened awareness of words to sharpen our perception of, though not as the final arbiter of, the phenomena. (Austin, p. 8; the qualification, "though not as the final arbiter of," is discussed in Note 15, below.)

The basic error of the Logical Atomists, then, was not in their view that language "marks" something about "realities," but only in their oversimple view that such "marking" must consist in "picturing." And because the relation between ordinary ways of talking and "the phenomena," "the realities," is not a simple "picturing" one, those early thinkers were led to suppose that ordinary language is imperfect, and so to search for an ideal language. Once we eliminate the naïve mistakes of Thesis 2, therefore, we can deny Thesis 1 and carry out the program of Logical Atomism as expressed in Thesis 3b — remembering, of course, that the word "facts" must not be taken in the way the Logical Atomists took it.[15]

(2) In Part I, Sections 89—107, of the *Philosophical Investigations,* Wittgenstein summarizes and analyzes the development of his own earlier views — the views of the *Tractatus,* and, therefore, of the tradition that developed from those views. He explains why he was led — mistakenly — to think of the task of philosophy as the construction or at least the outlining of

> a final analysis of our forms of language, and so a *single* completely resolved form of every expression. That is, as if our usual forms of expression were, essentially, unanalyzed; as if there were something hidden in them that had to be brought to light. . . . It can also be put like this: we eliminate misunderstandings by making our expressions more exact; but now it may look as if we were moving towards a particular state, a state of complete exactness;

> and as if this were the real goal of our investigation. (*Philosophical Investigations,* I, 91).

The essence of language, he thought then, "'is hidden from us'; this is the form our problem now assumes. We ask, '*What is* language?'" (I, 92) And this question — of what is the essential form of language — he wanted to answer by means of the apparatus of "logical form" and "atomic facts" and their accoutrements. "Thought, language, now appear to us as the unique correlate, picture, of the world." (I, 96) "Its essence, logic, presents an order, in fact the a priori order of the world." (I, 97) But

> the more narrowly we examine actual language, the sharper becomes the conflict between it and our requirement. (For the crystalline purity of logic was, of course, not a *result of investigation:* it was a requirement.) The conflict becomes intolerable; the requirement is now in danger of becoming empty. . . . Back to the rough ground! (I, 107).

In Sections 108—33, he continues explaining his criticisms of these views, and outlines his new procedures. "It was true to say," as he had said in the *Tractatus,* "that

> our considerations could not be scientific ones. . . . We must do away with all *explanation,* and description alone must take its place. And this description gets its power of illumination — i.e. its purpose — from the philosophical problems. These are, of course, not empirical problems; they are solved, rather, by looking into the workings of our language, and that in such a way as to make us recognize those workings: *in despite of* an urge to misunderstand them." (I, 109).

In order to bring out the "workings" of an expression, we point, among other things, to examples of its use in actual contexts, and particularly to uses which the philosopher's employment of the expression fails to cover, or with which his uses conflict; we try to show how his problem arises out of his peculiar use of the expression, his use of it in a peculiar context; to show what makes such a misunderstanding not only possible, but seemingly plausible and even appealing; and to show that, once one sees the workings of the expression, all temptation to misunderstand it — to understand it in the peculiar way the philosopher does — disappears, and with that, the philosopher's problem and his doctrine disappear. "The results of philosophy are the uncovering of one or another piece of plain nonsense and of bumps that the understanding has got by running its head up against the limits of language" (I, 119); and this will indeed mean "*complete* clarity. But this simply means," not the achievement of perfect formulation, but "that the philosophical problems should *completely* disappear." (I, 133). Not only will the philosopher's problems dissolve, but also his positive doctrines which are either the attempted answers to or the misguided sources of those problems.

Thus, as Austin continues in the tradition of Logical Atomism, so Wittgenstein carried on the therapeutic conception of philosophy which he advanced in his earlier work.[16] To what extent have these approaches been successful? The present essay has provided a context, a basis, in terms of which a thorough critical examination of them can be made.

## NOTES

* [Note added, 1982] The most unambiguous statement of this position by Russell that I know of is the following: "There is, I think, a discoverable relation between the stucture of sentences and the structure of the occurrences to which the sentences refer. I do not think the structure of non-verbal facts is wholly unknowable, and I believe that, *with sufficient caution, the properties of language may help us to understand the structure of the world.*" (*An Inquiry into Meaning and Truth*, London, George Allen and Unwin, 1948, p. 341; italics mine.) Although this passage was written rather late (1940) in his career, thought in the same vein occurs throughout his writings from at least the mid-1900's through the first World War and beyond. Logic he held to be "the essence of philosophy" (the title of Chapter II of *Our Knowledge of the External World*); and it played not only the negative role of being a tool for removing philosophical confusions engendered by the "form" of statements in ordinary language, but also the positive one of revealing the types of (forms of) facts there are. See, for example, *Our Knowledge of the External World*, London, George Allen and Unwin, 1949, pp. 60–69, which conclude that "The old logic put thought in fetters, while the new logic gives it wings. It has, in my opinion, introduced the same kind of advance into philosophy as Galileo introduced into physics, making it possible at last to see what kinds of problems may be capable of solution, and what kinds must be abandoned as beyond human powers. And where a solution appears possible, the new logic provides a method which enables us to obtain results that do not merely embody personal idiosyncrasies, but must command the assent of all who are competent to form an opinion." "The Philosophy of Logical Atomism" (1918) opens with

a promise "to set forth... a certain kind of logical doctrine, and on the basis of this a certain kind of metaphysic." (*Logic and Knowledge*, London, George Allen and Unwin, 1956, p. 178). Again, in that work, philosophy is not the mere removal of confusions through casting language into its true logical form, as is was for the author of the *Tractatus Logico-Philosophicus*; through the "structure" which logic provides of a "perfect language," something about the world, the "forms" or "structures" of facts, or at least the possible such forms or structures, can be revealed. "A logically perfect language... will show at a glance the logical structure of the facts asserted or denied. The language which is set forth in *Principia Mathematica* is intended to be a language of that sort." (*Ibid.*, pp. 197–8.) The interpretation offered here brings out the importance of Russell's early work on Leibniz in the development of his thinking: Leibniz's mistakes arose not from his attempt to draw metaphysical conclusions from logical considerations; they stemmed rather from the fact that he had an inadequate logic. (*A Critical Exposition of the Philosophy of Leibniz*, London, George Allen and Unwin, 1949; first published in 1900.)

1 In sharpening this distinction between the views of the *Tractatus* and Logical Atomism, this essay will be in disagreement with several current interpretations of Wittgenstein's early thought. Warnock, for example, in expounding the view of the *Tractatus*, claims that "This was in fact closely related to Russell's Logical Atomism; it could be called perhaps a more consistent, more thorough, and therefore more extreme working out of some of Russell's principles and ideas." (G. J. Warnock, *English Philosophy Since 1900*, London, Oxford, 1958, p. 64). Urmson, though he admits that Wittgenstein's early thought was probably different from Russell's (as Wittgenstein himself insisted it was), nevertheless maintains that "it was the sort of interpretation I have given" – of Wittgenstein as a Logical Atomist, – "right or wrong, which was accepted in the period under examination." (J. O. Urmson, *Philosophical Analysis*, Oxford, Clarendon, 1956, pp. ix–x). It will be an incidental purpose of the present essay to argue that the differences between the views of the *Tractatus* and those of the *Philosophical Investigations* are not as great as the Logical Atomist interpretation of the former would suggest; and to show that the influence of the purely therapeutic character of the *Tractatus* was greater than Urmson says it was.

2 G. Ryle, "Systematically Misleading Expressions," *Proceedings of the Aristotelian Society*, 1931–32; reprinted in A. G. N. Flew, *Logic and Language*, First Series, Oxford, Blackwell, 1952. (Page references are to Flew.)

3 Whitehead's metaphysics took events rather than facts as the ultimate building-blocks of reality; and perhaps this accounts for Ryle's otherwise unaccountable introduction here of the term "event," and also for his habitual alternation of "facts or states of affairs": possibly he wanted to surmount the whole problem of facts versus events by relying on an expression which would allow for either solution: "state of affairs" could be construed as *either* "fact" *or* "event," depending on the way philosophers finally decided the question. Ryle's attention would certainly have been called to Whitehead's views by Russell's frequent admiring references to them: Whitehead had explained points in terms of events, and Russell was fond of giving this analysis as a second example (in addition to his own theory of descriptions, which Ramsey pronounced a "paradigm of philosophy") of the technique of logical construction.

4 "That the elements of the picture are combined with one another in a definite way, represents that the things are so combined with one another. This connection of the elements of the picture is called its structure" (*Tractatus,* 2.15); "The logical picture of the facts is the thought" (*ibid.,* 3); "The thought is the significant proposition" (*ibid.,* 4); "To the configuration of the simple signs in the propositional sign corresponds the configuration of the objects in the state of affairs" (*ibid.,* 3.21).

5 The *Tractatus* itself does not employ "facts" (or anything else, for that matter) as criteria for discovering what is the correct formulation of propositions. Wittgenstein is there concerned to show what the logical structure of any possible language must be; and, though he holds that some forms of expression would exhibit the logical form of facts more clearly than others, he does not deal with the question of how we would tell which forms of expression are actually clearest. (Probably he would have held that we simply understand expressions, and need no criterion to tell us what they mean; this is a view also suggested by some passages in Ryle, and I will discuss it presently.) For his successors, however, who wanted to apply his views and eliminate some philosophical confusions, the problem could not fail to arise.

6 Yet another criterion of misleadingness is suggested by Ryle in this paper, *viz.,* Occam's Razor: a statement is misleading if it tempts us to multiply entities. This criterion is not necessarily connected with the "facts" and "paralogisms" criteria discussed above. I will not deal with it in this paper.

7 In this paper I will pass over the question whether the Logical Atomists believed that the analysis of language would lead to the discovery of the forms of new facts, or of new forms of facts.

8 *Tractatus* 5.5563 must be understood in the light of such passages as 4.014–4.015, and 4.002: ordinary language is (and must be) as much a picture of reality as any other language; but the "law of projection" which projects reality into language (as a "law of projection . . . projects the symphony into the language of the musical score") is an extremely complex one; hence "The silent adjustments to understand colloquial language are enormously complicated." It is to reveal the picturing relation more simply and clearly that philosophical translation is needed: the picturing relation is there, even in ordinary language (otherwise it could not be a language); hence that language is "logically completely in order."

9 Needless to say, traditional philosophers have looked on this doctrine as a conclusion drawn from a detailed analysis of ordinary ways of looking at things and talking about them – not as an assumption made in advance of their thinking. And the critics of the doctrine – to be discussed below – have not just baldly denied it, but claim that detailed analyses of the particular arguments from which philosophers have drawn this thesis, show these arguments to be in error; and that this fact suggests that perhaps the whole doctrine is in error.

10 Some theses of Ryle's own position have been passed over in this essay: (1) that misleadingness is a matter of degree; (2) that perhaps no sentence can ever be freed entirely of misleadingness (*i.e.,* that the ideal language can never be achieved); (3) that misleading statements fall into definite types or classes; and (4) that misleadingness is not relative to the reader or hearer of a sentence, but is an inherent property of it.

I have ignored these theses, despite their considerable intrinsic interest, because they are not of such central importance in the overall historical picture with which I am here concerned.

11 It is sometimes said that the "fundamental assumption" (or "mistake") of these early analytic philosophers lay in the crucial place which they conceived logic to have in the philosophical quest; or, again, that their basic doctrine (and error) was that meaning is naming. But these are merely the trailers of Thesis 2: the crucial importance of logic is asserted by 2i, the doctrine that meaning is naming, by 2ii. All of the theses which I have outlined are central to the positions with which I have been concerned; to maintain that any one of them, or any part of any one of them, is more fundamental than any other, is to falsify the historical picture.

12 When the next step is taken, and we begin looking for a criterion to tell us *both* what is the true form of the proposition *and* what is the true form of the fact, the differences between the two approaches begin to disappear. This fact has increased the tendency of interpreters to conflate the two fundamentally different initial positions stemming from Theses 3a and 3b.

13 Another very important tradition has refused to relinquish Thesis 1, holding that ordinary language must still be translated into another form. But even for this tradition, this other form is no longer considered to be "correct" in the sense of being a true or accurate representation of the world. Thus Carnap says that "the widely held opinion" that the language recommended by philosophers "must be proved to be 'correct' and to constitute a faithful rendering of 'the true logic'" has led to "pseudo-problems and wearisome controversies" (R. Carnap, *The Logical Syntax of Language*, London, Routledge and Kegan Paul, 1937, pp. xiv–xv). He maintains rather "that we have in every respect complete liberty with regard to the forms of language; that both the forms of construction for sentences and the rules of transformation . . . may be chosen quite arbitrarily" (*ibid.*, p. xv). The choice of a syntactical system is not to be made on the basis of the "correctness" or "incorrectness" of the system – of whether it corresponds to the "logical form of the facts" – but on the basis of purely pragmatic considerations. And although Carnap in this work explicitly applies his "Principle of Tolerance" only to "syntax," it was not long before the choice of "semantics," too, retreated from the high ground of correctness: the choice, say, between a "sense-datum language" and a "language of appearing" is not to be made on the basis of whether the expressions of one "reach out" to "objects" or "simples" in the world (as Wittgenstein would have had it in the *Tractatus*), but on the basis of "convenience." It was not too long a step from such views to the full-fledged "Logical Pragmatism" of Quine, White, and Goodman; or to the disturbing difficulties concerning "analyticity" and related concepts. But I will not discuss this line of development here; I wish only to point out that it still remains in the tradition of Thesis 1: ordinary language, common sense, is still "vague, cocksure, and self-contradictory," as Russell remarked in *An Outline of Philosophy* (p. 1); and the philosopher's task is still a reconstruction of ordinary language.

14 J. L. Austin, "A Plea for Excuses," *Proceedings of the Aristotelian Society*, Vol. LVII (1956–57), p. 8.

[15] Austin places two very important qualifications on the use of his technique: (1) we should restrict our application of it to areas of ordinary language which have not been corrupted by philosophical disputes (p. 8); (2) ordinary language must not be considered to be "the final arbiter" of "the phenomena": "ordinary language is *not* the last word: in principle it can everywhere be supplemented and improved upon and superseded. Only remember, it *is* the *first* word." (p. 11)

[16] New reasons are sometimes given in contemporary philosophy for fusing or confusing these two approaches. "What probably happened is this: in the process of dissolving philosophical problems it was gradually seen that certain of these problems arose because of systematic deviations from the ordinary logic of certain concepts. Soon the interest in the logic rather than the deviation became paramount; and philosophy reconstituted itself as a positive, quite autonomous logical activity which is important independently of its ability to clean up traditional mistakes." (M. Weitz, "Oxford Philosophy," *Philosophical Review*, Vol. LXII, 1953, p. 188.) In the light of this interpretation, one might ask whether there is any fundamental difference between the Austinian and the Wittgensteinian approaches. But although there is a sense in which it can be said that Wittgenstein's therapeutic measures are supposed to result in something "positive" (namely, an understanding of the logic of our language), he would never have subscribed to Austin's remarks, quoted above, about the results of philosophy. According to Wittgenstein, we explain our ordinary uses by looking at certain "general facts of nature"; extralinguistic facts are no more revealed by the study of ordinary concepts (uses) than by the examination of any other usable set of concepts. The difference between the two approaches is seen in the fact that, for Wittgenstein, the concepts we choose to examine are not (as for Austin) the ones with which philosophers have *not* dealt, but rather those with which philosophers *have* dealt.

CHAPTER 2

# MATHEMATICAL IDEALS AND METAPHYSICAL CONCEPTS

In his recent series on "Criticism in the History of Science: Newton on Absolute Space, Time, and Motion"—*Philosophical Review*, LXVIII (1959), 1-29; 203-227—Professor Stephen Toulmin has argued that the concepts of absolute space and time functioned, in Newton's scientific work, as "mathematical ideals"—as formal elements of an axiomatic system—rather than as empirical postulates.

The temper of empiricist philosophy since Mach has been to label such clearly nonempirical concepts as "absolute space" and "absolute time" as "metaphysical," and therefore and therewith to dismiss them as "meaningless." Toulmin's account is a suggestive addition to the growing number of attempts to rectify this unbalanced view of the logic of science. But although I agree with him that such concepts can be interpreted as playing a significant role as elements of a nonempirical side of a theory, I do not feel that he has done justice to this point. The expression "mathematical ideal" (instead of which, for reasons he does not explain, he often uses "dynamical ideal") not only is left too vague to explain the role of the concepts in question, but even is so misleading as to falsify that role and to produce important distortions in Toulmin's own account. Once we have clarified that role, it will appear that the employment of these concepts, even in this nonempirical yet nonmetaphysical way, need not be, and was not in classical physics, as guiltless of violating sound scientific attitude as Toulmin suggests it was.

The only explanation Toulmin offers of his notion of a "mathematical ideal" occurs on pages 16-18; the core of his discussion is contained in the following passage:

Suppose we make a pendulum clock as accurately as we can, and compare the rate of swing of the pendulum with the motion of the sun across the sky; we shall find that they do not proceed exactly in step, their relative rates varying between the equinoxes and the solstices. What are we to make of this discovery? If we use the "natural" or solar day as our measure of "vulgar time," we shall be obliged to say that the pendulum swings at a rate which varies slightly according to the season of the year, and that this variation persists however much we improve the construction of the pendulum. (This by itself might be a genuine phenomenon calling for a physical explanation.) But now suppose in addition that we carefully observe the motions of Jupiter's satellites, timing the instants at which they are eclipsed once again by solar time: in that case, we shall find a precisely similar seasonal variation. Yet who will want to concede that the rate of circulation of these satellites around Jupiter is affected by the season of the year—the season, that is, not of the jovial but of the terrestrial year? We clearly do better to regard the pendulum and the circumjovial satellites alike as moving more uniformly than the sun across the sky, hoping in due course to explain this variation in the sun's apparent motion as the more genuine phenomenon. . . . If we accept this alternative, however, that will mean abandoning the previous "vulgar" measure of time and taking in its place for theoretical purposes a dynamical ideal: namely, the common ideal toward which both terrestrial and celestial motions—pendulums and satellites—are found to approximate. In these so nearly uniform motions we see the closest physical realization of the "absolute, true and mathematical time" of Newton's axiomatic system [I, pp. 16-17].

According to this account the reasoning behind the rejection of the solar time scale is as follows. We discover the same kind of variation, relative to this time scale, in two entirely different sets of events (that one happens to be terrestrial, the other celestial, seems to me wholly irrelevant). Now usually the discovery of such precisely similar variations leads us to suspect the existence of a causal relationship between the events—in this case, according to Toulmin, we would have to say that the rate of revolution of the Jovian satellites is affected by the seasons of the terrestrial year[1]; and this, we are led to believe, would simply be a ridiculous hypothesis: "who will want to concede" its possibility? "We clearly do better" to abandon the standard of temporal measurement which leads to this absurd conclusion.

But surely there is more to the case than this: surely there are good reasons, which ought to be brought out, why we would not want to concede the possibility of a causal relation between the terrestrial seasons and the motions of the Jovian satellites. There is, after all, nothing logically impossible about entertaining such a hypothesis: the observed precise similarity of variations *could*, just as much as the observed irregularities in the swing of the pendulum, be admitted as "a genuine phenomenon calling for a physical explanation." Nor would the discovery of still further cases of events showing variations precisely similar to these make logically untenable the hypothesis of a causal relationship between all the events concerned.[2] That is, even after the discovery of the irregularities in the motions of the Jovian satellites relative to the sun's motion, and even after the discovery of precisely similar variations in other, apparently independent events, we could still continue to consider the sun's motion as temporally uniform, and so as suitable to serve as a standard of temporal measurement, and attempt to find some way of explaining the variations.

What, then, is the reason why, in the case given by Toulmin, the decision is made to reject the possibility of such interaction? Fundamentally, the reason is that, within the framework of current physical theory, there is no way to explain how such interaction might take place. Even this reason, of course, is not completely compelling: we might still be willing to admit that there might be some kind of "force" hitherto unrecognized that is responsible for the variations. And we shall see, below, that sometimes it might be convenient to take this approach. But in the case given by Toulmin, the reasoning might be summarized in the following procedural principle: *If, relative to a particular standard of temporal measurement, precisely similar variations are found in a number of temporal processes, and if current physical theory does not provide the possibility of there being a causal relationship between those processes, then we are to attribute those observed irregularities to the standard rather than to those processes.* When we discover the same variation in the motions of the Jovian satellites relative to the sun's motion as standard of temporal uniformity, as in the motion of the pendulum relative to the same standard, and when no way is found in current physical theory of relating those events, then we are, according to this principle, to consider those variations as due to our choice of standard, rather than as being "a genuine phenomenon calling for a physical explanation."

This procedural principle can also be stated, not as a principle of the rejection of time scales, but rather as a principle of the acceptance of new time scales: *So choose the standard of temporal measurement that no systematic irregularities will occur in nature which are not explained in terms of current physical theory.* I shall call this principle, in either of these forms, *P*. It should be noted that *P* can be rephrased to refer to standards of spatial rather than of temporal measurement, or to refer to standards of measurement in general.

Now Toulmin claims that the reasoning involved in his example is clarified by bringing in the expression "mathematical ideal," and that, understood as mathematical ideals, Newton's notions of absolute space and absolute time make scientific sense. But the exact role of these ideals in the reasoning about time scales is not clear. On the one hand, one gets the impression throughout the paper that the mathematical ideals, "absolute space" and "absolute time," serve as standards by which we decide which time scales are satisfactory and which are not. For example:

In dynamics as in geometry it can always be asked how accurately in fact particular figures or measures realize our ideals, whether of "right lines and circles" or of "absolute, true" time and space. Sensible standards and measures of position, velocity, and lapse of time will possess more or less in fact the characteristics which the variables $x$, $y$, $z$, $t$, and their derivatives possess in the theory from their own nature [I, 18].

On the other hand, in the example quoted above, he seems to be saying that we arrive at the necessity of using mathematical ideals *after* we have found at least one sensible standard to be unsatisfactory: in place of that sensible standard we take "for theoretical purposes a dynamical ideal." This statement would lead us to suppose that we decided that standard to be unsatisfactory on grounds other than its failure to conform to the mathematical ideal.

Toulmin nowhere makes clear what are "the characteristics which the variables $x$, $y$, $z$, $t$, and their derivatives possess in the theory" of Newton. Nor does he tell us how those formal characteristics function as "ideals" governing the choice of a particular sensible standard—if this is how he supposes them to function in that theory. If, on the contrary, he supposes them to be arrived at by a decision to reject some (or all) sensible standards, the process by which this takes place is left equally unexplained. But fortunately, it is unnecessary to try to determine which of these interpretations is really Toulmin's meaning; for in neither case is the expression "mathematical ideal" enlightening about the reasoning involved in his example.

We have seen that the logic of the choice made in Toulmin's example, about whether to reject a certain time scale, is brought out by the procedural principle $P$. That is, $P$ summarizes the reasons—provides the standard—for selecting and rejecting time scales. Now is there any useful sense in which $P$ can be said either to be, or to lead to the adoption of, a "mathematical ideal"? For if the mathematical ideal "absolute time" functions as a standard for determining whether a sensible standard of time measurement is satisfactory, then it should amount to $P$; and if $P$ leads to the adoption of the mathematical ideal "absolute time," then we should be able to understand how this comes about.

Can $P$ be said, in any useful sense, to be either "mathematical" or "ideal"? One might be led to call it "mathematical" by thinking along lines like these: we decide whether to change time scales by seeing whether we can account for the observed variations in terms of the laws of our physical theory; but these laws are derivable, in Newtonian physics, from certain axioms; and the fundamental variables of the system are defined through the ways in which they are related in these axioms. Hence, ultimately, it is the formal characteristics of the variables, as set down in the axioms, that lead us to decide what kind of time scale to look for.

But this reasoning would overlook some important points. First, the laws used to determine whether or not to attribute the variations to the time scale are not pure, uninterpreted mathematical formulae; they are interpreted, not merely "formal." Furthermore, it is not merely—as Toulmin, on this interpretation, suggests it is—the characteristics of the time variable that tell us what kind of time scale to try to find; it is the laws of the system as a whole. In fact, as I will point out later, time in Newtonian physics can have no causal efficacy; and since, in trying to decide on a time scale, we are concerned to see if certain observed variations can be explained physically, the characteristics of the time variable itself are not at all central to the problem of choosing a time scale.

I can therefore see no good argument for referring to *P*, the reasoning behind the selection and rejection of time scales, as "mathematical," or as dependent purely on the "formal characteristics" of the variable *t* and its derivatives. That reasoning makes use of the characteristics of the mathematical variables, since it is concerned with the physical laws; but it would be entirely misleading to say that that reasoning depends on nothing but those formal characteristics—and still worse to say that that reasoning depends on nothing but the formal characteristics of the time variable *t*.

Nor can we say, without careful qualification, that *P* functions as an "ideal" in terms of which we choose our standards of temporal measurement: if we call *P* an ideal, then we must remember that it is an ideal in the sense of being a standard for choosing time scales; it is *not* an ideal in the sense of being *itself* a time scale, but more perfect than ordinary ones, to which all ordinary time scales "approximate." And this brings us back to the fault in the other possible interpretation of Toulmin's "mathematical ideals"—that we decide to adopt them after we have reasoned according to *P*. For in deciding to consider the sun's motion as temporally irregular (that is, as no longer serving, without qualification, as our standard of temporal measurement), we are not "abandoning the previous 'vulgar' measure of time and taking in its place for theoretical purposes a dynamical ideal." What we are doing is taking as our standard another empirical one—albeit, no doubt, a highly complex one—which takes into consideration the various systematic irregularities which would result from taking *any one* of a number of candidates (pendulum, sun's motion, and so forth) as our sole standard, and so balances our particular standards as to cancel, as much as possible, the resultant systematic irregularities.[8] And we must be especially careful in speaking of "so nearly uniform motions" which are "the closest physical realization" of our ideal; for such locutions make it all too easy to slip into thinking of the ideal as a perfectly uniform, nonphysical motion, and to forget, as Toulmin seems sometimes to have forgotten, what kind of a thing the standard really is.

Thus it does not seem very profitable to speak of the reasons for abandoning a temporal scale, and for choosing a new one, in terms of a fulfillment of "mathematical ideals." For, by saying that the "formal characteristics" of the Newtonian mathematical system serve as "mathematical ideals" in terms of which we are, according to that system, to choose our empirical standards of temporal measurement, Toulmin obscures the reasoning behind the choice of temporal scale, and even misleads himself into a Platonic interpretation of such "ideals." Nor is it any better to say that, when we decide to abandon a time scale, we replace it by a perfect nonempirical one in the form of our "mathematical ideals."

In general, observance of *P* has great advantages for science. Science has always aimed at accounting for as much of experience as possible in terms of as few independent hypotheses as possible. And *P* is essentially a technique for avoiding the necessity of introducing any new independent hypotheses or modifying old ones. Looked at in this way, the reasoning summarized in *P* becomes just a special application of the principle of "simplicity" in science.

But it may nevertheless be true that, in the long run, a better—that is, a simpler, more usable—account of phenomena might be given if we admit the existence, say, of the variations observed in our example, even at the expense of introducing a new mechanical principle or of modifying some of our old ones; for such change might well make theoretically dispensable a number of other laws or concepts previously thought fundamental and independent. A truly undogmatic, scientific attitude must admit at least the possibility that "the rate of circulation of these satellites around Jupiter is affected by the seasons" of the terrestrial year.

But when faced with the possibility of adding to, dropping, or modifying some of its fundamental principles, science is naturally, and for good reasons, conservative. Usually, in fact, whether conditions require that *P* should or should not be observed is not entirely clear; and in such cases science adopts a "wait-and-see" attitude. The following discussion will illustrate three circumstances—often closely conjoined in practice—in which such a tentative attitude may be adopted.

(1) Although the observed irregularities may actually have been explained in terms of current theory, that explanation may not be wholly satisfactory; in such cases we leave open the possibility that we may yet have to modify our time scale. For example, the spectra of distant galaxies show a shift toward the red end of the spectrum: the emitted light has a longer wave length, a lower frequency, than normal; and this irregularity is systematic, in the sense that, statistically speaking, the dimmer (and therefore, again statistically speaking, the farther) the galaxy, the greater the shift to the red. Here, then, we have a case of "precisely similar" variations to account for. Now these variations can be explained in terms of the Doppler Effect: the galaxies are moving away from us, and the farther the galaxy, the greater the velocity of recession. But many prominent scientists have not considered this explanation entirely satisfactory for many reasons (for example, it involves an enormous extrapolation of the Doppler Effect from terrestrial cases to a cosmic scale; it requires the galaxies to possess, relative to one another, velocities which approach and, at great enough distances, can even be expected to surpass the velocity of light). And if we have to abandon this explanation, we may have to try attributing the irregularity to our present time and distance scales rather than to the motions of the galaxies themselves.

(2) In cases in which everything seems to indicate the advisability of a modification of time scale, there may be no clear way forthcoming of so choosing a new time scale, or of modifying our old one, that all unexplainable irregularities will be eliminated completely: all proposed new scales may, for instance, take care of the irregularities in question, but may introduce new ones of their own. In such cases again, the final decision as to a course of action is left open. And even

if the new time scale seems free of such irregularities, there is no way of telling immediately whether it is to be trusted or not; for we must always be prepared to discover, relative to any time scale we adopt, systematic irregularities which we had not noticed before. (There is, of course, no reason why this *must* happen.) Here again we can see the virtue of the careful, tentative approach in the employment of *P*.

(3) Even when a new time scale is available which does not appear to introduce any unexplainable systematic irregularities, that new scale may itself require serious modifications of current theory. This will clearly be the case if (as in the example of the Doppler Effect given above) an explanation in terms of current theory is rejected as unsatisfactory; but it may not always be immediately clear whether such modifications of theory will be necessitated or not. For example, to attribute the supposed expansion of the universe to our choice of time scale may entail abandonment of the hypothesis that the Doppler Effect holds throughout the universe, or else that the frequency of light does not change with time. However, at least one scientist has been willing to make such sacrifices: one of the primary purposes of E. A. Milne, in his construction of his "tau-time" scale, was to eliminate the necessity of assuming that the universe is expanding without introducing any other unexplained irregularities. But because Milne's approach—apart from the many difficulties in the details of his theory—requires the abandonment of some principles that have proved of great explanatory value in science, it has been viewed with extreme caution (sometimes, unfairly, with worse). And this is only right: it would be dangerous to throw over the cumulative advances of years in favor of a radically new and relatively untried system which is designed to take care of only a small part of the total number of facts which the older theory has proved capable of handling. And besides, it may be possible to remove all the difficulties in the current explanation of the red-shift; for such explanations, too, cannot always be presented immediately with all possible clarity. But the fact remains that Milne (for example) may be right, at least in his approach: such a radical change of time scale, even with its attendant modifications of current physical laws, might in the long run provide a better account of experience than we now have.

And so the problem of the selection of and adherence to a particular standard of temporal measurement is far more complicated than Toulmin has suggested. And once we have seen that the issue with which he is concerned involves a procedural principle regarding acceptance or rejection of such standards, and that *in some cases it might be convenient to violate that principle*, we can see that his apology for Newton's use of "absolute time" cannot be successful.

For in Newtonian physics it was an ultimate and unquestioned dogma that all physical events were to be accounted for solely in terms of the properties and interactions of masses; and it was for *this* reason, and not for the good reasons discussed above, that in Newtonian physics *P* was to be observed—and observed in all cases, without exception. Whenever a systematic irregularity was encountered which could not be accounted for in terms of the specified properties and interactions of masses, that irregularity was to be treated as due solely to our choice of reference frame. Space and time thus became for Newtonian physics convenient dumping-grounds to which could be relegated all observations which did not fit this fundamental dogma.[4]

Thus, even if Newton (or any other scientist) used a concept, which he called "absolute time," which amounted in practice to adherence to the procedural principle *P* as a fundamental and always-to-be-obeyed rule, such adherence would still be open to severe criticism. For though it might not be adherence to a meaningless metaphysical principle, nevertheless there can arise in science important cases in which adherence to the principle might be unwarranted in the light of the fact that rejection of it might produce a better system. Hence Toulmin's apology for Newton's work has failed to bring out the remaining anti-scientific element of the "absolute space and time" doctrine.

It would be futile to argue that metaphysicians who talked about "substance" were anticipating or struggling toward a scientific procedural principle but mistaking it for a proposition about "reality": this *may* be true of the "absolute time" doctrine, but may not be true of other such notions. Still, a parallel can be drawn between the old arguments that there "must" be a substratum of change, and later scientific conservation principles and principles about the continuity of physical processes; and it might be argued that those principles, too, are procedural, in the same way that *P*, which may be the meaningful core or part of the meaningful core of the "absolute time" doctrine, is a procedural principle. And again, though the "principle of the continuity of physical processes" seems to have been held as a dogma by classical physics, its hold has been greatly weakened by the quantum theory. Only in this century have scientists begun to realize the width and depth of the options open to them in the acceptance or rejection of such principles.

## NOTES

[1] Actually, all we would have to admit (and even that, only on the acceptance of Mill's criteria of causal explanation) would be that there is some interrelation of some kind between them: perhaps the motions of the Jovian satellites affect the seasons of the year; or perhaps some third factor is the cause of both variations.

[2] One might expect that, if the variations are due to our choice of standard, precisely similar variations should exist in *all* physical events. In practice, however, the determination that such systematic irregularities occur, or that they are "precisely similar," is not so simple; in the case given by Toulmin, for example, other systematic irregularities overlie the one attributed to the choice of temporal standard: *e.g.*, in the seventeenth century, it was noticed that the times of eclipse of the Jovian satellites showed a periodic deviation from the predicted times dependent on the distance of Jupiter from the earth. This systematic deviation was attributed to the finite velocity of light, and on this hypothesis Römer made his original measurement of that velocity (1675). It is not always so possible to disentangle the various observed deviations from the expected, however. Still, when the number of events begin to multiply in which precisely, or almost precisely, the same systematic variations of process occur, we are tempted to look rather to our standard than to the rest of the universe for an explanation of the variations. But it must be understood that this reasoning is not logically compelling.

[3] People are often puzzled by questions like this: how can we say, of an atomic maser clock—the best clock we have—that it is accurate to so many seconds per ten thousand years (doesn't that imply our comparing it with a more perfect clock?)? The answer, built into physical theory, is not that we compare its time measurement with a perfect but nonempirical time scale, but rather that, if we did not say this, then many other things—perhaps even all other things—would be found to happen with a slight, and in all cases exactly similar, irregularity. And we can use other particular standards, like pendulums, because we know, in each case, what qualifications to make in order that, within the limits of present empirical knowledge, no systematic irregularities will occur in nature that do not have a physical explanation. It should be pointed out, incidentally, that change or modification of time scale has, according to the analysis I have given, nothing to do with passing from a "vulgar" to a nonvulgar scale.

[4] It follows from this dogma that neither the irregularities nor any other phenomena could be said to be due to the spatial or temporal location, or to the spatial orientation, of the events concerned; only masses could have

causal efficacy. This requirement of spatial and temporal homogeneity and isotropy imposed further restrictions on what could count as an explanation, as well as on the concepts of space and time, in Newtonian physics. Any complete account of the Newtonian concepts of absolute space and time would have to exhibit those concepts as hybrids of many separate ideas: I would not like to claim that *P* is *all* that was meaningful in the Newtonian notion of the "absoluteness" of absolute time.

CHAPTER 3

# THE STRUCTURE OF SCIENTIFIC REVOLUTIONS

THIS important book[1] is a sustained attack on the prevailing image of scientific change as a linear process of ever-increasing knowledge, and an attempt to make us see that process of change in a different and, Kuhn suggests, more enlightening way. In attacking the "concept of development-by-accumulation," Kuhn presents numerous penetrating criticisms not only of histories of science written from that point of view, but also of certain philosophical doctrines (mainly Baconian and positivistic philosophies of science, particularly verification, falsification, and probabilistic views of the acceptance or rejection of scientific theories) which he convincingly argues are associated with that view of history. In this review, I will not deal with those criticisms or with the details of the valuable case studies with which Kuhn tries to support his views; rather, I will concentrate on certain concepts and doctrines which are fundamental to his own interpretation of the development and structure of science. His view, while original and richly suggestive, has much in common with some recent antipositivistic reactions among philosophers of science—most notably, Feyerabend, Hanson, and Toulmin—and inasmuch as it makes explicit, according to Kuhn, "some of the new historiography's implications" (p. 3), it is bound to exert a very wide influence among philosophers and historians of science alike. It is therefore a view which merits close examination.

Basic to Kuhn's interpretation of the history of science is his notion of a paradigm. Paradigms are "universally recognized scientific achievements that for a time provide model problems and solutions to a community of practitioners" (p. x). Because a paradigm is "at the start largely a promise of success discoverable in selected and still incomplete examples" (pp. 23-24), it is "an object for further articulation and specification under new or more stringent conditions" (p. 23); hence from paradigms "spring particular coherent traditions of scientific research" (p. 10) which Kuhn calls "normal science."

[1] *The Structure of Scientific Revolutions.* By Thomas S. Kuhn. (Chicago, University of Chicago Press, 1962. Pp. xiv, 172.) All page references, unless otherwise noted, are to this work.

Normal science thus consists largely of "mopping-up operations" (p. 24) devoted to actualizing the initial promise of the paradigm "by extending the knowledge of those facts that the paradigm displays as particularly revealing, by increasing the extent of the match between those facts and the paradigm's predictions, and by further articulation of the paradigm itself" (p. 24). In this process of paradigm development lie both the strength and weakness of normal science: for though the paradigm provides "a criterion for choosing problems that, while the paradigm is taken for granted, can be assumed to have solutions" (p. 37), on the other hand those phenomena "that will not fit the box are often not seen at all" (p. 24). Normal science even "often suppresses fundamental novelties because they are necessarily subversive of its basic commitments. Nevertheless, so long as those commitments retain an element of the arbitrary, the very nature of normal research ensures that novelty shall not be suppressed for very long" (p. 5). Repeated failures of a normal-science tradition to solve a problem or other anomalies that develop in the course of paradigm articulation produce "the tradition-shattering complements to the tradition-bound activity of normal science" (p. 6).

The most pervasive of such tradition-shattering activities Kuhn calls "scientific revolutions."

> Confronted with anomaly or with crisis, scientists take a different attitude toward existing paradigms, and the nature of their research changes accordingly. The proliferation of competing articulations, the willingness to try anything, the expression of explicit discontent, the recourse to philosophy and to debate over fundamentals, all these are symptoms of a transition from normal to extraordinary research [p. 90].
> Scientific revolutions are inaugurated by a growing sense . . . that an existing paradigm has ceased to function adequately in the exploration of an aspect of nature to which that paradigm itself had previously led the way [p. 91].

The upshot of such crises is often the acceptance of a new paradigm:

> Scientific revolutions are here taken to be those non-cumulative developmental episodes in which an older paradigm is replaced in whole or in part by an incompatible new one [p. 91].

This interpretation of scientific development places a heavy burden indeed on the notion of a paradigm. Although in some passages we are led to believe that a community's paradigm is simply "a set of recurrent and quasi-standard illustrations of various theories," and that these are "revealed in its textbooks, lectures, and laboratory exercises" (p. 43), elsewhere we find that there is far more to the paradigm than is contained, at least explicitly, in such illustrations. These "accepted examples of actual scientific practice . . . include law, theory, application, and instrumentation together" (p. 10). A paradigm consists of a "strong network of commitments—conceptual, theoretical, instrumental, and methodological" (p. 42); among these commitments are "quasi-metaphysical" ones (p. 41). A paradigm is, or at least includes, "some implicit body of intertwined theoretical and methodological belief that permits selection, evaluation, and criticism" (pp. 16-17). If such a body of beliefs is not implied by the collection of facts (and, according to Kuhn, it never is), "it must be externally supplied, perhaps by a current metaphysic, by another science, or by personal and historical accident" (p. 17). Sometimes paradigms seem to be patterns (sometimes in the sense of archetypes and sometimes in the sense of criteria or standards) upon which we model our theories or other work ("from them as models spring particular coherent traditions"); at other times they seem to be themselves vague theories which are to be refined and articulated. Most fundamentally, though, Kuhn considers them as not being rules, theories, or the like, or a mere sum thereof, but something more "global" (p. 43), from which rules, theories, and so forth are abstracted, but to which no mere statement of rules or theories or the like can do justice. The term "paradigm" thus covers a range of factors in scientific development including or somehow involving laws and theories, models, standards, and methods (both theoretical and instrumental), vague intuitions, explicit or implicit metaphysical beliefs (or prejudices). In short, anything that allows science to accomplish anything can be a part of (or somehow involved in) a paradigm.

Now, historical study does bear out the existence of guiding factors which are held in more or less similar form, to greater or less extent, by a multitude of scientists working in an area over a number of years. What must be asked is whether anything is gained by referring to such common factors as "paradigms," and whether such gains, if any, are offset by confusions that ensue because of such a way of speaking. At the very outset, the explanatory value of the notion of a paradigm is suspect: for the truth of the thesis that shared paradigms are (or are behind) the common factors guiding scientific research appears to be guaranteed, not so much by a close examination of actual historical cases, however scholarly, as by the breadth of definition of the term "paradigm." The suspicion that this notion plays a determinative role in shaping Kuhn's interpretation of history is strengthened by his frequent remarks about what *must* be the case with regard to science and its development: for example, "No natural history can be interpreted in the absence of at least some . . . belief" (pp. 16-17); "Once a first paradigm through which to view nature has been found, there is no such thing as research in the absence of any paradigm" (p. 79); "no experiment can be conceived without some sort of theory" (p. 87); "if, as I have already urged, there can be no scientifically or empirically neutral system of language or concepts, then the proposed construction of alternate tests and theories must proceed from within one or another paradigm-based tradition" (p. 145). Such views appear too strongly and confidently held to have been extracted from a mere investigation of how things *have* happened.

Still greater perplexities are generated by Kuhn's view that paradigms cannot, in general, be formulated adequately. According to him, when the historian tries to state the rules which scientists follow, he finds that "phrased in just that way, or in any other way he can imagine, they would almost certainly have been rejected by some members of the group he studies" (p. 44). Similarly, there may be many versions of the same theory. It would appear that, in Kuhn's eyes, the concepts, laws, theories, rules, and so forth that are common to a group are just not common enough to guarantee the coherence of the tradition; therefore he concludes that the paradigm, "the concrete scientific achievement" that is the source of that coherence, must not be identified with, but must be seen as "prior to the various concepts, laws, theories, and points of view that may be abstracted from it" (p. 11). (It is partly on the basis of this argument that Kuhn rejects the attempt by philosophers of science to formulate a "logic" of science in terms of precise rules.) Yet if it is true that all that can be said about paradigms and scientific development can and must be said only in terms of what are mere "abstractions" from paradigms, then it is difficult to see what is gained by appealing to the notion of a paradigm.

In Kuhn's view, however, the fact that paradigms cannot be described adequately in words does not hinder us from recognizing them: they are open to "direct inspection" (p. 44), and historians can "agree in their *identification* of a paradigm without agreeing on, or even attempting to produce, a full *interpretation* or *rationalization* of it" (p. 44). Yet the feasibility of a historical inquiry concerning paradigms is exactly what is brought into question by the scope of the term "paradigm" and the inaccessibility of particular paradigms to verbal formulation. For on the one hand, as we have seen, it is *too* easy to identify a paradigm; and on the other hand, it is not easy to determine, in particular cases treated by Kuhn, what the paradigm is supposed to have been in that case. In most of the cases he discusses, it is the theory that is doing the job of posing problems, providing criteria for selection of data, being articulated, and so forth. But of course the theory is not the paradigm, and we might assume that Kuhn discusses the theory because it is as near as he can get in words to the inexpressible paradigm. This, however, only creates difficulties. In the case of "what is perhaps our fullest example of a scientific revolution" (p. 132), for instance, what was "assimilated" when Dalton's theory (paradigm) became accepted? Not merely the laws of combining proportions, presumably, but something "prior to" them. Was it, then, the picture of matter as constituted of atoms? But contrary to the impression Kuhn gives, that picture was never even nearly universally accepted: from Davy to Ostwald and beyond there was always a very strong faction which "regarded it with misgiving, or with positive dislike, or with a constant hope for an effective substitute" (J. C. Gregory, *A Short History of Atomism* [London, 1931], p. 93), some viewing atoms as convenient fictions, others eschewing the vocabulary of atoms entirely, preferring to talk in terms of "proportions" or "equivalents." (It is noteworthy that Dalton was presented with a Royal Medal, not unequivocally for his development of the atomic theory, but rather "for his development of the Theory of Definite Proportions, usually called the Atomic Theory of Chemistry"; award citation, quoted in Gregory, p. 84.) No, it was certainly not atoms to which the most creative chemists of the century were "committed"—unless (contrary to his general mode of expression) Kuhn means that they were "committed" to the atomic theory because they—most of them—used it even though they did not believe in its truth. Further, what else was "intertwined" in this behind-the-scenes paradigm? Did it include, for instance, some inexpressible Principle of Uniformity of Nature or Law of Causality? Is this question so easy to answer—a matter of "direct inspection"—after all these years of philosophical dispute? One begins to doubt that paradigms are open to "direct inspection," or else to be amazed at Professor Kuhn's eyesight. (And why is it that such historical facts should be open to direct inspection, whereas scientific facts must always be seen "through" a paradigm?) But if there are such difficulties, how can historians know that they agree in their identification of the paradigms present in historical episodes, and so determine that "the same" paradigm persists through a long sequence of such episodes? They cannot, by hypothesis, compare their formulations. Suppose they disagree: how is their dispute to be resolved?

On the other hand, where do we draw the line between different paradigms and different articulations of the same paradigm? It is natural and common to say that Newton, d'Alembert, Lagrange, Hertz, Hamilton, Mach, and others formulated different versions of classical mechanics; yet certainly some of these formulations involved different "commitments"—for example, some to forces, others to energy, some to vectorial, others to variational principles. The distinction between paradigms and different articulations of a paradigm, and between scientific revolutions and normal science, is at best a matter of degree, as is commitment to a paradigm: expression of explicit discontent, proliferation of competing articulations, debate over fundamentals are all more or less present throughout the development of science; and there are always guiding elements which are more or less common, even among what are classified as different "traditions." This is one reason why, in particular cases, identification of "the paradigm" is so difficult: not just because it is hard to see, but because looking for the guiding elements in scientific activity is not like looking for a unitary entity that either is there or is not.

But furthermore, the very reasons for supposing that paradigms (nevertheless) exist are unconvincing. No doubt some theories are very similar—so similar that they can be considered to be "versions" or "different articulations" of one another (or of "the same subject"). But does this imply that there must be a common "paradigm" of which the similar theories are incomplete expressions and from which they are abstracted? No doubt, too, many expressions of methodological rules are not as accurate portrayals of scientific method as they are claimed to be; and it is possible that Kuhn is right in claiming that no such portrayal can be given in terms of any one set of precise rules. But such observations, even if true, do not compel us to adopt a *mystique* regarding a single paradigm which guides procedures, any more than our inability to give a single, simple definition of "game" means that we must have a unitary but inexpressible idea from which all our diverse uses of "game" are abstracted. It may be true that "The coherence displayed by the research tradition . . . may not imply even the existence of an underlying body of rules and assumptions" (p. 46); but neither does it imply the existence of an underlying "paradigm."

Finally, Kuhn's blanket use of the term "paradigm" to cover such a variety of activities and functions obscures important differences between those activities and functions. For example, Kuhn claims that "an apparently arbitrary element . . . is always a formative ingredient" (p. 4) of a paradigm; and, indeed, as we shall see shortly, this is a central aspect of his view of paradigms and scientific change. But is the acceptance or rejection of a scientific theory "arbitrary" in the same sense that acceptance or rejection of a standard (to say nothing of a metaphysical belief) is? Again, Newtonian and Hertzian formulations of classical mechanics are similar to one another, as are the Einstein, Whitehead, Birkhoff, and Milne versions of relativity, and as are wave mechanics and matrix mechanics. But there are significant differences in the ways in and degrees to which these theories are "similar"—differences which are masked by viewing them all equally as different articulations of the same paradigm.

There are, however, deeper ways in which Kuhn's notion of a paradigm affects adversely his analysis of science; and it is in these ways that his view reflects widespread and important tendencies in both the history and philosophy of science today.

Because a paradigm is

> the source of the methods, problem-field, and standards of solution accepted by any mature scientific community at any given time, . . . the reception of a new paradigm often necessitates a redefinition of the corresponding science. . . . And as the problems change, so, often, does the standard that distinguishes a real scientific solution from a mere metaphysical speculation, word game, or mathematical play. The normal-scientific tradition that emerges from a scientific revolution is not only incompatible but often actually incommensurable with that which has gone before [p. 102].

Thus the paradigm change entails "changes in the standards governing permissible problems, concepts, and explanations" (p. 105). In connection with his view that concepts or meanings change from one theory (paradigm) to another despite the retention of the same terms, Kuhn offers an argument whose conclusion is both intrinsically important and crucial to much of his book. This argument is directed against the "positivistic" view that scientific advance is cumulative, and that therefore earlier sciences are derivable from later; the case he considers is the supposed deducibility of Newtonian from Einsteinian dynamics, subject to limiting conditions. After summarizing the usual derivation, Kuhn objects that

> the derivation is spurious, at least to this point. Though the [derived statements] are a special case of the laws of relativistic mechanics, they are not Newton's Laws. Or at least they are not unless those laws are reinterpreted in a way that would have been impossible until after Einstein's work. ... The physical referents of these Einsteinian concepts are by no means identical with those of the Newtonian concepts that bear the same name. (Newtonian mass is conserved; Einsteinian is convertible with energy. Only at low relative velocities may the two be measured in the same way, and even then they must not be conceived to be the same.) ... The argument has still not done what it purported to do. It has not, that is, shown Newton's Laws to be a limiting case of Einstein's. For in the passage to the limit it is not only the forms of the laws that have changed. Simultaneously we have had to alter the fundamental structural elements of which the universe to which they apply is composed [pp. 100-101].

But Kuhn's argument amounts simply to an assertion that despite the derivability of expressions which are in every formal respect identical with Newton's Laws, there remain differences of "meaning." What saves this from begging the question at issue? His only attempt to support his contention comes in the parenthetical example of mass; but this point is far from decisive. For one might equally well be tempted to say that the "concept" of mass (the "meaning" of "mass") has remained the same (thus accounting for the deducibility) even though the *application* has changed. Similarly, rather than agree with Kuhn that "the Copernicans who denied its traditional title 'planet' to the sun ... were changing the meaning of 'planet' " (p. 127), one might prefer to say that they changed only the application of the term. The real trouble with such arguments arises with regard to the cash difference between saying, in such cases, that the "meaning" has changed, as opposed to saying that the "meaning" has remained the same though the "application" has changed. Kuhn has offered us no clear analysis of "meaning" or, more specifically, no criterion of change of meaning; consequently it is not clear why he classifies such changes as changes of meaning rather than, for example, as changes of application. This is not to say that no such criterion could be formulated, or that a distinction between change of meaning and change of application could not be made, or that it might not be very profitable to do so for certain purposes. One might, for example, note that there are statements that can be made, questions that can be raised, views that may be suggested as possibly correct, within the context of Einsteinian physics that would not even have made sense—would have been self-contradictory—in the context of Newtonian physics. And such differences might (for certain purposes) be referred to with profit as changes of meaning, indicating, among other things, that there are differences between Einsteinian and Newtonian terms that are not brought out by the deduction of Newtonian-like statements from Einsteinian ones. But attributing such differences to alterations of "meaning" must not blind one to any resemblances there might be between the two sets of terms. Thus it is not so much Kuhn's conclusion that is objectionable as, first, the fact that it is based, not on any solid argument, but on the feature of meaning dependence which Kuhn has built into the term "paradigm" (scientists see the world from different points of view, through different paradigms, and therefore see different things through different paradigms); and second, the fact that this feature leads him to a distorted portrayal of the relations between different scientific theories. For Kuhn's term "paradigm," incorporating as it does the view that statements of fact are (to use Hanson's expression) theory-laden, and as a consequence the notion of (in Feyerabend's words) meaning variance from one theory or paradigm to another, calls attention excessively to the differences between theories or paradigms, so that relations that evidently do exist between them are in fact passed over or denied.

The significance of this point emerges fully when we ask about the grounds for accepting one paradigm as better than another. For if "the differences between successive paradigms are both necessary and irreconcilable" (p. 102), and if those differences consist in the paradigms' being "incommensurable"—if they disagree as to what the facts are, and even as to the real problems to be faced and the standards which a successful theory must meet—then what are the two paradigms disagreeing about? And why does one win? There is little problem for Kuhn in analyzing the notion of progress within a paradigm tradition (and, indeed, he notes, such evolution is the source of the prevailing view of scientific advance as "linear"); but how can we say that "progress" is made when one paradigm replaces another? The logical tendency of Kuhn's position is clearly toward the conclusion that the replacement is not cumulative, but is mere change: being "incommensurable," two paradigms cannot be judged according to their ability to solve the same problems, or deal with the same facts, or meet the same standards. "If there were but one set of scientific problems, one world within which to work on them, and one set of standards for their solution, paradigm competition might be settled more or less routinely. . . . But . . . The proponents of competing paradigms are always at least slightly at cross-purposes" (pp. 146-147). Hence "the competition between paradigms is not the sort of battle that can be resolved by proofs" (p. 147), but is more like a "conversion experience" (p. 150). In fact, in so far as one can compare the weights of evidence of two competing paradigms—and, on Kuhn's view that after a scientific revolution "the whole network of fact and theory . . . has shifted" (p. 140), one must wonder how this can be done at all—the weight of evidence is more often in favor of the older paradigm than the new (pp. 155-156). "What occurred was neither a decline nor a raising of standards, but simply a change demanded by the adoption of a new paradigm" (p. 107). "In these matters neither truth nor error is at issue" (p. 150); indeed, Kuhn's view of the history of science implies that "We may . . . have to relinquish the notion, explicit or implicit, that changes of paradigm carry scientists and those who learn from them closer and closer to the truth" (p. 169).

Kuhn is well aware of the relativism implied by his view, and his common sense and feeling for history make him struggle mightily to soften the dismal conclusion. It is, for instance, only "often" that the reception of a new paradigm necessitates a redefinition of the corresponding science. Proponents of different paradigms are only "at least partially" at cross-purposes. Though they "see different things when they look from the same point in the same direction," this is "not to say that they can see anything they please. Both are looking at the world" (p. 149). It is only "in some areas" that "they see different things" (p. 149). But these qualifications are more the statement of the problems readers will find with Kuhn's views than the solutions of those problems. And it is small comfort to be told, in the closing pages of the book, that "a sort of progress will inevitably characterize the scientific enterprise" (p. 169), especially if that "progress," whether or not it is aimed toward final truth, is not at least an advance over past error. Nor will careful readers feel reassured when they are asked, rhetorically, "What better criterion [of scientific progress] than the decision of the scientific group could there be" (p. 169)? For Kuhn has already told us that the decision of a scientific group to adopt a new paradigm is not based on good reasons; on the contrary, what counts as a good reason is determined by the decision.

A view such as Kuhn's had, after all, to be expected sooner or later from someone versed in the contemporary treatment of the history of science. For the great advances in that subject since Duhem have shown how much more there was to theories that were supposedly overthrown and superseded than had been thought. Historians now find that "the more carefully they study, say, Aristotelian dynamics, phlogistic chemistry, or caloric thermodynamics, the more certain they feel that those once current views of nature were, as a whole, neither less scientific nor more the product of human idiosyncrasy than those current today" (p. 2). Yet perhaps that deep impression has effected too great a reaction; for that there is more to those theories than was once thought does not mean that they are immune to criticism —that there are not *good* reasons for their abandonment and replacement by others. And while Kuhn's book calls attention to many mistakes that have been made regarding the (good) reasons for scientific change, it fails itself to illuminate those reasons, and even obscures the existence of such reasons. We must, as philosophers of science, shape our views of the development and structure of scientific thought in the light of what we learn from science and its history. But until historians of science achieve a more balanced approach to their subject—neither too indulgently relativistic nor intransigently positivistic —philosophers must receive such presentations of evidence with extremely critical eyes.

Certainly there is a vast amount of positive value in Kuhn's book. Besides making many valid critical remarks, it does bring out, through a wealth of case studies, many common features of scientific thought and activities which make it possible and, for many purposes, revealing to speak of "traditions" in science; and it points out many significant differences between such traditions. But Kuhn, carried away by the logic of his notion of a paradigm, glosses over many important differences between scientific activities classified as being of the same tradition, as well as important continuities between successive traditions. He is thus led to deny, for example, that Einsteinian dynamics is an advance over Newtonian or Aristotelian dynamics in a sense more fundamental than can consistently be extracted from his conceptual apparatus. If one holds, without careful qualification, that the world is seen and interpreted "through" a paradigm, or that theories are "incommensurable," or that there is "meaning variance" between theories, or that all statements of fact are "theory-laden," then one may be led all too readily into relativism with regard to the development of science. Such a view is no more implied by historical facts than is the opposing view that scientific development consists solely of the removal of superstition, prejudice, and other obstacles to scientific progress in the form of purely incremental advances toward final truth. Rather, I have tried to show, such relativism, while it may seem to be suggested by a half-century of deeper study of discarded theories, is a *logical* outgrowth of conceptual confusions, in Kuhn's case owing primarily to the use of a blanket term. For his view is made to appear convincing only by inflating the definition of "paradigm" until that term becomes so vague and ambiguous that it cannot easily be withheld, so general that it cannot easily be applied, so mysterious that it cannot help explain, and so misleading that it is a positive hindrance to the understanding of some central aspects of science; and then, finally, these excesses must be counterbalanced by qualifications that simply contradict them. There are many other facets of Kuhn's book that deserve attention—especially his view that a paradigm "need not, and in fact never does, explain all the facts with which it can be confronted" (p. 18), and his suggestion that no paradigm ever could be found which would do so. But the difficulties that have been discussed here indicate clearly that the expanded version of this book which Kuhn contemplates will require not so much further historical evidence (p. xi) as—at the very least—more careful scrutiny of his tools of analysis.

CHAPTER 4

# THE PARADIGM CONCEPT

**The Structure of Scientific Revolutions.** THOMAS S. KUHN. Second edition. University of Chicago Press, Chicago, 1970. xii, 210 pp. Cloth, $6; paper, $1.50. International Encyclopedia of Unified Science, vol. 2, No. 2.

**Criticism and the Growth of Knowledge.** Proceedings of the International Colloquium on the Philosophy of Science, London, July 1965, vol. 4. IMRE LAKATOS and ALAN MUSGRAVE, Eds. Cambridge University Press, New York, 1970. viii, 282 pp. Cloth, $11.50; paper, $3.45.

Since its publication in 1962, Thomas Kuhn's *The Structure of Scientific Revolutions* has become one of the most popular attempts of all time to interpret the nature of science. It has proved an important step in a movement away from the positivistic empiricism that has held sway, among both philosophers and working scientists, for well over two generations. Writers in many disciplines have adopted the book's fundamental notion of "paradigm" in analyses of their subject matter and controversies. The book has had an impact also on a wide body of laymen, even, on occasion, being cited as authority by spokesmen of the New Left.

The thesis of the original edition was that "particular coherent traditions of scientific research" (p. 10), which Kuhn called "normal science," are unified by and emerge from "paradigms." Paradigms are "universally recognizable scientific achievements that for a time provide model problems and solutions to a community of practitioners" (p. x). Kuhn conceived of a paradigm as not identifiable with any body of theory, being more "global" (p. 43) and generally incapable of complete formulation. He held it to include "law, theory, application, and instrumentation together" (p. 10), consisting of a "strong network of commitments, conceptual, theoretical, instrumental, and methodological" (p. 42), and even "quasi-metaphysical" (p. 41); it is, he claimed, "the source of the methods, problem-field, and standards of solution accepted by any mature scientific community at any given time" (p. 102), permitting "selection, evaluation, and criticism" (p. 17). "Normal science" consists of working within and in the light of the paradigm, making it more and more explicit and precise, actualizing its initial promise "by extending the knowledge of those facts that the paradigm displays as particularly revealing, by increasing the extent of the match between those facts and the paradigm's predictions, and by further articulation of the paradigm itself" (p. 24). In the course of such articulation, however, "anomalies" arise which, after repeated efforts to resolve them have failed, give birth to the kind of situation in which a scientific revolution can take place:

> Confronted with anomaly or with crisis, scientists take a different attitude toward existing paradigms, and the nature of their research changes accordingly. The proliferation of competing articulations, the willingness to try anything, the expression of explicit discontent, the recourse to philosophy and to debate over fundamentals, all these are symptoms of a transition from normal to extraordinary research. . . . Scientific revolutions are inaugurated by a growing sense . . . that an existing paradigm has ceased to function adequately in the exploration of an aspect of nature to which that paradigm itself had previously led the way (pp. 90–91).

New candidates for fundamental paradigm are introduced; ultimately one may become accepted, often necessitating "a redefinition of the corresponding science" (p. 102). Kuhn emphasized that scientific revolutions are "non-cumulative developmental episodes in which an older paradigm is replaced in whole or in part by an incompatible new one" (p. 91).

Kuhn's views diverge radically from those dominant since Mach and Ostwald, developed in the outlook of the Vienna Circle and its intellectual associates, and paralleled in the views of Bridgman and Frank and a host of more recent thinkers. Whereas those views tended, at least in their heydays, to separate sharply "fact" (or observation" or "operation") from "interpretation" – thus claiming to preserve the "objectivity" of science – Kuhn emphasizes the dependence of what counts as a "fact," a "problem," and a "solution of a problem" on presuppositions, theoretical or otherwise, explicit or implicit. Likewise, he attacks traditional "development-by-accumulation" views of science – views according to which science progresses linearly by accumulation of theory-independent facts, older theories giving way successively to wider, more inclusive ones. In these respects, Kuhn's book has had an undeniably healthy influence on discussions of the nature of science, bringing them to a closer inspection of science and more in line with what recent scholarship has revealed about its history.

Despite these beneficial effects, however, Kuhn's views as expressed in the first edition have faced severe criticism. Two main types of objections have been raised. Those of the first type revolve around ambiguities in the notion of "paradigm." For that term, although at the outset it is applied to "a set of recurrent and quasi-standard illustrations of various theories," which are "revealed in . . . textbooks, lectures, and laboratory exercises" (p. 43), ultimately appears, as the reader may have gathered from the passages quoted above, to cover anything and everything that allows the scientist to do anything. The assertion that a scientific tradition is paradigm-governed then appears to become a tautology, and all the wealth of Kuhn's historical analysis becomes irrelevant. On the other hand, the term is so vague that, in particular cases, it is difficult to identify what is supposed to be the paradigm. (This problem is, of course, compounded by Kuhn's insistence that the paradigm is not, and in general cannot be, completely expressed.) Furthermore, the vagueness of the term makes the distinction between "normal" and "revolutionary" science seem a matter more of degree than of kind, as Kuhn claims: expression of explicit discontent, proliferation of competing articulations, debate over fundamentals are all more or less present throughout the development of science. And similarly for the distinction, drawn with uncompromising sharpness by Kuhn, between different "traditions" in science; far from there being such sharp discontinuities, there are always guiding factors which are more or less common, even among what are somewhat artifically classified as different "traditions." Finally, "commitment" to such guiding factors does not in general seem to be as rigid as Kuhn suggests.

The second major type of objection against Kuhn's first-edition view has to do with the relativism in which it apparently eventuates. In emphasizing the determinative role of background paradigms, and attacking the notion of theory- (or paradigm-) independent "facts" (or any such independent factors or standards whatever), Kuhn appears to have denied the possibility of reasonable judgment, on objective grounds, in paradigm choice; there can be no good reason for accepting a new paradigm, for the very notion of a "good reason" has been made paradigm-dependent. And certainly, though in some passages Kuhn denied this implication of his view, in most he gloried in it: "the competition between paradigms is not the sort of battle that can be resolved by proofs" (p. 147), but is more like a "conversion experience" (p. 150); "What occurred [in a paradigm change] was neither a decline nor a raising of standards, but simply a change demanded by the adoption of a new paradigm" (p. 107); "In these matters neither proof nor error is at issue" (p. 150); "We may. . . have to relinquish the notion, explicit or implicit, that changes of paradigm carry scientists and those who learn from them closer and closer to the truth" (p. 169). Objectivity and progress, the pride of traditional interpretations of science, have both been abandoned. Indeed, Kuhn's relativism did not stop here: for not only is there no means of rationally assessing two competing paradigms; there is no way of comparing them at all, so different is the world as seen through them (or – in an alternative formulation that is in many ways more consonant with Kuhn's general thesis – so different are the worlds they define). "The normal-scientific tradition that emerges from a scientific revolution is not only imcompatible but often actually incommensurable with that which has gone before" (p. 102). Kuhn carried this view to the point of holding that if the same terms continue to be used after a scientific revolution (like "mass" after the replacement of the Newtonian by the Einsteinian "paradigm") those terms have different meanings.

In this new edition, Kuhn has altered little of the original text; however, he has added a 36-page "Postscript" (p. 174 ff.) reviewing and attempting to meet criticisms that were made of the first edition. This discussion is supplemented by opening and closing essays by Kuhn in *Criticism and the Growth of Knowledge*, a collection of papers – the others are by Paul Feyerabend, Imre Lakatos, Margaret Masterman, Karl Popper, Stephen Toulmin, John Watkins, and L. P. Williams – discussing Kuhn's ideas in relation to those of Popper. (There is good reason to compare the two; for despite the differences that emerge from the discussions in that book, Popper's contention that there is no rationale in the introduction of new "conjectures" in science, but only in the exposure of such conjectures to tests potentially falsifying them, and Kuhn's insistence, at least in the first edition, and despite a number of contradictory statements, that there is no rationale in the introduction of a new paradigm, but only in the attempt to "articulate" the paradigm and make it deal successfully with "anomalies," are basically similar. There is room here to discuss only Kuhn's contributions to this volume; the paper by Lakatos, however, may be recommended as being particularly important and provocative.)

It is important to recognize the extent – and the significance – of Kuhn's withdrawal from his original position. With regard to the concept of paradigm, Kuhn now wishes to distinguish two different senses of the term.

> On the one hand, it stands for the entire constellation of beliefs, values, techniques, and so on shared by the members of a given community. On the other, it denotes one sort of element in that constellation, the concrete puzzle-solutions which, employed as models or examples, can replace explicit rules as a basis for the solution of the remaining puzzles of normal science (p. 173).

For the former, broader sense Kuhn suggests the name "disciplinary matrix," distinguishing four components of such matrices (pp. 182–86): "symbolic generalizations," "metaphysical paradigms," "values," and "exemplars," the "concrete puzzle-solutions" referred to above. All these elements were lumped together in the first edition; "they are, however, no longer to be discussed as though they were all of a piece" (p. 182). This distinction, however, is of little help to those who found the earlier concept of "paradigm" obscure. Contrary to Kuhn's complaint, few critics failed to see that the *primary* sense of "paradigm" had to do with the "concrete puzzle-solution." The difficulty was, rather, that Kuhn never adequately clarified how the remaining factors covered by that term were related to (embodied in) the concrete examples in such a way that the whole outlook ("paradigm" in the broader sense) of the tradition would be conveyed to students through such examples. Nor did he clarify the ways in which, through the concrete examples, this general paradigm determined the course of scientific research and judgment. Yet it was precisely the unity, and the controlling status, of paradigms that constituted the appeal and the challenge of Kuhn's original view: the contention that there was a coherent, unified viewpoint, a single overarching *Weltanschauung*, a disciplinary *Zeitgeist*, that determined the way scientists of a given tradition viewed and dealt with the world, that determined what they would consider to be a legitimate problem, a piece of evidence, a good reason, an acceptable solution, and so on. (The affinities of Kuhn's view with 19th-century Idealism run deep.) Does he now hold that this "constellation" that makes up the disciplinary matrix is just a loosely associated assemblage, each of whose components has its own separate and separable function? (And Kuhn offers precious little discussion of those functions.) Certainly Kuhn's emphasis here is on the distinction between the components rather than on any unity underlying them; but if this is his new view, then – especially when it is coupled with his apparent abandonment (to be discussed below) of the controlling status of the paradigm – Kuhn will have abandoned what was, however obscure, one of the most provocative and influential aspects of his earlier view. Perhaps this would be – if the remaining elements of his new position prove consistent with this view – for the best. For it could then be argued that he has moved in the direction of a salutary concern with the details of scientific reasoning – for example, with specific ways in which specific background presuppositions may influence scientific judgment and activity – rather than with sweeping but vague generalities that are ultimately tautological. But in any case it would not be the old Kuhn. (It should be remarked that Kuhn still, in spite of his critics' attacks, maintains the sharp distinction between "revolutionary" and "normal" science; indeed, the latter and its characteristic activity of "puzzle-solving" – a notion which Kuhn uses far too lightly – acquire in his essays in *Criticism and the Growth of Knowledge* an even more central cole.)

But it is in his attempt to meet the charge of relativism that Kuhn's most striking retreats from his original extreme position occur. Now, what counts as a scientific problem is not determined, at least completely, by the paradigm: "Most of the puzzles of normal science are directly presented by nature, and all involve nature indirectly" (*Criticism*, p. 263); there is, apparently, a paradigm-independent objective world (nature) which presents problems that a paradigm must solve. Further, paradigms no longer, apparently, determine, at least completely, what counts as a good reason:

> It should be easy to design a list of criteria that would enable an uncommitted observer to distinguish the earlier from the most recent theory time after time. Among the most useful would be: accuracy of prediction, particularly of quantitative prediction; the balance between esoteric and everyday matter; and the number of different problems solved. . . . Those lists are not yet the ones required, but I have no doubt that they can be completed. If they can, then scientific development is, like biological, a unidirectional and irreversible process. Later scientific theories are better than earlier ones for solving puzzles in the often quite different environments to which they are applied. That is not a relativist's position, and it displays the sense in which I am a convinced believer in scientific progress (pp. 205–06).

No, that is not a relativist's position; but it is a far cry from Kuhn's first-edition attack on the view of scientific change as a linear process of ever-increasing knowledge (to say nothing of its view that there is no such thing as an "uncommitted observer"), and its defense of the view that what happens in a scientific revolution is "neither a decline nor a raising of standards, but simply a change demanded by the adoption of a new paradigm." It is, in fact, for better or for worse, a long step toward a more conventional position in the philosophy of science – one that makes a distinction between the "given" and the "interpretation" (or "theory") and holds that the latter are adequate to the extent that they account for the former.

It appears, then, that Kuhn now believes that the conceptual guiding factors in scientific research are more diverse and complicated in their functioning, and that there are objective factors that are independent of and exercise some constraint on them ("nature cannot be forced into an arbitrary set of conceptual boxes" – *Criticism*, p. 263). Such sober retrenchment is not, however, consistent with Kuhn's simultaneous adherence to many of his old views. Despite his claim that his view does not imply "either that there are no good reasons for being persuaded [in favor of a new paradigm] or that those reasons are not ultimately decisive for the group" (p. 179), he still tells us,

> What it should suggest, however, is that such reasons function as values and that they can thus be differently applied, individually and collectively, by men who concur in honoring them. If two men disagree, for example, about the relative fruitfulness of their theories,

or if they agree about that but disagree about the relative importance of fruitfulness and, say, scope in reaching a choice, neither can be convicted of a mistake. Nor is either being unscientific (pp. 199–200).

But if there are, as Kuhn suggests here and elsewhere, no constraints on what one can assert in the name of "values," it seems gratuitous to speak of *reasons* in such contexts. And yet this seems to be the sort of thing Kuhn intends when he speaks of "good reasons" for adopting a new paradigm (for example, after telling us that his view does not imply "that the reasons for choice are different from those usually listed by philosophers of science: accuracy, simplicity, fruitfulness, and the like" [p. 199], he declares that "such reasons function as values" in the sense just discussed). It is a viewpoint as relativistic, as antirationalistic, as ever.

Particularly unhelpful is Kuhn's reply to the charge that his view of paradigm "incommensurability" implies that competition or communication between different paradigms is impossible. This is partly due to his residual ambiguity regarding the extent to which paradigms determine meanings and views of "nature": for in the absence of a clear idea of the extent of that determination, it is impossible to be clear about the extent to which meanings determined by one paradigm can be expressed in the language of another. This ambiguity in turn destroys the effectiveness of his suggestion that Quine's views on translation can help alleviate the difficulty: for Quine's views (briefly, that "radical" translation is indeterminate in that it depends on some "analytic hypothesis" which is highly arbitrary, though subject to some constraints) are not obviously consistent with Kuhn's first-edition view of paradigm determination of meanings, hypotheses, and standards. Finally, Kuhn's strange view of neural stimuli and processes and their relation to meanings and knowledge muddies the situation still further: on the one hand, we read that "people do not see stimuli; our knowledge of them is highly theoretical and abstract" (p. 192); but on the other hand – when he is trying to face the problem of incommensurability – he says that "the stimuli that impinge on [the adherents of two different paradigms] are the same" (p. 210). It is thus unclear whether what we consider to be stimuli is paradigm-independent or is relative to our paradigm (for Kuhn does call our knowledge of them "theoretical"). Beyond these contributions to confusion, Kuhn's discussion (pp. 202–04; *Criticism*, p. 266 ff.) fails utterly to come to grips with the issue. He now admits (in denial of complete paradigm-determination of meanings) to a great deal of overlap of meanings, and this, he claims, helps to circumscribe the areas of communication breakdown between adherents of different paradigms. But how are mutual understanding and comparison of adequacy to be achieved with regard to *those* areas, once located? His answer is simply that competing scientists proceed to observe one another and "may in time become very good predictors of each other's behavior. Each will have learned to translate the other's theory and its consequences into his own language and simultaneously to describe in his language the world to which that theory applies" (p. 202). But this begs the question, amounting merely to an assertion that such translation is possible. Kuhn has not succeeded in showing how he can retain paradigm incommensurability in the sense of the first edition while allowing cross-paradigm communication and comparison.

In summary, then, Kuhn appears to have retreated from his earlier position in just those respects in which it was most suggestive, important, and influential, and to have retained aspects which many have felt were the most objectionable features of his earlier view. Finally, the consistency of what he has retained with his apparent departures from his former view is certainly open to question. And it is far from being unambiguously clear what his current view really is. He seems to want to say that there are paradigm-independent considerations which constitute rational bases for introducing and accepting new paradigms; but his use of the term "reasons" is vitiated by his considering them to be "values," so that he seems not to have gotten beyond his former view after all. He seems to want to say that there is progress in science; but all grounds of assessment again apparently turn out to be "values," and we are left with the same old relativism. And he seems unwilling to abandon "incommensurability," while trying, unsuccessfully, to assert that communication and comparison are possible.

These issues come to a head in Kuhn's proposals as to what must be done if a complete understanding of science is to be obtained, and what the character of that understanding will be once obtained. For the fundamental question is, Do scientists (at least sometimes, even in "revolutionary" episodes) proceed as they do because there are objective reasons for doing so, or do we call those procedures "reasonable" merely because a certain group sanctions them? Despite the ambiguities and inconsistencies of many of his remarks, Kuhn's tendency is clearly toward the latter alternative. Though occasionally tentative ("Some of the principles deployed in my explanation of science are irreducibly sociological, at least at this time" – *Criticism*, p. 237), in most passages he asserts his view categorically: "The explanation [of scientific progress] must, in the final analysis, be psychological or sociological. . . . I doubt that there is another sort of answer to be found" (*Criticism*, p. 21). "Whatever scientific progress may be, we must account for it by examining the nature of the scientific group, discovering what it values, what it tolerates, and what it disdains. That position is intrinsically sociological" (*Criticism*, p. 238). We must study scientific communities not as one of several steps in clarifying the nature of science (in attempting, say, to separate the irrational from the rational components as a prelude to analyzing the latter); it is the *only* step. What the community says is rational, scientific, is so; beyond this, there is no answer to be found. An alternative to this view is to think of sociology as able to bring to our attention the kinds of biases which scientists should learn to avoid, as interferences, hindrances to good scientific judgment. For Kuhn, however, such biases are an integral, and indeed the central, aspect of science. The point I have tried to make is not merely that Kuhn's is a view which denies the objectivity and rationality of the scientific enterprise; I have tried to show that the arguments by which Kuhn arrives at his conclusion are unclear and unsatisfactory.

CHAPTER 5

# MEANING AND SCIENTIFIC CHANGE

## THE REVOLT AGAINST POSITIVISM

In the past decade, a revolution – or at least a rebellion – has occurred in the philosophy of science. Views have been advanced which claim to be radically new not only in their doctrines about science and its evolution and structure, but also in their conceptions of the methods appropriate to solving the problems of the philosophy of science, and even as to what those problems themselves are. It will be the primary purpose of this paper to examine some of the tenets of this revolution, in order to determine what there is in them of permanent value for all people who wish to understand the nature of science.

But before proceeding to this study, it will be worthwhile to examine some of the sources of these new views; and the first thing to do in this regard will be to summarize (at considerable risk of oversimplification) some of the main features of the approach to the philosophy of science against which these new approaches are in part reacting.[1]

The mainstream of philosophy of science during the second quarter of this century – the so-called "logical empiricist" or "logical positivist" movement and related views – was characterized by a heavy reliance on the techniques of mathematical logic for formulating and dealing with its problems. Philosophy of science (and, indeed, philosophy in general) was pronounced to be "the logic of science," this epithet meaning to attribute to the subject a number of important features. First, philosophy of science was to be conceived of on the analogy of formal logic: just as formal logic, ever since Aristotle, has been supposed to be concerned with the "form" rather than with the "content" of propositions and arguments, so also philosophy of science was to deal with the "form" – the "logical form" – of scientific statements rather than with their "content," with, for example, the logical structure of *all possible* statements claiming to be scientific laws, rather than with any particular such statements; with the logical skeleton of *any possible* scientific theory, rather than with particular actual scientific theories; with the logical pattern of any possible scientific explanation, rather than with particular actual scientific explanations; with the logical relations between evidence-statements and theoretical conclusions, rather than with particular scientific arguments. Of course, the philosophical conclusions arrived at were supposed, in principle, to be tested against actual scientific practice, but the actual work of the philosopher of science was with the construction of adequate formal representations of scientific expressions in general, rather than with the details of particular current scientific work (and much less with past scientific work).[2]

Alternatively, the analogy between logic and "the logic of science" can be drawn in another manner which is in some ways even more revealing. Just as modern logicians make a distinction between logic proper – particular systems of logic, formulated in an "object language" – and metalogic, which consists of an analysis of expressions (like "true," "provable," "is a theorem") which are applied to statements and sequences of statements expressed in the object language, so also "the logic of science" can be seen as concerning itself primarily with the analysis of expressions which are applied to actual scientific terms or statements – which are used in talking about science (expressions like "is a law," is meaningful," "is an explanation," "is a theory," "is evidence for," "confirms to a higher degree than").

On the basis of either analogy, some conclusions can be drawn which will be of importance for our later discussion. First, since philosophy of science, so conceived, does not deal with particular scientific theories, it is immune to the vicissitudes of science – the coming and going of particular scientific theories, – for those changes have to do with the content of science, whereas the philosopher of science is concerned with its structure; not with specific mortal theories, but with the characteristics of any possible theory, with the meaning of the word "theory" itself. It also follows that the philosopher of science, insofar as he is successful, will provide us with a *final* analysis of the expressions which he analyzes; in giving us the characteristics of, for instance, all possible explanations, he is *a fortiori* giving us the formal characteristics of all future explanations. It is thus assumed that a revealing account can be given of such terms as "explanation" which will hold true always: although particular scientific explanations may change from theory to theory, nevertheless that which is *essential* to being an explanation – those features of such accounts which make them deserve the title "explanation" – can be laid down once and for all; and furthermore, those essential characteristics can be expressed in purely logical terms, as characteristics of the form or structure of explanation.

Besides conceiving the philosophy of science along the lines of formal logic as a model, the "logical empiricist" tradition also *used* the techniques of modern mathematical logic in approaching their problems. Thus, fatal objections were raised against proposed views because of some flaw in the logical formulation of the position; and such difficulties were to be overcome not by abandoning the safe ground of formulation in terms of the already well-developed mathematical logic, but rather by giving a more satisfactory reformulation in terms of that logic. Again, scientific theories were conceived of as being, or as most easily treated as being, axiomatic (or axiomatizable) systems whose connection with experience was to be achieved by "rules of interpretation," the general characteristics of which could again be stated in formal terms. The conclusions of philosophy of science were therefore supposed to be applicable only to the most highly developed scientific theories, those which had reached a stage of articulation and sophistication which permitted treating them as precisely – and completely – formulated axiomatic systems with precise rules of interpretation. (Whether any scientific theory has ever achieved such a pristine state of completeness, or whether it even makes sense to talk about precision in such an absolute sense in connection with scientific concepts and theories, is questionable.) Hence, an examination of the history of science was considered irrelevant to the philosophy of science. This concentration on perfected (even idealized) systems was part of what was embodied in the slogan, "There is no logic of discovery." Insofar as the development of science was considered at all, it tended to be looked upon as a process of ever-increasing accumulation of knowledge, in which previous facts and theories would be incorporated into (or reduced to) later theories as special cases applicable in limited domains of experience.

All this, in summary, constituted the "logical" aspect of logical empiricism. The "empiricist" aspect consisted in the belief, on the part of those philosophers, that all scientific theory must, in some precise and formally specifiable sense, be grounded in experience, both as to the meanings of terms and the acceptability of assertions. To the end of showing how the meanings of terms were grounded in experience, a distinction was made between "theoretical terms" and "observation terms," and a central part of the program of logical empiricism consisted of the attempt to show how the former kind of terms could be "interpreted" on the basis of the latter. Observation terms were taken to raise no problems regarding their meanings, since they referred directly to experience. As to the acceptability of assertions, the program was to show how scientific hypotheses were related to empirical evidence verifying or falsifying them (or confirming or disconfirming them); and if there were any other factors (such as "simplicity") besides empirical evidence influencing the acceptability of scientific hypotheses, those other factors, if at all possible, should be characterized in formal terms as rigorously as the concept of verification (or confirmation).

The views which have been presented to date within the general logical empiricist framework have not met with unqualified success. Although analyses of meaning, of the difference between theoretical and observation terms and of the interpretation of the former on the basis of the latter, of lawlikeness, of explanation, of acceptability of theories, etc., have been developed in considerable detail, they have all been subjected to serious criticism. Continuing efforts have been made to adjust and extend those analyses to meet the criticisms – and, after all, the logical empiricist programs are not self-contradictory *enterprises*, so that the hope can always be held out that they will yet be carried through to success. But because of the multitude of difficulties that have been exposed, many philosophers think that an entirely new approach to the problems of the philosophy of science is required.[3]

In addition to such criticisms of specific views, however, objections have also been raised against the general logical empiricist approach of trying to solve the problems of the philosophy of science by application of the techniques of, and on analogy with, formal logic. For in its concentration on technical problems of logic, the logical empiricist tradition has tended to lose close contact with science, and the discussions have often been accused of irrelevancy to real science. Even if this criticism is sometimes overstated, there is surely something to it, for in their involvement with logical details (often without more than cursory discussion of any application to science at all), in their claim to be talking only about thoroughly developed scientific theories (if there are any such), and in their failure (or refusal) to attend at all to questions about the historical development of actual science, logical empiricists have certainly laid themselves open to the criticism of being, despite their professed empiricism, too rationalistic in failing to keep an attentive eye on the facts which constitute the subject matter of the philosophy of science.

Such disenchantment with the general mode of approach that has been dominant in the philosophy of science since at least the early days of the Vienna Circle has been reinforced by developments in other quarters. Many proponents of the "rebellion" against logical empiricism have been heavily influenced by the later philosophy of Ludwig Wittgenstein,[4] which was itself partly a reaction against the attempt to deal, through the "ideal language" of logic, with all possible cases. Wittgenstein warned that a great many functions of language can be ignored if language is looked upon simply as a calculus, and philosophers of science have found application for this warning by pointing out functions of, say, scientific laws which could not be noticed by looking at them solely in terms of their logical form.[5]

Other thinkers have been influenced in turning to a new, nonpositivistic approach to philosophy of science by developments in science itself. This is particularly the case with Paul Feyerabend, whose work departs not only from a reaction against contemporary empiricism, but also from his opposition to certain features of the Copenhagen Interpretation of quantum theory.[6] Feyerabend attacks as dogmatic the view of the Copenhagen Interpretation according to which all future developments of microphysical theory will have to maintain certain features of the present theory, or will otherwise fall into formal or empirical inconsistency. He characterizes this view as being opposed to the spirit of true empiricism; but, as we shall see shortly, he finds the same sort of dogmatism inherent in contemporary (and past) versions of empiricism also, particularly in current analyses of the nature of scientific explanation and of the reduction of one scientific theory to another.

But by far the most profound influence shaping the new trends in the philosophy of science has come from results attained by the newly professionalized discipline of the history of science. I have already mentioned that the logical empiricist tradition has tended to ignore the history of science as being irrelevant to the philosophy of science, on the ground that there could be no "logic of discovery," the processes by which scientific discovery and advance are achieved being fit subject matter for the psychologist and the sociologist, but hardly for the logician. I also noted that, insofar as logical empiricists considered the history of science at all, they tended to look on it as largely a record of the gradual removal of superstition, prejudice, and other impediments to scientific progress in the form of an ever-increasing accumulation and synthesis of knowledge – an interpretation of the history of science which Thomas Kuhn has called "the concept of development-by-accumulation."[7] This interpretation, coupled with the logical empiricists' exclusive concern with "completely developed" theories, led them to ignore as unworthy of their attention even the ways in which incomplete theories ultimately eventuated in "completely developed" (or more completely developed) ones. But in the years since the pioneering historical research of Pierre Duhem early in this century, the history of science has come a long way from the days when most writers on the subject were either themselves confirmed positivists or else scientists, ignorant of the details of history, who read the past as a record of great men throwing off the shackles of a dark inheritance and struggling toward modern enlightenment. The subject has developed high standards of scholarship, and much careful investigation has brought out features of science which seemed clearly to conflict with the positivist portrayal of it and its evolution. Many older theories that were supposedly overthrown and superseded – Aristotelian and medieval mechanics, the phlogiston and caloric theories – have been found to contain far more than the simple-minded error and superstition which were all that was attributed to them by earlier, less scholarly and more positivistic historians of science. Indeed, those theories have been alleged to be as deserving of the name "science" as anything else that goes by that name. On the other hand, previous pictures of the work of such men as Galileo and Newton have been found riddled with errors, and the "Galileo-myth" and the "Newton-myth," products of an excessively Baconian and positivistic interpretation, have been mercilessly exposed.[8] Newton made hypotheses after all, and rather alarmingly nonempirical ones at that; and, it is suggested, he had to make them. Galileo, now often demoted to a status little above that of press agent for the scientific revolution, did not base his views on experiments, and even when he performed them (which was more rarely and ineffectively than had previously been supposed), he did not draw conclusions from them, but rather used them to illustrate conclusions at which he had already arrived – ignoring, in the process, any deviations therefrom.

Further, the *kind* of change involved in the history of science has been found (so the story continues) not to be a mere process of accumulation of knowledge, synthesized in more and more encompassing theories. Contemporary historians of science have emphasized again and again that the transition from Aristotelian to seventeenth-century dynamics required not a closer attention to facts (as older histories would have it), but rather, in the words of Herbert Butterfield, "handling the same bundle of data as before, but placing them in a new system of relations with one another by giving them a different framework, all of which virtually means putting on a different kind of thinking-cap."[9] Such words as "virtually" tend to be dropped as deeper and more sweeping conclusions are drawn. The underlying philosophy of the sixteenth- and seventeenth-century scientific revolution has been held to have been strongly infused, not with Baconian empiricism, but rather – irony of ironies! – with Platonic rationalism.[10] Such conclusions have been generalized still further: while experiment plays far less of a role than many philosophers have supposed in the great fundamental scientific revolutions, certain types of presuppositions, not classifiable in any of the usual traditional senses as "empirical," play a crucial role. The most pervasive changes in the history of science are to be characterized, according to these writers, in terms of the abandonment of one set of such presuppositions and their replacement by another. It is no wonder that Thomas Kuhn begins his influential book, *The Structure of Scientific Revolutions*, with the words, "History, if viewed as a repository for more than anecdote or chronology, could produce a decisive transformation in the image of science by which we are now possessed."[11] And it is no wonder, either, that many of the leaders in presenting this new image – Kuhn, Alexandre Koyré – have been historians of science. Nor is it any accident that many philosophers dissatisfied with current logical empiricist approaches to science – Paul Feyerabend,[12] N. R. Hanson,[13] Robert Palter,[14] Stephen Toulmin[15] – have found inspiration for their views in the work of contemporary historians of science, and have even, in some cases, made original contributions to historical research.

The view that, fundamental to scientific investigation and development, there are certain very pervasive sorts of presuppositions, is the chief substantive characteristic of what I have called the new revolution in the philosophy of science (although the authors concerned do not usually use the word "presupposition" to refer to these alleged underlying principles of science). Of course, there have been presupposition analyses of science before, but the present movement (if it can be called that) is different from its predecessors in certain important respects. Any consistent body of propositions, scientific or not, contains "presuppositions" in one sense, namely, in the sense of containing a (really, more than one) subset of propositions which are related to the remainder of the propositions of the set as axioms to theorems. But these new sorts of presuppositions are alleged to be related to scientific methods and assertions not simply (if at all) as axioms to theorems, but in some other, deeper sense which will be discussed in the course of this paper. For most writers, these presuppositions are not what are ordinarily taken to be fundamental scientific laws or theories or to contain the ordinary kind of scientific concepts; they are more fundamental even than that – more "global,"[16] as Kuhn says. Even when they are called "theories," as by Feyerabend, it turns out (as we shall see) that the author does not really mean that word in any usual sense; and even when the author speaks of a certain scientific law as having the character of a fundamental presupposition – as Toulmin describes the law of inertia – he reinterprets that law in an entirely novel way.

Again, in opposition to what might be called a "Kantian" view, the presuppositions are held to vary from one theory or tradition to another; indeed, what distinguishes one theory or tradition from another ultimately is the set of presuppositions underlying them. Hence, although these writers hold that *some* presuppositions always have been made and (at least according to some authors) must always be made, there is no single set which must always be made. In defending these views, as has been suggested above, the authors make extensive appeal to cases from the history of science.

More positively, different writers characterize these "presuppositions," as I have called them, in different ways – but, as we shall see, with much in common, despite significant differences. Koyré speaks of a "philosophic background"[17] influencing the science of a time; Palter, too, speaks of "'philosophic' principles which tend to diversify scientific theories."[18] Toulmin, in *Foresight and Understanding*, calls them "ideals of natural order" or "paradigms," and describes them as "standards of rationality and intelligibility"[19] providing "fundamental patterns of expectation."[20] "We see the world through them to such an extent that we forget what it would look like without them"; [21] they determine what questions we will ask as well as "giving significance to [facts] and even determining what are 'facts' for us at all."[22] Finally, "Our 'ideals of natural order' mark off for us those happenings in the world around us which do require explanation, by contrasting them with 'the natural course of events' – i.e., those events which do not."[23] He suggests that "These ideas and methods, and even the controlling aims of science itself, are continually evolving";[24] and inasmuch as what counts as a problem, a fact, and an explanation (among other things) changes with change of ideal, it follows that we cannot hope to gain an understanding of these basic features of science by merely examining logical form; we must examine the content of particular scientific views. "In studying the development of scientific ideas, we must always look out for the ideas and paradigms men rely on to make Nature intelligible."[25]

Kuhn's *The Structure of Scientific Revolutions* presents a view which is in many respects similar to that of Toulmin. Analyzing the notion of "normal science" as a tradition of workers unified by their acceptance of a common "paradigm," Kuhn contrasts normal science with scientific revolutions: "Scientific revolutions are ... non-cumulative episodes in which an older paradigm is replaced in whole or in part by an incompatible new one."[26] Kuhn considers his paradigms as being not merely rules, laws, theories, or the like, or a mere sum thereof, but something more "global,"[27] from which rules, theories, and the like can be abstracted, but to which no mere statement of rules, theories, and so forth can do justice. A paradigm consists of a "strong network of commitments – conceptual, theoretical, instrumental, and methodological";[28] among these commitments are "quasi-metaphysical"[29] ones. A paradigm is, or at least includes, "some implicit body of intertwined theoretical and methodological belief that permits selection, evaluation, and criticism";[30] it is "the source of the methods, problem-field, and standards of solution accepted by any mature scientific community at any given time."[31] Even what counts as a fact is determined by the paradigm. Because of this pervasive paradigm-dependence, "the reception of a new paradigm often necessitates a redefinition of the corresponding science .. , And as the problems change, so, often, does the standard that distinguishes a real scientific solution from a mere metaphysical speculation, word game, or mathematical play. The normal-scientific tradition that emerges from a scientific revolution is not only incompatible but often actually incommensurable with that which has gone before."[32] Thus, a paradigm entails "changes in the standards governing permissible problems, concepts, and explanations"[33] – changes that are so fundamental that the meanings of the terms used in two different paradigm traditions are "often actually incommensurable," incomparable.

It thus appears that there are at least the following theses held in common by a number of proponents of the "new philosophy of science" (including, as we shall see, Feyerabend):

(a) A *presupposition theory of meaning*: the meanings of all scientific terms, whether "factual" ("observational") or "theoretical," are determined by the theory or paradigm or ideal of natural order which underlies them or in which they are embedded. (This thesis is in opposition to the traditional view of logical empiricism to the effect that there is an absolute, theory-independent distinction between "theoretical terms" and "observation terms," the latter having the same meanings, or at least a core of common meaning, for all (or at least for competing) scientific theories, and against which different theories are judged as to adequacy. It also opposes the attempt to distinguish, in a final manner, "meaningful" ("verifiable," "confirmable," or perhaps "falsifiable") statements from "meaningless" ("metaphysical") ones.

(b) A *presupposition theory of problems* that will define the domain of scientific inquiry, *and of what can count as an explanation* in answer to those problems. (Most obviously, this thesis is directed against the attempt of Hempel and others to give a "deductive-nomological and statistical" analysis of the concept of scientific explanation.)

(c) A *presupposition theory of the relevance of facts to theory, of the degree of relevance* (i.e., of the relative importance of different facts), *and, generally, of the relative acceptability or unacceptability of different scientific conclusions* (laws, theories, predictions). (This thesis is directed primarily against the possibility, or at least the value as an interpretation of actual scientific procedure, of a formal "inductive logic" in Carnap's sense.)

It will be the purpose of this essay to examine critically some aspects of this revolutionary philosophy of science, especially what I have called the "presupposition theory of meaning," although, in later parts of the paper, something will be said also about other facets of these new ideas. I will focus my critical examination on one particular view, that presented by Paul Feyerabend in a number of papers, especially in his "Explanation, Reduction, and Empiricism," "Problems of Microphysics," and "Problems of Empiricism." After discussing his views as presented in those papers, I will consider his recent attempt, in a paper entitled "On the 'Meaning' of Scientific Terms," to clarify his position. At the end of this discussion of Feyerabend's work, I will compare my criticisms of him with criticisms which I have raised previously against Kuhn.[34] This comparison will enable us to see not only some deeply underlying mistakes (or rather excesses) of the "new philosophy of science," but also, through an examination of a case study in the development of science, some of the not insignificant elements of positive value in it.

Feyerabend bases his position on an attack on two principles following from the theory of explanation which is "one of the cornerstones of contemporary philosophical empiricism."[35] These two principles are (1) *the consistency condition*: "Only such theories are ... admissible in a given domain which either *contain* the theories already used in this domain, or which are at least *consistent* with them inside the domain"; (2) *the condition of meaning invariance*: "meanings will have to be invariant with respect to scientific progress; that is, all future theories will have to be framed in such a manner that their use in explanations does not affect what is said by the theories, or factual reports to be explained."[36]

In opposition to these two conditions, Feyerabend argues (1) that scientific theories are, and ought to be, inconsistent with one another, and (2) that "the meaning of every term we use depends upon the theoretical context in which it occurs. Words do not 'mean' something in isolation; they obtain their meanings by being part of a theoretical system."[37] This dependence of meaning on theoretical context extends also to what are classified as "observation terms"; such terms, like any others, depend for their meanings on the theories in which they occur. The meanings of theoretical terms do not depend (as they were alleged to by the logical empiricist tradition) on their being interpreted in terms of an antecedently understood observation-language; on the contrary, Feyerabend's view implies a reversal

> in the relation between theory and observation. The philosophies we have been discussing so far [i.e., versions of empiricism] assumed that observation sentences are meaningful *per se*, that theories which have been separated from observations are not meaningful, and that such theories obtain their interpretation by being connected with some observation language that possesses a stable interpretation. According to the point of view I am advocating, the meaning of observation sentences is determined by the theories with which they are connected. Theories are meaningful independent of observations; observational statements are not meaningful unless they have been connected with theories. . . . It is therefore the *observation sentence* that is in need of interpretation and *not* the theory.[38]

What, then, of the traditional empiricist view that a theory must be tested by confrontation with objective (theory-independent) facts and that one theory is chosen over another because it is more adequate to the facts – facts which are *the same* for both theories? Such factual confrontation, Feyerabend tells us, will not work for the most fundamental scientific theories.

> It is usually assumed that observation and experience play a theoretical role by producing an observation sentence that by virtue of its meaning (which is assumed to be determined by the nature of the observation) may *judge* theories. This assumption works well with theories of a low degree of generality whose principles do not touch the principles on which the ontology of the chosen observation language is based. It works well if the theories are compared with respect to a background theory of greater generality that provides a stable meaning for observation sentences. However, this background theory, like any other theory, is itself in need of criticism.[39]

But the background theory cannot be criticized on its own terms; arguments concerning fundamental points of view are "invariably *circular*. They show what is implied in taking for granted a certain point of view, and do not provide the slightest foothold for a possible criticism."[40] How, then, are such theories to be criticized? The theory-dependence of meanings, together with the fact that each theory specifies its own observation-language, implies, according to Feyerabend, that "each theory will have its own experience."[41] This, however, does not prevent the facts revealed by one theory from being relevant to another theory. This means, in Feyerabend's eyes, that in order to criticize high-level background theories, "We must choose a point outside the system or the language defended in order to get an idea of what a criticism would look like."[42] It is necessary to develop alternative theories.

> Not only is the description of every single fact dependent on *some* theory ... , but there also exist facts that cannot be unearthed except with the help of alternatives to the theory to be tested and that become unavailable as soon as such alternatives are excluded.[43]
>
> Both the relevance and the refuting character of many decisive facts can be established only with the help of other theories that, although factually adequate, are not in agreement with the view to be tested.... Empiricism demands that the empirical content of whatever knowledge we possess be increased as much as possible. Hence, *the invention of alternatives in addition to the view that stands in the center of discussion constitutes an essential part of the empirical method.*[44]

An adequate empiricism itself therefore requires the detailed development of as many different alternative theories as possible, and "This ... is the methodological justification of a plurality of theories."[45]

Since meanings vary with theoretical context, and since the purpose of such theoretical pluralism is to expose facts which, while relevant to the theory under consideration, cannot be expressed in terms of that theory, and would not ordinarily be noticed by upholders of that theory (or speakers of that language), it follows that we cannot be satisfied with alternatives that are "created by arbitrarily denying now this and now that component of the dominant point of view."[46] On the contrary, "Alternatives will be the more efficient the more radically they differ from the point of view to be investigated."[47] In fact, "It is ... better to consider conceptual systems all of whose features deviate from the accepted points of view,"[48] although "failure to achieve this in a single step does not entail failure of our epistemological program."[49] Thus "the progress of knowledge may be by replacement, which leaves no stone unturned, rather than by subsumption.... A scientist or a philosopher must be allowed to start completely from scratch and to redefine completely his domain of investigation."[50]

There are a number of difficulties with these views, both as to interpreting what, exactly, they are supposed to assert, and – when one can arrive at an interpretation – as to whether they are adequately defended or, even if not, whether they are correct.

First, it is not clear whether Feyerabend believes that it is impossible ever to change a theoretical context (to change a theory) without violating the conditions of meaning invariance and consistency – so that the older empiricist viewpoint cannot be correct – or whether, while those conditions *can*, in some cases at least, be satisfied, it is inadvisable or undesirable to do so. On the one hand, we are led to believe that the theory-dependence of meanings is a necessary truth, that since the meaning of *every* term depends on its theoretical context, therefore a change of theory *must* produce a change of meaning of every term in the theory. But on the other hand, we learn that the two conditions *are* "adopted by some scientists":

> The quantum theory seems to be the first theory after the downfall of the Aristotelian physics that has been quite explicitly constructed, at least by some of the inventors, with an eye both on the consistency condition and on the condition of meaning invariance. In this respect it is very different indeed from, say, relativity, which violates both consistency and meaning invariance with respect to earlier theories.[51]

That is to say, the Copenhagen Interpretation of quantum theory, restated by Feyerabend as a "physical hypothesis," holds that the terms "space," "time," "mass," etc., are used by quantum theory in their classical senses; and Feyerabend declares himself "prepared to defend the Copenhagen Interpretation as a physical hypothesis and I am also prepared to admit that it is superior to a host of alternatives."[52] Thus, Feyerabend alleges that this view is evidence for the *possibility* of upholding meaning invariance. If, however, meanings *must* vary with theoretical context, and if – as surely must be admitted under any reasonable interpretation of the expression "difference of theoretical context" – those classical terms occur in a different theoretical context when they occur in quantum-theoretical contexts, then they should have meanings which are *different* from their meanings in classical physics. In short, in Feyerabend's own terms we are hard put to understand his contention, in *Problems of Microphysics*, that the Copenhagen Interpretation (restated as a physical hypothesis), while it is overly dogmatic in barring theories which are inconsistent with it and whose terms differ in meaning from its own, is nevertheless a satisfactory scientific theory.

These difficulties concerning the general thesis of the theory-dependence of meanings have implications for the more specific view that there is no core of observational meaning which is common to all theories and which provides the basis for testing and comparing them. *Can* there be no observational core? Or is it merely *undesirable* to maintain one? Despite the suggestions conveyed by Feyerabend's statements about the relations between theories and meanings, we find that "it is completely up to us to have knowledge by acquaintance and the poverty of content that goes with it or to have hypothetical knowledge, which is corrigible, which can be improved, and which is informative."[53] Again, he tells us that "the ideal of a purely factual theory . . . was first realized by Bohr and his followers . . ."[54] – "factual" because everything in quantum theory, on Bohr's view, is to be expressed in "purely observational" terms, the classical terms "space," "time," "mass," etc. being taken (strangely!) as "purely observational."

Again, although we are told that "the meaning of *every* term we use depends upon the theoretical context in which it occurs"[55] – suggesting that the slightest alteration of theoretical context alters the meaning of every term in that context – Feyerabend introduces, at numerous points, qualifications which appear to contradict this thesis. Thus, "High-level theories ... *may not* share a single observational statement,"[56] although one would suppose that, if they are really different theories, all their terms would be different in meaning, so that it is difficult to see how they *could* share *any* statement. Similar difficulties arise with regard to the qualifications made in such remarks as the following:

> Statements that are empirically adequate and are the result of observation (such as "here is a table") *may* have to be reinterpreted ... because of changes in sometimes very remote parts of the conceptual scheme to which they belong.[57]
>
> ... the methodological unit to which we must refer when discussing questions of test and empirical content is constituted by a *whole set of partly overlapping, factually adequate, but mutually inconsistent theories.*[58]

The root of such difficulties is, of course, the lack of sufficient explanation and detailed defense which Feyerabend offers of his doctrine of the theory-dependence of meanings. We are given no way of deciding either what counts as a part of the "meaning" of a term or what counts as a "change of meaning" of a term. Correspondingly, we are given no way of deciding what counts as a part of a "theory" or what counts as a "change of theory." Hence, it is not clear what we should say when confronted with proposed objections to Feyerabend's analysis. We may be confronted, for example, with cases of theoretical changes which seem too minor to affect the meanings of the expressions concerned (much less terms "far removed" from the area of change): the addition of an epicycle; a change in the value of a constant; a shift from circular to elliptical orbits;[59] the ascription of a new property to some type of entity. Yet such cases might not be accepted by Feyerabend as counting against him; he might consider such changes as not really being changes of theory (perhaps they are only changes *in* theory, but at what point, exactly, do such changes become major enough to constitute changes of theory – i.e., to affect meanings?). Or, alternatively, perhaps he would consider that the mere difference itself *constitutes* a change of meaning of all terms in the theory – so that the doctrine that "meanings change with change of context" becomes a tautology.

It seems sensible to ask whether every change constitutes a change of meaning, but what Feyerabend would say about this is unclear. Much the same must be said about the question of whether every change constitutes a change of theory. What, on Feyerabend's view, is the appropriate reply to objections such as the following: Do mere extensions or applications of a theory make a difference to the "theoretical context," and so to the meanings, of the terms involved? Do alternative axiomatizations constitute different theoretical contexts, so that the meanings of the expressions axiomatized change with reaxiomatization? And do logical terms, like "and" and "if-then," change their meanings under alteration of theory? Presumably, one would want to answer such questions in the negative; but Feyerabend does not deal with such points, and his statements about the relation between meaning changes and changes of theory leave much to be desired. (Remember: "The meaning of every term we use depends upon the theoretical context in which it occurs.")

Further, what counts as part of a theory? Did Kepler's mysticism determine the meanings of the terms used in his laws of planetary motion? And did the meanings of those laws change when they were removed from that context and incorporated in the Newtonian theory? Or to consider a more difficult question: Are Newton's conceptions of "absolute space" and "absolute time" relevant parts of the theoretical context of his mechanical theory, or are they essentially irrelevant? Where does one draw the line? These difficulties might at first appear rather minor; one might want to reply, "But we can at least point to clear examples of theories, and this is all Feyerabend needs to make his point clear enough." This impression disappears, however, and the difficulty takes on crucial importance, when one looks closely at what Feyerabend means in talking about "theories." The usual idea, made familiar to us by logicians, is that a theory is a set of statements formulable in a language, in which language alternatives (e.g., the denial) to the theory can also be expressed. Perhaps this is true for Feyerabend's "lower-level" theories (although this is not clear), but it certainly does not do justice to his conception of higher-level background theories. On the contrary, such theories are *presupposed by* a language, and in terms of that language, alternatives to the background theory are absurd, inconceivable, self-contradictory. A theory is "a way of looking at the world";[60] it is really a philosophical point of view, a metaphysics, although it need not be so precise or well formulated; superstitions also count as theories. Thus, we have the following (the only) explanation of what he means by a "theory":

> In what follows, the term "theory" will be used in a wide sense, including ordinary beliefs (e.g., the belief in the existence of material objects), myths (e.g., the myth of eternal recurrence), religious beliefs, etc. In short, any sufficiently general point of view concerning matter of fact will be termed a "theory."[61]

It is this breadth allowed to what can count as a theory that makes it difficult – even impossible – to say, in cases like that of Kepler's mysticism and Newton's absolutes, whether they are to be considered, on Feyerabend's view, as part of the theoretical context.[62] (Was Kepler perhaps holding two *different*, mutually independent theories in adhering to his laws of planetary motion on the one hand and to his mysticism on the other? But Feyerabend has given us no criterion for distinguishing theories – no "principle of individuation" of theories – and so this possibility is of no help either.)

Still more difficulties arise: how is it possible to reject *both* the consistency condition *and* the condition of meaning invariance? For in order for two sentences to contradict one another (to be inconsistent with one another), one must be the denial of the other; and this is to say that what is denied by the one must be what the other asserts; and this in turn is to say that the theories must have some common meaning. Perhaps Feyerabend has in mind some special sense of "inconsistent" (although he claims not to be abandoning the principle of noncontradiction), or else of "meaning"; but in the absence of any clarification, it is difficult to see how one could construct a theory which, while differing in the meanings of all its terms from another theory, can nevertheless be inconsistent with that other theory. It is no wonder that Feyerabend, like Kuhn, often uses the word "incommensurable" to describe the relations between different background theories.[63]

This brings us to what I believe is the central difficulty in Feyerabend's philosophy of science. He tells us that the most desirable kinds of theories to have are ones which are *completely* different from the theory to be criticized – which "do not share a single statement" with that theory, which "leave no stone unturned." Yet – even if we agree to pass over any feelings of uneasiness we may have about what such an absolute difference would be like – how could two such theories be relevant to one another? How is criticism of a theory possible in terms of facts unearthed by another if meaning depends on, and varies with, theoretical context, and especially if there is *nothing* common to the two theories? Facts, after all, on Feyerabend's view, are not simply "unearthed" by a theory; they are *defined* by it and *do not exist* for another theory. ("Each theory will possess its own experience, and there will be no overlap between those experiences.") Even if two sentences in two different theories are written in the same symbols, they will have different meanings. How, then, can evidence for or against a theory be forthcoming because of another theory which does not even talk the same language – and in a much stronger sense than that in which French and English are different languages, since, for Feyerabend's two radically different high-level theories, presumably, translation – even inaccurate translation – appears to be impossible in principle?

But even if facts unearthed by one high-level theory *could* be relevant to the testing of some other, completely different theory, it is hard to see how such relevant criticism could be effective. For why should it not be possible to reinterpret the fact unearthed by the alternative theory so that it either is no longer relevant or else supports our theory? Feyerabend's own words lend credence to this: "Observational findings can be reinterpreted, and can perhaps even be made to lend support to a point of view that was originally inconsistent with them."[64] And he himself asks the crucial question: "Now if this is the case, does it not follow that an objective and impartial judge of theories does not exist? If observation can be made to favor any theory, then what is the point of making observations?"[65]

How, then, does Feyerabend answer this question? What are "the principles according to which a decision between two different accounts of the external world can be achieved,"[66] when those two accounts are high-level background theories which are so radically different as to leave no stone unturned? He lists three such principles. "The first [procedure] consists in the invention of a still more general theory describing a common background that defines test statements acceptable to *both* theories."[67] But this third theory is still a different theory, and even though it contains a subset of statements which *look* exactly like statements in the two original theories, the meanings of those statements in the new metatheory will still be different from the meanings of the corresponding statements in either of the two original theories. In fact, the meanings will be *radically* different; for any term in the metatheory will have, as part of the theoretical context which determines its meaning, not only the set of statements corresponding to statements in one of the two original theories, but also a set of statements corresponding to statements in the other, radically different original theory. The context of any term in the new metatheory will thus be radically different from the context in which a corresponding term occurred in one of the two original theories, and so its meaning will be radically different. Thus, the same problems arise concerning the possibility of comparing the metatheory with either of the two original theories as arose with regard to the possibility of comparing the two original theories with one another.

"The second procedure is based upon an internal examination of the two theories. The one theory might establish a more direct connection to observation and the interpretation of observational results might also be more direct."[68] I confess that I do not understand this, since each theory defines its own facts or experience, and what could be more direct than this?

Feyerabend's third procedure for choosing between two different high-level theories consists of "taking the pragmatic theory of observation seriously."[69] He describes this theory as follows:

> A statement will be regarded as observational because of the *causal context* in which it is being uttered, and *not* because of what it means. According to this theory, "this is red" is an observation sentence, because a well-conditioned individual who is prompted in the appropriate manner in front of an object that has certain physical properties will respond without hesitation with "this is red"; and this response will occur independently of the *interpretation* he may connect with the statement.[70]

According to the pragmatic theory, then,

> observational statements are distinguished from other statements not by their meaning, but by the circumstances of their production. . . . These circumstances are open to observation and . . . we can therefore determine in a straightforward manner whether a certain movement of the human organism is correlated with an external event and can therefore be regarded as an indicator of this event.[71]

This theory provides, according to Feyerabend, a way of choosing between even radically different high-level background theories:

> It is bound to happen, then, at some stage, that the alternatives do not share a single statement with the theory they criticize. The idea of observation that we are defending here implies that they will not share a single observation statement either. To express it more radically, each theory will possess its own experience, and there will be no overlap between these experiences. Clearly, a crucial experiment is now impossible. It is impossible not because the *experimental device* would be too complex or expensive, but because there is no universally accepted *statement* capable of expressing whatever emerges from observation. *But there is still human experience as an actually existing process*, and it still causes the observer to carry out certain actions, for example, to utter sentences of a certain kind. Not every interpretation of the sentences uttered will be such that the theory furnishing the interpretation predicts it in the form in which it has emerged from the observational situation. Such a combined use of theory and action leads to a selection even in those cases where a common observation language does not exist . . . the theory – an acceptable theory, that is – has an inbuilt syntactical machinery that *imitates* (but does not *describe*) certain features of our experience. This is the *only* way in which experience judges a general cosmological point of view. Such a point of view is not removed because its observation *statements* say that there must be certain experiences that then do not occur. . . . It *is* removed if it produces observation *sentences* when observers produce the *negation* of these sentences. It is therefore

> still judged by the predictions it makes. However, it is not judged by the truth or falsehood of the prediction-statements – this takes place only after the general background has been settled – but by the way in which the prediction sentences are ordered by it and by the agreement or disagreement of this *physical* order with the *natural* order of observation sentences as uttered by human observers, and therefore, in the last resort, with the natural order of sensations.[72]

It turns out that there is, after all, something that is theory-independent and against which we can compare and test theories: it is "human experience as an actually existing process," which causes the well-conditioned observer to utter a sequence of noises (observation sentences). That this takes place can be determined "in a straightforward manner" (i.e., independently of theory); it is only when we assign meanings to the sequence of noises uttered by the observer that we bring in theoretical considerations. The human organism emits results of experiments or experience (in the form of sequences of noises) which must be interpreted in the light of theory, just as other scientific instruments produce pointer-readings which must then be interpreted in the light of theory. Theories are to be compared and judged, not by reference to their meanings (for those are necessarily different) but by reference to the common domain of "features of experience" which they are concerned to "imitate" or "order": the theory, if it is an acceptable one, "has an inbuilt syntactical machinery" which "produces observation sentences"; and the theory is to be "removed" not when "its observation statements say that there must be certain experiences that then do not occur. . . . It is removed if it produces observation sentences when observers produce the negation of these sentences."

We thus have come back to an older empiricism: there is, after all, something common to all theories, in terms of which they can be compared and judged; only, what is objective, independent of theory, given, is not an observation-language but something nonlinguistic; for Feyerabend's observation-sentences, being mere uninterpreted noises, are no more "linguistic" than is a burp. We place an interpretation on this "given" only when we read meanings into those utterances; and to read in meanings is to read in a theory. Hence, in the light of the pragmatic theory of observation, we must give a conservative interpretation to Feyerabend's more radical declarations – e.g., that "the given is out," that each theory "possesses its own experience." The given is indeed still "in," and there is human observation, experience, which is the same for all theories: it is not theory-independent observation, but a theory-independent observation *language* that Feyerabend is set against.[73]

One can wonder, among other things, whether the view that statements made by human beings pop out as conditioned responses, as the word "Ouch!" sometimes pops out when one is stuck with a pin, is not a drastic oversimplification. More important for present purposes is the question as to whether Feyerabend has shown that theories can really be judged against one another despite the theory-dependence of meanings. The answer, it seems to me, is clearly that he has not; nothing has been said by the pragmatic theory of observation to remove the fatal objection of Feyerabend's own words: "Observational findings can be reinterpreted, and can perhaps even be made to lend support to a point of view that was originally inconsistent with them"; and Feyerabend has still given no reason why the qualification "perhaps" is included in this statement. Knowledge by acquaintance – raw, meaningless "human experience" (including uninterpreted "observation statements"), after all, according to Feyerabend, exhibits a complete "poverty of content": such experience tells us nothing whatever; uninterpreted observation statements convey no information whatsoever and, therefore, cannot convey information which would serve as a basis for "removing" a theory. They can do so only when they are assigned meanings and, thereby, are infused with a theoretical interpretation. This "poverty of content," therefore, not only leaves open the possibility of interpretation, but even *requires* that interpretation be made in order to allow judgment of theories. It is not any help to say that theories must at least "imitate" the "order" of experiences ("and ultimately the order of sensations"). For scientific theories often, as a matter of fact, *alter* that order rather than imitate it; and in many cases, some of the elements of experience are declared irrelevant. So "interpretation," rather than "imitation," takes place even with regard to the alleged "order" of experience or sensations. And with the liberty – no, rather with the license – which Feyerabend grants us for interpreting experience, for assigning meanings to observation statements, we must conclude that, with regard either to single "experiences" (or observation statements) or allegedly ordered sets of them, anything goes: we are always able to interpret experience so that it supports, rather than refutes, our theory. The truth of the matter is thus that Feyerabend's kind of experience is altogether *too weak*, in its pristine, uninterpreted form, to serve as grounds for "removal" of any theory; and his view of meaning is *too strong* to preclude the possibility of *any* interpretation whatever of what is given in experience.[74]

I have confined the above remarks to those sorts of high-level background theories which "leave no stone unturned." One might suppose that the situation is less serious with less radically different theories. There, at least, there are some similarities, and perhaps relevance can be established and comparison made of the two theories on the basis of those similarities. One might suppose, for example, that a slight amendment of Feyerabend's position, introducing the notion of degrees of likeness of meaning, might answer the question of how theories all of whose terms must differ in meaning can yet, in some cases at least, be mutually relevant, since the relevance could be established through the likenesses, despite the differences. This view would also remove our difficulties, discussed above, with Feyerabend's description of some background theories as, e.g., "partially overlapping." *Prima facie*, this seems a promising move to make, although the notion of "degrees of likeness of meaning" may well introduce complications of its own; and in any case, making this move, as will become clear in what follows, would be tantamount to confessing that Feyerabend's technical notion of "meaning" is an unnecessary obstruction to the understanding of science. In any case, however, it is not a move that Feyerabend himself makes.[75] We have seen that he admits only three ways of comparing and judging two high-level theories: by constructing a metatheory, by examining the relative "directness" of their connection with experience, or via their common domain of experience. Different high-level theories, even those which are "partially overlapping," are apparently not comparable in spite of their similarities; Feyerabend's general tendency is to look on the similarities as rather unimportant, superficial, inessential. And this is only what we would expect if likeness and difference of meaning are not a matter of degree, for if difference of meaning makes *all* the difference, then any two theories must be incommensurable, incomparable, despite any (superficial, inessential) similarities. Thus, all our dire conclusions regarding theories which have nothing in common are extended even to theories which do not turn every stone.

We are thus left with a complete relativism with regard not only to the testing of any single theory by confrontation with facts, but also to the relevance of other theories to the testing of that theory. Feyerabend's attempts "to formulate a methodology that can still claim to be *empirical*,"[76] as well as his efforts to justify a "methodological pluralism," have ended in failure.

In a recent short paper, "On the 'Meaning' of Scientific Terms," Feyerabend has attempted to reply to some criticisms of his views which had been raised by Achinstein and which are similar to some of the questions raised above concerning the interpretation of Feyerabend's views on meaning variance and the dependence of meaning on theoretical context. In that paper, Feyerabend admits that certain changes, although they count as changes of theory, do not involve a change of meaning. He cites as an example a case of two theories, $T$ (classical celestial mechanics) and $\bar{T}$ (like classical celestial mechanics except for a slight change in the strength of the gravitational potential). $T$ and $\bar{T}$, he declares,

are certainly different theories – in our universe, where no region is free from gravitational influence, no two predictions of $T$ and $\bar{T}$ will coincide. Yet it would be rash to say that the transition $T \rightarrow \bar{T}$ involves a change of meaning. For though the *quantitative values* of the forces differ almost everywhere, there is no reason to assert that this is due to the action of different *kinds of entities.*[77]

It thus appears that Feyerabend wants to say that two theories are different theories if they assign different quantitative values to the factors involved ("almost everywhere"); and the meanings of terms involved are different if they have to do with different kinds of entities. He makes his notion of "change of meaning" (and, conversely, of "stability of meaning") explicit in the following passage:

A diagnosis of *stability of meaning* involves two elements. First, reference is made to rules according to which objects or events are collected into classes. We may say that such rules determine concepts or kinds of objects. Secondly, it is found that the changes brought about by a new point of view occur *within* the extension of these classes and, therefore, leave the concepts unchanged. Conversely, we shall diagnose a *change of meaning* either if a new theory entails that all concepts of the preceding theory have extension zero or if it introduces rules which cannot be interpreted as attributing specific properties to objects within already existing classes, but which change the system of classes itself.[78]

At first glance, this discussion does seem to introduce some clarification, although at the price of adopting what seems an unreasonably extreme notion of "difference of theory" (after all, a slight refinement in the value of a fundamental constant will lead to widespread differences in quantitative predictions, and so, on Feyerabend's criterion, to a new, "different" theory). However, closer inspection reveals that the improvement achieved is by no means substantial. Consider the analysis of "change of meaning" (and, correlatively, of "stability of meaning"). This analysis depends on the notion of being able to collect "entities" ("objects or events") into classes, and this in turn rests on being able to refer to "rules" for so collecting them. If the changes occur only within the extensions of these classes ("kinds of entities," "objects or events"), the meanings have not changed; if the new theory changes the whole system of classes (or "entails that all concepts of the preceding theory have extension zero"), the meanings have changed.

However, first, in order to apply this criterion, the rules of classification must be unique and determinate, allowing an unambiguous classification of the "entities" involved. Otherwise, we might not be able to determine whether the system of classes, or merely the extension of the previous classes, has changed. Furthermore, there may be two different sets of rules and consequent systems of classification, according to one of which a change of meaning has taken place, while the other implies that the meaning has not changed. Indeed, this would seem to be generally the case: one can, in scientific as in ordinary usage, collect entities into classes in a great variety of ways, and on the basis of a great variety of considerations ("rules"); and which way of classifying we use depends largely on our purposes and not simply on intrinsic properties of the entities involved by means of which we are supposed to fit them unambiguously into classes. Are mesons different "kinds of entities" from electrons and protons, or are they simply a different subclass of elementary particles? Are the light rays of classical mechanics and of general relativity (two theories which Feyerabend claims are "incommensurable") different "kinds of entities" or not? Such questions can be answered *either* way, depending on the kind of information that is being requested (this is to say that the questions, as they stand, are not clear), for there are differences as well as similarities between electrons and mesons, as between light rays in classical mechanics and light rays in general relativity. They can be given a simple answer ("different" or "the same") only if unwanted similarities or differences are stipulated away as inessential. And even if we agree to Feyerabend's (rather arbitrary) decision "not to pay attention to any *prima facie* similarities that might arise at the observational level, but to base our judgment [as to whether change or stability of meaning has occurred] on the principles of the theory only,"[79] the spatiotemporal frameworks of classical mechanics and general relativity are still comparable with respect to their possession of certain kinds of mathematical properties – metrical and topological ones (both theories have something to do with "spaces" in a well-defined mathematical sense). And the question must still arise – and is equally useless and answerable only by stipulation – as to whether the spatiotemporal frameworks involved share the same *kinds* of properties and are the same *kinds* of entities ("spaces") or whether these properties are not "specific"[80] enough to count toward making those frameworks the same "kinds of entities."[81]

Under any interpretation, it is hard to see how *any* theory would entail that *all* the concepts of a rival theory have extension zero or would change the whole system of classes.[82] Even theories having to do with very different subjects – e.g., geological theories of the structure and evolution of the earth on the one hand and physical theories of waves and their transmission on the other – have something in common. (Theories of the structure and evolution of the earth in fact depend intimately on the ways in which earthquake waves are transmitted through different kinds of material.) Of course, one *can* say, in examples like this, that the physical theory is part of the "borrowed background" of the geological theory rather than being *part of* the geological theory. But this again simply throws us back to the question, asked earlier in regard to Feyerabend's views, of what is and what is not supposed to be included in a "theory."

## MEANINGS AND THE ANALYSIS OF SCIENCE

We have seen that Feyerabend's interpretation of science eventuates in a complete relativism, in which it becomes impossible, as a consequence of his views, to compare any two scientific theories and to choose between them on any but the most subjective grounds. In particular, his "pragmatic theory of observation," which constitutes his main effort to avoid this disastrous conclusion, does not succeed in doing so for, inasmuch as all meanings are theory-dependent, and inasmuch as theories can be shaped at will, and inasmuch, finally, as all observational data (in his sense) can be reinterpreted to support any given theoretical framework, it follows that the role of experience and experiment in science becomes a farce. In trying to assure freedom of theorizing, Feyerabend has made theory-construction too free; in depriving observation statements of any meaning whatever (independent of theories), he has deprived them also of any power of judgment over theories: they must be interpreted by reading meaning into them, and thus reading theory into them; and we are at liberty to interpret them as we will – as irrelevant, or as supporting evidence. By granting unlimited power of interpretation, on the one hand, over that which allows limitless possibilities of interpretation on the other, Feyerabend has destroyed the possibility of comparing and judging theories by reference to experience. And by holding that all meanings vary with theoretical context, and by implying that a difference of meaning is *a fortiori* a complete difference, an "incommensurability," he has destroyed the possibility of comparing them on any other grounds either.

In the first section of this chapter, I called attention to the very great similarities between Feyerabend's views and those of a number of other recent writers whom I grouped together, on the basis of those similarities, as representatives of a new approach to the philosophy of science. Among those writers is Thomas Kuhn. There are differences, of course, between Kuhn's views and those of Feyerabend. For example, while Feyerabend insists on the desirability of developing a large number of mutually inconsistent alternative theories at all stages of the history of science, Kuhn claims that, both as a matter of desirability and as a matter of fact through most of its actual development, science is "normal," in the sense that there is one dominant point of view or "paradigm" held in common by all the members of the tradition; it is only on the very exceptional and rare occasions of scientific revolutions that we find the development of competing alternatives. However, it is not in the differences, but rather in the similarities between their views that I am interested here.

In view of these similarities, it is only to be expected that Kuhn's and Feyerabend's interpretations of science may be open to many of the same objections. This is indeed the case. In an earlier paper reviewing Kuhn's book, *The Structure of Scientific Revolutions*, I made a number of criticisms of his views which are in fact remarkably like those which I tried to bring out in connection with Feyerabend.[83] Kuhn's notion of a "paradigm," like Feyerabend's notion of a "theory," becomes so broad and general in the course of his discussion that we are often at a loss to know what to include under it and what to exclude. Again, neither author gives us a criterion for determining what counts as a part of the meaning of a term, or what counts as a change of meaning, even though these notions are central to their portrayals of science. They share other criticisms as well; most important for present purposes, however, is the fact (which I tried to establish for Kuhn in my review of his book, and for Feyerabend in this chapter) that both views result in relativism: the most fundamental sorts of scientific change are really complete replacements; the most fundamental scientific differences are really utter incompatibilities. It will be instructive for us to compare the source of this relativism in these two writers, because the trouble, as I think could be shown, is shared by a large number of current writers representative of what I have called "the new philosophy of science," and is, I think, the major pitfall facing that view.

What are the grounds, in Kuhn's view, for accepting one paradigm as better, more acceptable, than another? He manages without difficulty to analyze the notion of progress within a paradigm tradition – i.e., within normal science. There, "progress" consists of further articulation and specification of the tradition's paradigm "under new or more stringent conditions."[84] The trouble comes when we ask how we can say that "progress" is made when one paradigm is replaced, through a scientific revolution, by another. For according to Kuhn, "the differences between successive paradigms are both necessary and irreconcilable";[85] those differences consist in the paradigms' being "incommensurable": they disagree as to what the facts are, and even as to the real problems to be faced and the standards which a successful theory must meet. A paradigm change entails "changes in the standards governing permissible problems, concepts, and explanations";[86] what is metaphysics for one paradigm tradition is science for another, and *vice versa*. It follows that the decisions of a scientific group to adopt a new paradigm cannot be based on good reasons of any kind, factual or otherwise; quite the contrary, what counts as a good reason is determined by the decision. Despite the presence in Kuhn's book of qualifications to this extreme relativism (although, as in Feyerabend, these qualifications really only contradict his main view), the logical tendency of his position is clearly toward the conclusion that the replacement of one paradigm by another is not cumulative, but is mere change: being "incommensurable," two paradigms cannot be judged according to their ability to solve the same problems, deal with the same facts, or meet the same standards. For problems, facts, and standards are all defined by the paradigm, and are different – *radically*, incommensurably different – for different paradigms.

How similar this is to the logical path that leads to relativism in the case of Feyerabend! It is, in fact, fundamentally the same path: meanings, whether of factual or of any other sorts of terms, are theory-(paradigm-) dependent and, therefore, are different for different theories (paradigms); for two sets of meanings to be different is for them to be "incommensurable"; if two theories (paradigms) are incommensurable, they cannot be compared directly with one another. Neither Kuhn nor Feyerabend succeeds in providing any extra-theoretical basis (theory-independent problems, standards, experiences) on the basis of which theories (paradigms) can be compared or judged indirectly. Hence, there remains *no* basis for choosing between them. Choice must be made without any basis, arbitrarily.

When their reasoning (and the objections thereto) is summarized in this way, it becomes obvious that the root of Kuhn's and Feyerabend's relativism, and of the difficulties which lead to it, lies in their rigid conception of what a difference of meaning amounts to – namely, absolute incomparability, "incommensurability." Two expressions or sets of expressions must either have precisely the same meaning or else must be utterly and completely different. If theories are not meaning-invariant over the history of their development and incorporation into wider or deeper theories, then those successive theories (paradigms) cannot *really* be compared at all, despite apparent similarities which must therefore be dismissed as irrelevant and superficial. If the concept of the history of science as a process of "development-by-accumulation" is incorrect, the only alternative is that it must be a completely noncumulative process of replacement. There is never any middle ground and, therefore, it should be no surprise that the rejection of the positivistic principles of meaning invariance and of development-by-accumulation leave us in a relativistic bind, for that is the only other possibility left open by this concept of difference of meaning. But this relativism, and the doctrines which eventuate in it, is not the result of an investigation of actual science and its history; rather, it is the purely logical consequence of a narrow preconception about what "meaning" is. Nor should anyone be surprised that the root of the trouble, although not easy to discern until after a long analysis, should turn out to be such a simple point, for philosophical difficulties are often of just this sort.

Having, then, found the place where Kuhn and Feyerabend took a wrong turn and ended by giving us a complete relativism with regard to the development of science, can we provide a middle ground by altering their rigid notion of meaning? For example, can we say that meanings can be similar, comparable in some respects even while also being different in other respects? For by taking this path, we could hope to preserve the fact that, e.g., Newtonian and relativistic dynamics *are* comparable – something Feyerabend and Kuhn deny – even while being more fundamentally different than the most usual logical empiricist views make them. Thus, we could hope, by this expedient, to avoid the excesses *both* of the positivistic view of the development of science as a process of development-by-accumulation (and systematization), characterized by meaning invariance, *and* of the view of the "new philosophy of science" that different theories, at least different fundamental theories (paradigms), are "incommensurable."

Whether this is a wise path to take depends on how we interpret this new concept of degrees of likeness (or difference) of meanings. For if we still insist on some distinction between what, in the use of a term, is and what is not a part of the meaning of the term, then we expose ourselves to the danger of relegating some features of the use of a term to the "less important" status of not being "part of the meaning." Yet those very features, for some purposes, may prove to be the very ones that are of central importance in comparing two uses, for relative importance of features of usage must not be enshrined in an absolute and a priori distinction between essential and inessential features. It thus seems wiser to allow *all* features of the use of a term to be equally potentially relevant in comparing the usage of the terms in different contexts. But this step relieves the notion of meaning of any importance whatever as a tool for analyzing the relations between different scientific "theories." If our purpose is to compare the uses of two terms (or of the same term in different contexts), and if *any* of their similarities and differences are at least potentially relevant in bringing out crucial relations between the uses – the actual relevance and importance being determined by the problem at hand rather than by some intrinsic feature of the uses (their being or not being "part of the meaning") – then what is the use of referring to those similarities and differences as similarities and differences of "meaning" *at all*? Once more, introducing the term "meaning," and even admitting degrees of meaning, suggests that there may be similarities and differences which are not "part of the meaning" of the terms, and this in turn might suggest that those features are, in some intrinsic, essential, or absolute sense less important than features which *are* "parts of the meaning." For the purpose of seeking out central features of scientific theories, and of comparing different theories, then, it seems unnecessary to talk about meanings, and on the other hand, that notion is potentially misleading. Worse still, we have already seen how that notion, which is made so fundamental in the work of Feyerabend and Kuhn, has *actually* been an obstruction, misleading those authors into a relativistic impasse.

All this is not to say that we *cannot*, or even that we *ought not*, use the term "meaning," even often if we like – so long as we do not allow ourselves to be misled by it, as Kuhn and Feyerabend were misled by it, or as we are liable to be misled by talk about "degrees of likeness of meaning." Nor is it to say that we could not formulate a precise criterion of meaning, which would distinguish between what is, and what is not, to count as part of the meaning, and which would also serve to specify what would count as a change of meaning. Nor is it to say that for some purposes it might not be very valuable to formulate such a precise criterion. All that has been said is that, *if* our purpose is to understand the workings of scientific concepts and theories, and the relations between different scientific concepts and theories – if, for example, our aim is to understand such terms as "space," "time," and "mass" (or their symbolic correlates) in classical and relativistic mechanics, and the relations between those terms as used in those different theories – then there is *no need* to introduce reference to meanings. And in view of the fact that that term *has* proved such an obstruction to the fulfillment of this purpose, the wisest course seems to be to avoid it altogether as a fundamental tool for dealing with this sort of problem.

Both the thesis of the theory-dependence of meanings (or, as I called it earlier – more accurately, as we have seen – the presupposition theory of meaning), and its opponent, the condition of meaning invariance, rest on the same kind of mistake (or excess). This does not mean that there is not considerable truth (as well as distortion) in both theses. There are, for example, as I have argued elsewhere,[87] statements that can be made, questions that can be raised, views that may be suggested as possibly correct, within the context of Einsteinian physics that would not even have made sense – would have been self-contradictory – in the context of Newtonian physics. And such differences, both naturally and, for many purposes, profitably, can be referred to as changes of meaning, indicating, among other things, that there are differences between Einsteinian and Newtonian terms that are not brought out by the deduction of Newtonianlike statements from Einsteinian ones. But attributing such differences to alterations of "meaning" must not blind one – as it has blinded Kuhn and Feyerabend – to any resemblances there might be between the two sets of terms.

It is one of the fundamental theses of Kuhn's view of science that it is impossible to describe adequately in words any paradigm; the paradigm, "the concrete scientific achievement" that is the source of the coherence of a scientific tradition, must not be identified with, but must be seen as "prior to the various concepts, laws, theories, and points of view that may be abstracted from it."[88] Yet why, simply because there are differences between views or formulations of views held by members of what historians classify as a "tradition" of science, *must* there be a single, inexpressible view held in common by all members of that tradition? No doubt some theories are very similar – so similar that they can be considered to be "versions" or "different articulations" of one another (or of "the same subject"). But this does not imply, as Kuhn seems to believe, that there must be a common paradigm of which the similar theories are incomplete and imperfect expressions and from which they are abstracted. There need not be, unifying a scientific "tradition," a single inexpressible paradigm which guides procedures, any more than our inability to give a single, simple definition of "game" means that we must have a unitary but inexpressible idea from which all our diverse uses of "game" are abstracted. It would appear that Kuhn's view that, in order for us to be able to speak of a "scientific tradition," there must be a single point of view held in common by all members of that tradition, has its source once again in the error of supposing that, unless there is absolute identity, there must be absolute difference. Where there is similarity, there must be identity, even though it may be hidden; otherwise, there would be only complete difference. If there are scientific traditions, they must have an identical element – a paradigm – which unifies that tradition. And since there are differences of formulation of the various laws, theories, rules, etc. making up that tradition, the paradigm which unifies them all must be inexpressible. Since what is visible exhibits differences, what unites those things must be invisible.

Again, then, Kuhn has committed the mistake of thinking that there are only two alternatives: absolute identity or absolute difference. But the data at hand are the similarities and differences; and why should these not be enough to enable us to talk about more, and less, similar views and, for certain purposes, to classify sufficiently similar viewpoints together as, e.g., being in the same tradition? After all, disagreements, proliferation of competing alternatives, debate over fundamentals, both substantive and methodological, are all more or less present throughout the development of science; and there are always guiding elements which are more or less common, even among what are classified as different "traditions." By hardening the notion of a "scientific tradition" into a hidden unit, Kuhn is thus forced *by a purely conceptual point* to ignore many important differences between scientific activities classified as being of the same tradition, as well as important continuities between successive traditions. This is the same type of excess into which Feyerabend forced himself through his conception of "theory" and "meaning." Everything that is of positive value in the viewpoint of these writers, and much that is excluded by the logic of their errors, can be kept if we take account of these points.

## IMPETUS AND INERTIAL DYNAMICS: A CASE STUDY

In Part I of this paper, mention was made of the influence of the history of science on recent philosophical interpretations of science. Of all the historical episodes contributing to such philosophical interpretation, none has had greater impact than the Impetus Theory, first brought to the attention of scholars by Pierre Duhem. This theory was advanced in the fourteenth century (and even earlier, as we shall see) in order to meet certain objections to Aristotle's account of projectile motion. Inasmuch as Feyerabend appeals to the Impetus Theory as an example of a theory which, despite appearances to the contrary, is "incommensurable" with Newtonian conceptions of inertia and momentum, and also inasmuch as the Impetus Theory has by now been rather (but by no means completely) well documented and discussed, it provides an excellent case study to illustrate the comparability of scientific concepts and the continuity of their evolution from one "tradition" to another.[89]

Aristotelian physics distinguished sharply between celestial motion – the type of motion characteristic of the heavenly spheres – and terrestrial motion. All celestial motion was held to be circular, the spheres revolving around the center of the universe, carrying the moon, planets, sun, and stars around with them; such motion in a perfect circle was the "natural motion" of the element ether, of which the incorruptible heavenly spheres were made. Sublunar (or terrestrial) motion, however, could be either "natural" or "violent." The four elements – earth, water, air, and fire – tended "by nature" to move rectilinearly, along a radius of the universe, toward a "natural place" in the sublunar domain, and to come to rest upon reaching that natural place. On the other hand, bodies composed of (mixtures of) those elements could also be moved "violently." It is in connection with this latter kind of motion that the focal weakness of Aristotelian physics arose and out of the criticisms of which the inertial conception ultimately evolved. However, the infection of this weakness spread also, as we shall see, to the cases of natural sublunar and of celestial motion; but for the present, we shall for the sake of simplicity ignore these cases.

The basic principle of Aristotelian dynamics, accepted throughout the Middle Ages, was that a force is required to maintain a body in motion: *omne quod movetur ab alio movetur* (everything that moves is moved by something else). Furthermore, this principle was usually (although by no means always) interpreted in a way that can be summarized, in modern quantitative terms, as saying that the greater the force, the greater the velocity. (That is, force is directly proportional to velocity; also, it was conceived of as being inversely proportional to the total resistance offered, e.g., by the medium, this latter proportionality being closely intertwined with the Aristotelian conception of the impossibility of a void. This quantitative formula can be said to hold only if the resistance is less than the force, for otherwise there would be no motion at all. It is also not certain, as Dijksterhuis points out,[90] whether Aristotle would have accepted the conclusion following from this formula that a body a hundred times as heavy as another must take one hundredth the time to fall a given distance.)

For the case of natural sublunar motions – the case which ultimately came to center around the problem of freely falling bodies – the character of this force was a matter of considerable debate; we shall, as was said above, pass over this problem. For the case of violent motion – the motion of projectiles – the force must in any case be external, and that external force must be in contact with the body throughout the motion. Upon removal of the force, the body will (if it is in its natural place) come to rest immediately or (if it is not in its natural place) immediately assume its natural motion. (Aristotelian principles did not allow the combined action of two different forces to produce a "resultant" motion.)

This view, however, faces some rather obvious difficulties, particularly as to why projectiles in fact do continue to move "violently" after being released by the propelling agent. Aristotle himself mentioned two ways of accounting for such continuation of violent motion, in both of which the air serves as the continually operating contact mover. According to the first, the theory of *antiperistasis*, a "mutual replacement" takes place between the projectile and the air, the air pushed out in front of the path by the moving projectile pushing other air out of place and, finally, air rushing in behind the projectile to avoid the creation of a vacuum; this air pushes the projectile on. Aristotle seems to have rejected this view, apparently preferring a second view, according to which the air moved, along with the projectile, by the original mover receives, because of its special nature, the power to act as a mover; this power, communicated in decreasing amounts to the air beyond, continues to carry the projectile along by contact action until the power or force is exhausted.

Already in the sixth century, the Byzantine philosopher John Philoponus rejected both theories on the basis of both rational and empirical considerations; he preferred a view according to which "some incorporeal motive force is imparted by the projector to the projectile, and ... the air set in motion contributes either nothing at all or else very little to this motion of the projectile."[91] This view, which anticipates the fourteenth-century Impetus Theory, influenced a number of Arabic writers, notably Avicenna and his followers (theory of *mail*). However, we shall consider here only the Impetus Theory as presented by the Fourteenth Century Parisian, Jean Buridan. After rejecting, on the basis of empirical evidence (*experientie*) both of the accounts of the continuation of projectile motion discussed by Aristotle, Buridan declares that:

> It seems to me that it ought to be said that the motor in moving a moving body impresses in it a certain impetus or a certain motive force of the moving body [which impetus acts] in the direction toward which the mover was moving the moving body, either up or down, or laterally, or circularly. And by the amount the motor moves that moving body more swiftly, by the same amount it will impress in it a stronger impetus. It is by that impetus that the stone is moved after the projector ceases to move. But that impetus is continually decreased by the resisting air and by the gravity of the stone,

which inclines it in a direction contrary to that in which the impetus was naturally predisposed to move it . . . by the amount more there is of matter, by that amount can the body receive more of that impetus and more intensely. . . .

From this theory also appears the cause of why the natural motion of a heavy body downward is continually accelerated. For from the beginning only the gravity was moving it. Therefore, it moved more slowly, but in moving it impressed in the heavy body an impetus. This impetus now [acting] together with its gravity moves it. Therefore, the motion becomes faster; and by the amount it is faster, so the impetus becomes more intense. Therefore, the movement evidently becomes continually faster.

. . . impetus is a thing of permanent nature, distinct from the local motion in which the projectile is moved. . . . And it also is probable that just as that quality (the impetus) is impressed in the moving body along with the motion by the motor; so with the motion it is remitted, corrupted, or impeded by resistance or a contrary inclination.[92]

A quantitative measure of impetus is at least suggested in this passage: it appears to be equal to the product of the quantity of matter and the velocity of the moved body. Notice also that the impetus is not self-corrupting (although it was according to many adherents of the Impetus Theory); it (and consequently the velocity of the body it pushes) is reduced by contrary or resisting agencies. Furthermore, Buridan extends his impetus account of motion from the case of projectile motion to that of freely falling bodies and thus to natural motions in the sublunar domain. He also proposes an extension of it to the natural motion of the celestial spheres:

Since the Bible does not state that appropriate intelligences move the celestial bodies, it could be said that it does not appear necessary to posit intelligences of this kind, because it would be answered that God, when He created the world, moved each of the celestial orbs as He pleased, and in moving them He impressed in them impetuses which moved them without his having to move them any more except by the method of general influence whereby He concurs as a co-agent in all things which take place. . . . And these impetuses which He impressed in the celestial bodies were not decreased nor corrupted afterwards, because there was no inclination of the celestial bodies for other movements. Nor was there resistance which would be corruptive or repressive of that impetus.[93]

Immediately upon unearthing the Impetus Theory, Duhem and others noticed the striking parallels between that theory and the principle of inertia, and also between that theory and the concept of momentum – equals mass ("quantity of matter") times velocity – in classical mechanics. Since Buridan's impetus would be preserved except for the degenerative effects of counter-agencies, only two main steps appeared necessary to achieve the concept of inertia.[94] First, the conception of the possibility of motion in a vacuum and, second, the abandonment of the idea of impetus as an internal force distinct from the body and maintaining it at constant velocity. The Impetus Theory itself contributed to the first of these steps by showing the medium to be unnecessary for the continuation of motion. The second could be seen as the product of a gradual transition, aided perhaps by increasing nominalistic tendencies. With those alterations, the Impetus Theory could be seen as transformed into the principle of inertia, that "Every body continues in its state of rest, or of uniform motion in a right [i.e., straight] line, unless it is compelled to change that state by forces impressed on it." The shift from Aristotelian to Newtonian physics could then be seen as a transition from the view that force is the cause of *motion* (i.e., of velocity) to the modern view that force is the cause of *change of motion* (i.e., of acceleration). The Impetus Theory thus stands as a transitional phase between these two traditions – still in the tradition of Aristotelian physics in its view that impetus is a cause of motion, but heading toward the modern view in making impetus an internal and incorporeal force rather than an external, corporeal one.

Historical evidence, although somewhat sketchy, was also available for the influence of the Impetus Theory on the development of the seventeenth-century scientific revolution. Galileo's early work was definitely in the tradition of the Impetus Theory, and he seems never to have escaped completely from the view that a constant force is required to maintain a body in motion. (Newton, too, was not perfectly clear in his conception of inertia; he sometimes spoke of a *vis inertiae* – not to be confused with the modern conception of "inertial force" – which he conceived of as an internal force maintaining the body in its state of uniform motion.) And a steady, although not always progressive, evolution seemed traceable from Buridan and his equally brilliant successor, Nicole Orêsme, to Newton, although complete details are still lacking. Among the major steps along this road were Galileo's view that bodies, once in "horizontal" motion, would continue to move at a uniform rate along a great circle on the surface of the earth and Descartes' conception that true inertial motion would be realized were it not for the fact that space is filled with (or identical with) incompressible matter, thus requiring all motion to be circular after all. Less famous, but equally revealing of the growing tendency to think in modern terms, were such developments as Benedetti's restriction of impetus to having rectilinear effects and the clear abandonment by Galileo's pupil, Baliani, of the notion of impetus as a cause of motion and his view that motion continues of its own accord.

It is not hard to see that the Impetus Theory (in the form in which it was presented by Buridan, at lease) made it possible to think in certain ways that had previously been precluded or at least discouraged by the Aristotelian system. (1) By removing the need for air as the agency responsible for continued motion, it helped make motion in a vacuum conceivable, and thus helped pave the way for thinking about the idealized case of a body moving in the absence of impeding forces. (2) It shifted attention away from the forces causing motion and toward those impeding motion. (3) It made plausible the notion of the composition of forces. (4) It was headed in the direction of concentration on quantitative features of motion. (It is noteworthy in this connection that Orêsme was one of the first, and perhaps the first, person to use graphing methods for the representation of intensive qualities.) (5) By treating celestial and terrestrial motion, and natural and violent motion, in terms of a single kind of cause (impetus), it helped lead toward a unified account of all motion.[95] (6) It provided counterarguments to two of Aristotle's proofs that the earth could not rotate and thus helped pave the way for Copernicus. Aristotle had held that, if the earth rotated, a terrific wind would be set up in the direction opposite to the rotation. The Impetus Theory's answer was that the earth communicates an impetus to the air, which impetus carries the air along with the rotating earth. Aristotle had further argued that, if the earth rotated from west to east, then an object dropped from a height would be left behind by the earth moving out from under it, so that the object would hit the ground at a point due west of the point directly underneath the dropping hand. The answer in terms of the Impetus Theory was as follows: the motion of the falling object is a composite of a downward motion caused by gravity, and an eastward-directed motion due to the impetus conveyed to the body by the dropping hand, which is being carried around by the rotating earth. The composition of the two forces will make the body hit the ground at the point directly beneath the dropping hand despite the rotation of the earth. (Although there is no definite evidence that Copernicus was acquainted with these arguments of the Impetus Theory, he may well have been exposed to them either at the University of Cracow or the University of Padua where he had studied and where the Impetus Theory was taught.)

Despite all this evidence of similarities and continuities between medieval and classical concepts, there has in recent years been a reaction against the claim that the concept of impetus is related to the concepts of inertia and momentum. Significant differences have been pointed out between these concepts,[96] so that some writers have been led to claim that impetus and classical mechanics are not nearly so closely alike as earlier writers had supposed and perhaps are entirely different. Feyerabend follows this reaction in claiming that the two theories are "incommensurable," what happened having been a "complete replacement . . . and a corresponding change in the meanings of all descriptive terms"[97] involved.

Feyerabend admits, quite correctly,[98] that the "inertial law" of the Impetus Theory (which he formulates as, "The impetus of a body in empty space which is not under the influence of any outer force remains constant"[99]) is not in quantitative disagreement with anything asserted by Newton's mechanics: "It is correct that the measure of [momentum] is identical with the measure that has been suggested for the impetus."[100] But such quantitative agreement, as usual, is not enough to satisfy Feyerabend, or even to make any difference at all, as far as he is concerned, to the comparability of the two concepts.

> It would be very mistaken if we were, on that account, to identify impetus and momentum. For whereas the impetus is supposed to be something that pushes the body along, the momentum is the result rather than the cause of its motion. Moreover, the inertial motion of classical mechanics is a motion which is supposed to occur by itself, and without the influence of any causes.[101]

But the point made here concerning the difference between impetus and momentum – that one is a cause, the other a result, of motion – seems slim grounds indeed for concluding that the two concepts have nothing at all in common. To put through an argument of this sort, one must first establish that it is relevant to discussing the meaning of a term whether the reference is the cause or the result of something else. (Does putting the cart before the horse change the meanings of "cart" and "horse"?) And, having established that, one must further show that any difference in meaning implies complete difference, "incommensurability," of the concepts. We have seen, however, that Feyerabend gives neither a clear criterion of what is supposed to be part of the meaning of a term, nor a convincing argument that differences of meaning make all the difference to the comparability of concepts.

Similar remarks apply to Feyerabend's alleged differences between impetus and inertia. The two theories differ with regard to what is considered to be the "natural state" of a body. For the Impetus Theory (and, indeed, this constitutes its primary affinity with the Aristotelian tradition), the natural state is the state of rest (in the natural place of the body). In such a place, no force is acting; it is only for bodies in motion that a force is required to maintain the motion. For the Newtonian view, on the other hand (according to Feyerabend), "it is the state of being at rest or in uniform motion which is regarded as the natural state,"[102] and thus force is the cause of deviation from a state of rest *or uniform motion* rather than simply from rest. But in the first place, as was pointed out earlier in this part of the present paper, Newton himself is not perfectly clear as to whether inertial motion requires a cause (*vis inertiae*), and so Feyerabend's discussion of "Newton's physics" is misleading on this point. It is true that the inertial conception became clearer in succeeding years and that the notion of a cause of inertial motion is unnecessary to classical mechanics; but Feyerabend's notion of theory gives us no way of distinguishing, here or elsewhere, between what, in Newton's thought, was necessary to his scientific theory and what was not. And furthermore, if we can ask whether the notion of an internal cause of inertial motion is necessary to Newton's physics, why can we not ask also the same question about the necessity of the concept of impetus as a cause of (uniform rectilinear) motion?

A more balanced appraisal can be outlined as follows. The Aristotelian-Scholastic system, through an intricately connected web of concepts and propositions, laid down certain patterns of thought, principles, for example, in terms of which certain sorts of questions could be raised and answered (e.g., that what needed to be explained was, in general, motion; and that motion was to be explained in terms of contact action only – a principle, incidentally, which it shared with later and otherwise far different "traditions"). And in terms of such principles, certain things appeared to be self-contradictory (vacuum, actual infinite) or physically impossible (rotation of the earth). In laying down such principles of investigation, it simultaneously set up obstacles, or limitations, both by theoretical argument and by suggestion, to thinking in certain other ways. In this sense, we can speak of the Aristotelian view as having involved certain "presuppositions" specifying (for example) what could and what could not count as an explanation. To this extent, Kuhn and Feyerabend have made an important point. But these "presuppositions" were not mysterious, invisible, behind-the-scenes "paradigms" (Kuhn) or "high-level background theories" (Feyerabend), but were involved in the straightforward scientific statements themselves, even though there were disagreements about details (and even about fundamentals), and even though the ways in which they restricted thought, or the importance of those restrictions, could not be seen so easily. (However, these things *were* seen by some people – Philoponus, Buridan, Orêsme, among others – who were taught in that tradition.)

The Impetus Theory, even though it did not go all the way toward thinking outside the Aristotelian limits, did open the door to doing so. This is not to say that the Impetus Theory was a *new* theory, distinct in all respects from its Greek and Scholastic predecessors. On the contrary, it constituted, in its inception, only an adjustment in one seemingly minor area of the total Aristotelian system, the account of projectile motion. It remained within the Aristotelian tradition in holding that without a continually acting force, there can be no continually proceeding motion. But to say all this is not to deny, either, its affinities with later scientific concepts: we can see clearly, from the above discussion, that there are a large number of such resemblances and continuities. Impetus is not yet inertia, nor is it momentum, but it is, in the ways described above, a visible move away from fundamental Aristotelian conceptions (e.g., it allows the sustaining force in projectile motion to be internal and incorporeal) and in the definite direction of classical mechanics. In our fear of falling into "precursoritis" – the mistake of supposing that all great achievements in science were anticipated long before – we must not fall into the opposite error of thinking that every great achievement in science is a "complete replacement" (in Feyerabend's sense) of the old by the new, a thoroughgoing "revolution" (in Kuhn's sense) – that it was not led up to by a succession of developments that can be described as reasonable.

Yet, as such thinkers as Feyerabend, Koyré, and Butterfield have pointed out, that reasonableness did not consist merely in the discovery of new facts, or even – merely – in a closer attention to facts already known. It is true, as Feyerabend notes, that the Impetus Theory can give the same quantitative results as the Newtonian. On the other hand, it must not be concluded from this (as Koyré and others have concluded) that the seventeenth-century scientific revolution was therefore "nonempirical," "rationalistic," or "Platonic": the distinction between "empirical" and "nonempirical" has not been clarified sufficiently by philosophers for either horn of the distinction to serve as an adequate tool for the interpretation of scientific change. And still less should the conclusion be drawn that the choice between impetus and inertial dynamics must be (and must have been in the sixteenth and seventeenth centuries) arbitrary (or a matter of convention). In a sense that still remains to be analyzed satisfactorily by philosophers of science, Galileo's inclined plane experiments were relevant as "factual evidence" for the modern inertial concept. The greater "simplicity" or "economy" of the latter view must be taken into account also.

No doubt many of the characteristics outlined here of the transition from Aristotelian to classical physics via the Impetus Theory are found also in later "scientific revolutions."[103] One must not generalize too hastily, however:

it was not too long ago that "medieval" was synonymous with "unscientific," and there may well be very important differences between the case considered here and later "revolutions," e.g., the transition from classical to relativistic dynamics. Conclusions about such matters should be drawn only in the light of careful examination of actual cases: as Dijksterhuis has commented, "The History of Science forms not only the memory of science, but also its epistemological laboratory."[104] There are two possible ways of interpreting this remark: the history of science (and we must include its product, contemporary science) can be looked on as a laboratory either in the sense that an understanding of science can come through an investigation of the logic of its development or in the sense that conceptions of the aims and methods of science may well have been forged and evolved in the practice thereof, just as has the content of science. We can profit by keeping in mind both of these possible interpretations.

In approaching the investigation of science, however, we must beware of analytical tools which, whether they have been employed throughout a long philosophical tradition or in less formal discussions of science, have never been adequately clarified. Such are the distinctions between what is and what is not part of the meaning of a term, and between stability and change of meaning, which have been examined in this essay. Such also are distinctions – some of which have been touched on here – between "meaningful" and "meaningless," "scientific" and "unscientific," and "empirical" and "nonempirical" ("metaphysical"). An adequate vocabulary for talking about science must, of course, be developed if understanding of science is to be achieved. But the vocabulary we adopt must not be laid down in advance of detailed examination of cases, as a set of logical categories which the cases must fit; this was a fault too often with philosophies of science in the past. Nor must our terms be employed without adequate analysis, and in such a way that those fundamental tools themselves determine the outcome of our investigation, as we have found to have been the case with some more recent attempts to interpret science.

## NOTES

[1] What follows is not meant to be a description of views to all of which any one thinker necessarily adheres, but rather a distillation of points of view which are widespread. Perhaps the writers who come closest to the characterizations given herein are Rudolf Carnap and Carl Hempel, at least in some of their works, although even they might not accept all the doctrines outlined here. The summary does, however, seem to me to

represent trends in a great many writings on such subjects as the verifiability theory of meaning, explanation, lawlikeness, counterfactual conditionals, theoretical and observational terms, induction, correspondence rules, etc. Conversely, many writers whose work fits, at least to some degree, the descriptions given here might object to the label "logical empiricist."

[2] There were, of course, some notable exceptions to this account – the work of Carnap and Reichenbach on relativity and quantum theory, for example.

[3] There have been a number of varieties of efforts to develop new approaches: conspicuous among them, in addition to the views to be discussed in this essay, are the work of Nelson Goodman (*Fact, Fiction, and Forecast*, Cambridge: Harvard University Press, 1955), and of those philosophers who have attempted to develop new sorts of logics ("modal," for example), in the hope that they will prove more suitable for dealing with philosophical problems.

[4] L. Wittgenstein, *Philosophical Investigations*, trans. G. E. M. Anscombe (New York: Macmillan, 1953).

[5] But Paul Feyerabend, whose views will be discussed in this essay, has not been particularly influenced by this approach and has opposed himself to some of its main features. But he is against overconcentration on formalisms; for example, he says, "Interesting ideas may . . . be invisible to those who are concerned with the relation between existing formalisms and 'experience' only." ("On the 'Meaning' of Scientific Terms," *J. of Philosophy* 62 (1965), 268.

[6] P. Feyerabend, "Problems of Microphysics," *Frontiers of Science and Philosophy*, ed. R. Colodny (Pittsburgh: U. of Pittsburgh Press, 1962), pp. 189–283.

[7] T. S. Kuhn, *The Structure of Scientific Revolutions* (Chicago: U. of Chicago Press, 1962), p. 2.

[8] For example, see E. J. Dijksterhuis, *The Mechanization of the World Picture* (Oxford: Clarendon Press, 1961), Pt. IV, Chaps. 2, sec. C and 3, sec. L.

[9] H. Butterfield, *The Origins of Modern Science* (New York: Macmillan, 1958), p. 1.

[10] See, for example, E. A. Burtt, *The Metaphysical Foundations of Modern Physical Science* (New York: Harcourt, Brace, 1925); A Koyré, "Galileo and Plato," *J. of the History of Ideas* 4 (1943), 400–28, reprinted in *Roots of Scientific Thought*, eds. P. P. Wiener and A. Noland (New York: Basic Books, 1957), pp. 147–75; A. R. Hall, *From Galileo to Newton* (New York: Harper & Row, 1963). For criticism of the view that the scientific revolution (and Galileo's philosophy of science in particular) was "Platonic," see L. Geymonat, *Galileo Galilei* (New York: McGraw-Hill, 1965); T. McTighe, "Was Galileo a Platonist?" (to appear in the proceedings of the 1964 Notre Dame Conference on Galileo, edited by E. McMullin and published by the U. of Notre Dame Press); and D. Shapere, "Descartes and Plato," *J. of the History of Ideas* 24 (1963), 572–76 (specifically concerned with Koyré's views).

[11] Kuhn, *Scientific Revolutions*, p. 1.

[12] See especially his "Explanation, Reduction, and Empiricism," in *Minnesota Studies in the Philosophy of Science*, eds. H. Feigl and G. Maxwell, *Scientific Explanation, Space, and Time* (Minneapolis: U. of Minnesota Press, 1962), III, 28–97; and "Problems of Empiricism," in *Beyond the Edge of Certainty*, ed. R. Colodny (Englewood Cliffs, N.J.: Prentice-Hall, 1965), pp. 145–260. Feyerabend's views will be discussed in detail later in this essay.

[13] N. R. Hanson, *Patterns of Discovery* (Cambridge: Cambridge U. Press, 1958); *The*

*Concept of the Positron* (Cambridge: Cambridge U. Press, 1963). Hanson has also expressed his views in a large number of articles.

[14] See especially his article, "Philosophic Principles and Scientific Theory," *Philosophy of Science* **23** (1956), 111–35.

[15] Especially in his *The Philosophy of Science* (New York: Hutchinson, 1953), and *Foresight and Understanding* (Bloomington: Indiana U. Press, 1961), and in his article, "Criticism in the History of Science: Newton on Absolute Space, Time, and Motion," *Philosophical Review* **68** (1959), 1–29, 203–27. For a critical examination of this article, see my "Mathematical Ideas and Metaphysical Concepts," *Philosophical Review* **69** (1960), 376–85.

[16] Kuhn, *Scientific Revolutions*, p. 43.

[17] A. Koyré, "Influence of Philosophic Trends on the Formulation of Scientific Theories," *The Validation of Scientific Theories*, ed. P. Frank (Boston: Beacon, 1954), p. 192.

[18] Palter, "Philosophic Principles and Scientific Theory," *Philosophy of Science* **23** (1956), p. 116.

[19] Toulmin, *Foresight and Understanding*, p. 56.

[20] *Ibid.*, p. 47.

[21] *Ibid.*, p. 101.

[22] *Ibid.*, p. 95.

[23] *Ibid.*, p. 79.

[24] *Ibid.*, p. 109.

[25] *Ibid.*, p. 81.

[26] Kuhn, *Scientific Revolutions*, p. 91.

[27] *Ibid.*, p. 93.

[28] *Ibid.*, p. 42.

[29] *Ibid.*, p. 41.

[30] *Ibid.*, pp. 16–17.

[31] *Ibid.*, p. 102.

[32] *Ibid.*, p. 102.

[33] *Ibid.*, p. 105.

[34] D. Shapere, "The Structure of Scientific Revolutions," *Philosophical Review* **73** (1964), pp. 383–94.

[35] Feyerabend, "Problems of Empiricism," p. 163.

[36] *Ibid.*, p. 164.

[37] *Ibid.*, p. 180.

[38] *Ibid.*, p. 213.

[39] *Ibid.*, p. 214.

[40] *Ibid.*, p. 150.

[41] *Ibid.*, p. 214.

[42] *Ibid.*, p. 151.

[43] *Ibid.*, p. 175.

[44] *Ibid.*, p. 176.

[45] *Ibid.*, p. 150.

[46] *Ibid.*, p. 149.

[47] *Ibid.*, p. 214.

[48] *Ibid.*, p. 254, n. 150.

49 *Ibid.*, p. 254, n. 150.
50 *Ibid.*, p. 199.
51 *Ibid.*, p. 167.
52 Feyerabend, "Problems of Microphysics," p. 201.
53 Feyerabend, "Problems of Empiricism," p. 259, n. 163.
54 *Ibid.*, p. 162.
55 *Ibid.*, p. 180.
56 *Ibid.*, p. 216; italics mine.
57 *Ibid.*, p. 180; italics mine.
58 *Ibid.*, p. 175.
59 An objection of this sort is raised by P. Achinstein, "On the Meaning of Scientific Terms," *J. of Philosophy* **61** (1964), 497–509. Feyerabend's "On the 'Meaning' of Scientific Terms," discussed later, is a reply to Achinstein's paper.
60 Feyerabend, "Explanation, Reduction, and Empiricism," p. 29.
61 Feyerabend, "Problems of Empiricism," p. 219, n. 3.
62 Thus, as the positions of the new approach and the older logical empiricist movement are reversed with respect to the relations between theory and observation, so also are their difficulties. For logical empiricism, observation terms were basic and it was "theoretical terms" that had to be interpreted; and many of the difficulties of that movement have revolved around the question of what counts as an "observation term." For the "new philosophy of science," on the other hand, which takes the notion of "theory" (or, for other writers than Feyerabend, some corresponding notion like "paradigm") as basic, difficulties arise concerning what counts as a theory.
63 In a footnote to "Problems of Empiricism," Feyerabend gives a definition of "incommensurable": "Two theories will be called incommensurable when the meanings of their main descriptive terms depend on mutually inconsistent principles" (p. 277, n. 19). In what language are these "principles" themselves formulated? Presumably (as we have seen), in order for them to be inconsistent with one another, they must be formulated, or at least formulable, in a common language. But if they are formulable in a common language, then in what way are the "main descriptive terms" of the theories "dependent" on them in such a way that *those* terms are not even translatable into one another? The characterization of incommensurability given in "On the 'Meaning' of Scientific Terms" does not seem to differ from that given in "Problems of Empiricism," and so does not help answer these objections.
64 Feyerabend, "Problems of Empiricism," p. 202.
65 *Ibid.*, p. 202.
66 *Ibid.*, p. 216.
67 *Ibid.*, pp. 216–17.
68 *Ibid.*, p. 217.
69 *Ibid.*, p. 217.
70 *Ibid.*, p. 198.
71 *Ibid.*, p. 212.
72 *Ibid.*, pp. 214–15.
73 One is tempted now to go back and say that Feyerabend's references to the "overlapping" of theories are just slips of the pen – that it is not *theories* that, strictly speaking, have any overlap by virtue of which they can be compared, but only their domain of experiences. If this is a proper reinterpretation of Feyerabend's position, it

only serves to emphasize how very radically (and peculiarly) he conceives of difference of meaning, as constituting a *complete* "incommensurability." This reinterpretation will in any case, however, not help Feyerabend, for reasons to be explained below; "experiences," in his sense, cannot provide a basis for comparison ("overlap") either.

[74] Further possible questions about this facet of Feyerabend's philosophy of science appear – as we might expect – as revivals of old problems about traditional phenomenalism: whether, for example, it is possible to observe "in a straightforward manner," without any importation of theoretical presuppositions, the "causal context" in which a statement is uttered; whether the "order" which is to be "imitated" by the theory does not itself presuppose an interpretation of experience; and whether the judgment that a certain theory is successful in imitating experience is itself a product of interpretation – whether, that is, we cannot still, despite the pragmatic theory, so interpret our experience that it will always support our theory.

[75] Perhaps one reason Feyerabend would object to making likeness of meaning a matter of degree is that, if relevance were to be established in terms of similarities, the conclusion might be drawn that two theories are more relevant to the testing of one another the more similar they are; and this would contradict his deep-rooted view that a theory is more relevant to the testing of another theory the more different the two are.

[76] Feyerabend, "Problems of Empiricism," p. 149.

[77] Feyerabend, "On the 'Meaning' of Scientific Terms," p. 267.

[78] *Ibid.*, p. 268. Feyerabend's criterion of change of meaning has some consequences that seem paradoxical, to say that least. If a new theory entails that *one* concept of the preceding theory has extension zero, apparently no meaning change has taken place. If *all but one* of the classes of the preceding theory have extension zero, again no meaning change has taken place. And if the extensions of all classes are changed radically, but not so much that the previous extensions are zero, again no meaning change has taken place.

[79] Feyerabend, "On the 'Meaning' of Scientific Terms," p. 270.

[80] *Ibid.*, p. 268.

[81] Feyerabend admits that his criteria require supplementation: "It is important to realize that these two criteria lead to unambiguous results only if some further decisions are first made. Theories can be subjected to a variety of interpretations . . ." (p. 268). But his ensuing discussion does nothing to take care of the difficulties raised here.

[82] In any case, it is unclear how a new theory can "entail" that concepts of another theory have extension zero if the latter concepts do not even occur in the new theory.

[83] See Note 34, above. The discussion of Kuhn's view which follows is based on that paper.

[84] Kuhn, *Scientific Revolutions*, p. 23.

[85] *Ibid.*, p. 102.

[86] *Ibid.*, p. 105.

[87] "The Structure of Scientific Revolutions." See also my "The Causal Efficacy of Space," *Philosophy of Science* **31** (1964), 111–12.

[88] Kuhn, *Scientific Revolutions*, p. 11.

[89] Documents and commentary concerning the material to be discussed in what follows are found in M. Clagett, *The Science of Mechanics in the Middle Ages* (Madison: U. of Wisconsin Press, 1959). Excellent discussions are also found in Butterfield, *The Origins*

*of Modern Science*; Dijksterhuis, *The Mechanization of the World Picture*; A. R. Hall, *The Scientific Revolution* (Boston: Beacon, 1954); Kuhn, *The Copernican Revolution* (Cambridge: Harvard U. Press, 1957).

90 E. J. Dijksterhuis, "The Origins of Classical Mechanics," *Critical Problems in the History of Science*, ed. M. Clagett (Madison: U. of Wisconsin Press, 1959), p. 167.

91 Quoted in *The Science of Mechanics in the Middle Ages*, p. 509.

92 J. Buridan, "Questions on the Eight Books of the Physics of Aristotle," in *The Science of Mechanics in the Middle Ages*, pp. 534–37.

93 *Ibid.*, p. 536.

94 Other steps were necessary in addition to these two main ones – e.g., the notion of circular impetus had to be abandoned in favor of rectilinear. It is noteworthy that Buridan's example of a case of circular impetus, a rotating wheel, is repeated by Newton.

95 The tendency of the theory was not, however, always in the direction of unification: not only did the distinction between different kinds of impetus (circular, rectilinear) obstruct the unity of the impetus concepts, but also, in at least one thinker (Thomas Bricot), impetus suffered the same sad fate as was dealt in the later Middle Ages to the doctrine of substantial forms: every different kind of initial propellant produced a different kind of impetus.

96 Chiefly as a result of the critical work of Anneliese Maier.

97 Feyerabend, "Explanation, Reduction, and Empiricism," p. 59.

98 Correctly, that is, if Buridan really would have agreed to this quantitative measure of impetus; but this is not completely certain. *Cf.* Dijksterhuis, *The Mechanization of the World Picture*, pp. 182–83.

99 Feyerabend, "Explanation, Reduction, and Empiricism," p. 54. This statement holds only for the nonself-degenerative kind of impetus accepted by Buridan.

100 Feyerabend, "Explanation, Reduction, and Empiricism," p. 56.

101 *Ibid.*

102 *Ibid.*

103 Compare the discussion of Aristotelian physics and the Impetus Theory given here with the discussion of Newtonian mechanics in my "The Causal Efficacy of Space."

104 Dijksterhuis, "The Origins of Classical Mechanics," p. 182.

CHAPTER 6

# NOTES TOWARD A POST-POSITIVISTIC INTERPRETATION OF SCIENCE, PART I

## POSITIVISM AND ITS OPPONENTS: A DIAGNOSIS

### 1. *Philosophical Functions of the "Theoretical-Observational" Distinction*

The distinction between "impressions" and "ideas," formulated during the development of British empiricism in the seventeenth and eighteenth centuries, was introduced primarily to fulfill two fundamental purposes. For, in maintaining that all ideas are based on impressions, the distinction embodied the views that (1) all meaningful concepts (terms) obtain their meanings from experience and (2) all meaningful propositions (statements) are to be judged true or false, acceptable or unacceptable, by reference to experience. Thus the distinction proposed—or, more accurately, promised—to make possible the solution of two crucially important philosophical problems: the problem of the *meaning of terms* (concepts) and the problem of the *acceptability of statements* (propositions).[1] Corresponding to these two positive aims of the distinction were the negative or critical ones of the elimination of meaningless or metaphysical terms and statements, and of false or unacceptable meaningful statements: for the precise way in which "ideas" were supposed to be based on "impressions" was held to provide, at the same time, criteria for the rejection of all those ideas or alleged ideas which are not so grounded.

As a lineal descendant of this empiricist distinction, formulated specifically with reference to the interpretation of scientific meaning and knowledge, the positivistic distinction between "observation" and "theory" has, with appropriate alterations, inherited

*I am indebted for discussions leading to many of the ideas in this paper to the students in my graduate seminar at the University of Chicago. I also wish to thank Professors Kenneth Schaffner and Sylvain Bromberger for valuable help on many points.*

the functions of its predecessor. True, the tendency in this century has been for the newer distinction to be stated (as above) as being concerned with terms and statements, thus avoiding, hopefully, the misleading psychological overtones of the word "ideas" (or even "concepts" and "propositions"). And, correlatively with this non-psychological approach, the newer distinction has been stated in terms of the definability or reducibility of theoretical terms, and of the justification for accepting or rejecting a theoretical statement, rather than, with Hume, in terms of the origin of such terms and statements. In addition, views of the exact manner in which theoretical terms and statements are supposed to be "based on" observation have also deviated in radical ways from what the classical empiricists probably would have wanted to say, even had they formulated their theses non-psychologically. Nevertheless, the newer, scientifically oriented distinction continues to serve the purposes of characterizing relationships between meaningful (or scientific) terms and true or acceptable statements on the one hand, and experience on the other, while at the same time providing criteria for the elimination of those terms and statements which do not satisfy such relationships.

But, in twentieth-century philosophy of science, the distinction between theory and observation has been utilized to serve a further purpose which was not served by the more general classical distinction between ideas and impressions. For, in its concern with theories rather than with ideas, the modern distinction has had to deal with *rival sets* of terms and statements in a way that Hume, for instance, did not. Hume did not explicitly discuss competing sets of ideas organized into the kinds of systems which in science are called theories. And it is here that philosophy of science in the twentieth century has added a new dimension to the traditional discussions (although, as we shall see, the newness and importance of this dimension have only recently become apparent and pressing).

For twentieth-century philosophy of science, and particularly the positivistic tradition, has utilized the theoretical-observational distinction not only for the purpose of analyzing the meaning and acceptability of *single* theories. It has also applied the distinction to the analysis of the basis on which *different* theories may be said to be in competition, one being chosen as more adequate than its competitors. The fact that two different theories "deal with (at least some of) the same observations" ("overlap in their observational vocabularies") provides a basis for comparing the *meanings* of the terms and statements of the two theories as well as the *relative acceptability* of the two theories.

Nevertheless, the problem of the comparability of theories has tended to play a subsidiary role in twentieth-century philosophy of science—that is, as will soon become clear, until quite recently. For it has usually been tacitly assumed (until recently) that, because the meanings of the theoretical terms of any one theory are at least partially determined by the observation terms with which they are correlated, and because there is a common pool of observation (observation terms) from which different theories can draw (i.e., with which their theoretical terms can be correlated), it follows as a matter of course that two theories may be compared with regard to their meanings insofar as they contain or are correlated with the same observational vocabulary. And one theory would be judged more acceptable than another if, for example, its degree of confirmation were higher than that of the other. In the case of meanings and of acceptability, then, the observational vocabulary provided an objective ("theory-independent") basis for interpreting and judging single theories; and the solution of any problems concerning the comparison of different theories would follow as a matter of course once the exact details of the basis for interpreting and judging single theories were laid out.

But in the last several years the problem of comparability (with respect to both meaning and acceptability) has increased in seriousness until it would not be an exaggeration to say that it is now one of the central problems in the philosophy of science, around which the treatment of a large number of other problems revolves. This shift of emphasis has been due primarily to two factors: (1) the failure of successive efforts to clarify the theoretical-observational distinction and to make it serve the purposes for which it was introduced; and (2) problems raised by the presentation of radical alternatives to that distinction in its positivistic forms. I will discuss these two factors briefly in turn.

2. *Difficulties of the Theoretical-Observational Distinction*

Attempts to solve the problems of meaning and acceptability by use of the distinction between theoretical and observational terms and statements have encountered severe difficulties. The distinction itself has proved resistant to clear and ultimately helpful formulation. Interpretations of the notions of "observation" and "theory" have proved highly suspect, and unobjectionable analyses of the meaning relations that are supposed to hold between observation terms and theoretical terms, and of the relations that are supposed to hold between evidence statements (observation statements) and theoretical ones, have not been forthcoming. Efforts to modify the basic approach—for instance, by considering the distinction to be a matter of *degree* rather than a sharp difference of *kind*, or by considering it to be a difference between kinds of *uses of terms* rather than a difference between kinds of *terms*, although such approaches have perhaps not yet had full opportunity for development and critical scrutiny—do not seem promising, for these types of approaches still require the very criteria for distinguishing observation and theory that they have sought to avoid.

The literature on these topics is well known, and the objections are familiar; and in any case the present essay will be concerned in general with very different kinds of objections and with an attitude toward the distinction which is different from that embodied in the usual objections. However, there is one by now familiar objection that must be reviewed briefly here, inasmuch as it marks out a transition to a view of science which stands in radical opposition to that of the empiricist-positivist tradition and has, both by its own freshness and its own failures, helped to bring about the shift of emphasis (noted above) in the problems of the philosophy of science.

The objection in question holds that, with regard to their meanings, the "observation terms" that serve as the basis for the scientist's work are *not*—as the empiricist-positivist tradition generally has tended to view them—completely free from "theory." What the scientist considers to be language appropriate for the presentation of empirical evidence seems not to be anything like the neutral "observational vocabulary" of the philosophers —not like the "red patch-here-now" of the sense datum or phenomenalistic analysis of observation, nor like the pointer readings of the operationalists, or even like the ordinary tables and chairs referred to by an everyday thing language. Far from it: according to this criticism, not only is the *relevance* of observations at least partly dependent on theory; even *what counts* as an observation, and the *interpretation* or *meaning* of observation terms, is at least partly so dependent. All "observation terms" in science are, in this view, at least to some extent "theory dependent" or "theory laden" in a sense which is passed over by the usual ways of making the distinction. Data are not "raw"; there are no "brute facts."

### 3. *The Approach of Feyerabend and Kuhn*

On the basis of this point, as well as of the other difficulties, mentioned earlier, in formulating the distinction between "observation" and "theory" so that it will do the jobs set for it, a number of writers have concluded that the meanings of observation terms are not dependent merely in part on their theoretical contexts but are *wholly* so dependent. "The meaning of every term we use," Paul Feyerabend declares, "depends upon the theoretical context in which it occurs. Words do not 'mean' something in isolation; they obtain their meanings by being part of a theoretical system." [2] In particular, according to Feyerabend, this holds for so-called observation terms.

> The philosophies we have been discussing so far [i.e., versions of empiricism] assumed that observation sentences are meaningful *per se,* that theories which have been separated from observations are not meaningful, and that such theories obtain their interpretation by being

connected with some observation language that possesses a stable interpretation. According to the point of view I am advocating, the meaning of observation sentences is determined by the theories with which they are connected. Theories are meaningful independent of observations; observational statements are not meaningful unless they have been connected with theories. . . . It is therefore the *observation sentence* that is in need of interpretation and *not* the theory.[3]

Advocates of this view do not stop with the claim that observation terms are theory dependent; the background point of view (theory, "paradigm" [4]) also is, according to Thomas Kuhn, "the source of the methods, problem-field, and standards of solution accepted by any mature scientific community at any given time. . . . And as the problems change [with change of fundamental point of view or "paradigm"], so, often, does the standard that distinguishes a real scientific solution from a mere metaphysical speculation, word game, or mathematical play." [5] Such basic shifts of viewpoint entail "changes in the standards governing permissible problems, concepts, and explanations";[6] and after such a change, "the whole network of fact and theory . . . has shifted." [7] What counts as a "fact"—the meanings of observation terms—is different, "incommensurable," from one theory (or at least one fundamental theory or paradigm) to another, and so is what counts as a real problem, a correct method, a possible explanation, an acceptable explanation, and nonsense or metaphysics.

Indicating ways in which the meanings of "observation terms" depend on a background of theory, the position advanced by these writers has certainly brought out forcefully the inadequacies of usual formulations of the distinction between theory and observation. Their view has also suggested that the analysis of many other supposedly "metascientific" concepts, such as "explanation," cannot be divorced completely from consideration of substantive developments in science. But in their extreme view that all such concepts are completely theory dependent they have made different fundamental theories "incommensurable," incomparable, and have failed to account for the fact that different theories—or different usages of the same terms or symbols—do in many cases exhibit a continuity in the development of science. Again, it is not necessary to repeat in detail the multitude of criticisms that have been leveled against the Feyerabend-Kuhn approach; for our purposes here, it is enough to note that, for all the value and suggestiveness of those writers, their views have not been formulated in a way which resolves the major problems of contemporary philosophy of science but only makes them more glaring.

### 4. *The Problem of Comparability*

In arguing for the successively stronger theses (*a*) that there is no observational vocabulary which serves as the common basis for comparison of theories, (*b*) that there is no separable component of the scientist's evidence talk which is common to all theories or even to more than one (fundamental) theory ("each theory will have its own experience," [8] according to Feyerabend), and (*c*) that what counts as an "observation term" and the meanings of such terms are wholly dependent on theory rather than vice versa, the Feyerabend-Kuhn approach has moved the problem of the comparability of theories from the periphery to the foreground of contemporary philosophy of science. The traditional empiricist-positivist approach, it will be remembered, accounted for the comparability of theories by maintaining that two rival theories, despite their differences, can be compared because they both talk about (deal with, try to take account of, explain, organize, systematize) at least some of the same observations. That view, which had long suffered from grave objections anyway, now faces the most serious challenge. But a satisfactory answer is not necessarily to go to the other extreme and hold, with Feyerabend and Kuhn, that science is not "objective," that it does not make real progress, that it is always relative to a background framework which itself is purely arbitrary and immune to rational criticism.

Thus the problem of comparability assumes far greater importance than before. Other problems, hitherto largely ignored, are also brought into prominence by this refocusing of the philosophy of science. For example, the positivistic tradition has made a sharp distinction between "scientific" terms—terms occurring within science, like "space," "time," "mass," "electron"—and "metascientific" expressions—terms or expressions used in talking about science, like "is a theory," "is a law," "is evidence for," "is an explanation." And, for that tradition, the "metascientific" expressions could and should be given an analysis which is independent of any particular scientific theories, laws, evidence, explanations. For positivism, to give such analyses was, indeed, one of the main duties of the philosopher of science. However, the Feyerabend-Kuhn approach has argued, as we have seen, that fundamental scientific revolutions (changes of "paradigm" or "high-level background theories") alter something (everything?) not only about the admittedly substantive content of science but even about what counts as, for example, an explanation (the meaning of "explanation"). The question is thus raised as to whether even allegedly "metascientific" concepts like "explanation" are not to at least some extent "theory dependent" and, therefore, whether, or to what extent, they can be used to talk about different theories. Indeed, the very possibility or significance of the general distinction between "metascientific" and "scientific" concepts becomes suspect, and, with it, a large part of the positivistic conception of the program and method of the philosophy of science.

The problem of comparability now appears in a new and deeper guise: for, with regard to both concepts allegedly occurring "within science" and concepts used in "talking about science," it becomes necessary to inquire whether there are any such concepts, or separable components of such concepts, which are common to more than one theory (or perhaps even *necessarily* common to *all* theories, at least to all those of certain types). And, if there are (as seems *prima facie* to be the case, despite Kuhn and Feyerabend) any common concepts or components of concepts, what gain if any is there, for the understanding of science, in analyzing them? Must we, in order to point out concepts that are (or perhaps must be) common to several or perhaps all scientific theories, or to point out metascientific concepts applicable to any scientific theory, point to something so abstract as to be rather empty and unilluminating? And, having found such concepts, must we face the possibility that future developments in science will force us, in order to maintain them as "common" or "applicable," to make them still more general—and empty? The whole problem of "comparability" thus becomes not simply *whether* or not there are similarities between different "theories" (regarding either the terms occurring in them or the terms used in talking about them) but rather, even granting the existence of such similarities, the extent to which pointing out those similarities and analyzing them is significant or illuminating for the attempt to understand science.

### 5. *Further Comments on the Relations between the Positivistic and Feyerabend-Kuhn Approaches*

The problem of comparability of theories, as it is exposed in the debate between the positivistic and the Feyerabend-Kuhn camps, may be expressed in the following way: How are we to give an account of the scientific enterprise according to which observations (experience, data, evidence) will be both *independent* of theory (any theory?) and *relevant* thereto? The difficulty is that, *prima facie,* a tension exists between these two requirements, for independence seems to demand that the meanings of observation terms be totally pure of any theoretical infusion, whereas relevance seems to demand that they be permeated, at least to some extent, by theory. And, when the problem is put in this way, it appears that, whereas the empiricist-positivist tradition in its concern with the objectivity of science has overemphasized independence, writers like Kuhn and Feyerabend have dwelt on relevance to the exclusion of independence. But an adequate interpretation of science must do justice to both features, and the problem is to steer a safe course between these two demands.

Nevertheless, this way of putting the comparability problem still relies on the notions of theory and observation for its formulation. And, in view of the difficulties which have plagued these notions, we must consider whether an attack on the comparability problem from this point of view is advisable. We should note particularly that the notion of theory is today in a worse state than ever. Consider first the view of (high-level background) theories advanced by Feyerabend, which is paralleled in many respects in Kuhn's notion of paradigms. Elsewhere I have argued that their approach, although it takes as the fundamental determinant of scientific meaning and acceptability the notion of a theory ("high-level background theory" or "paradigm"), fails to give an adequate analysis of that concept.[9] For it is unclear what is to be included in and excluded from Kuhn's "paradigms" and Feyerabend's "high-level background theories." In order to give his background theories sufficient pervasiveness and scope, Feyerabend declares that "in what follows, the term 'theory' will be used in a wide sense, including ordinary beliefs (*e.g.*, the belief in the existence of material objects), myths (*e.g.*, the myth of eternal recurrence), religious beliefs, *etc.* In short, any sufficiently general point of view concerning matter of fact will be termed a 'theory.' "[10] Similarly, Kuhn's "paradigm" consists of a "strong network of commitments—conceptual, theoretical, instrumental, and methodological,"[11] including "quasi-metaphysical"[12] ones.

But the problem of what is supposed to be a "part of a theory" is not confined to such questions as whether Kepler's mysticism was an integral part of some underlying "theory" he held, or whether absolute space was an integral part of Newton's science. In the second part of this paper we will several times encounter cases where it is not clear whether *admittedly scientific* propositions should be judged "part" of a certain theory or not. The positivistic analysis of scientific theories as interpreted axiomatic systems is of no help whatever in such cases. For considering a theory in this way presupposes that we already know *which* propositions *are* part of the theory, in order to axiomatize them; but the problem is to decide which propositions *are* to be considered members of the set to be axiomatized. Here again, philosophers of science are indebted to the Kuhn-Feyerabend view, even for its deficiencies, because the exposure of the difficulties in its notion of theories has brought out the essential triviality (with respect to the present problem) of the positivistic view. There is today no completely—one is almost tempted to say remotely—satisfactory analysis of the notion of a scientific theory.

It should be emphasized that the inadequacies of attempted analyses of the concepts of theory and observation do not imply that those terms ought to be eliminated from legitimate scientific discourse, or that they will not have a place in a full interpretation of science. They are, after all, terms which have perfectly good and common scientific uses. What has failed, so far, are not those uses but rather the *technical* distinction, which, while supposedly doing justice to those actual uses, was really introduced to perform certain philosophical functions.

### 6. *The Generation of Artificial Problems by the Theoretical-Observational Distinction*

The distinction between "theoretical" and "observational," and the relations between them, have proved extraordinarily difficult to formulate, particularly in a way which would deal successfully with the specific problems for the sake of which the distinction was introduced—namely, as we have seen, the problems of meaning, acceptability, and comparability. But from the point of view of the present paper there are yet further deficiencies of the distinction which are even more important to recognize, inasmuch as they will be found to suggest fresh and constructive steps to be taken in the attempt to understand science. First of all, it seems reasonable to ask whether some problems that have arisen within the context of discussions relying on the theoretical-observational distinction are *created, at least in part, by the limitations of that technical distinction and the roles for which it was introduced. If so, then the problematic character of those created "problems" must be reconsidered in the light of the failures of the background against which they arose.* That this is in fact true in at least one case may be argued as follows.

One of the most notorious problems arising within the context of discussions employing the notions of theory and observation as analytical tools is the so-called problem of the ontological status of theoretical entities, or the question of whether a "realistic" interpretation of scientific theories can be upheld. Can such terms as "electron"—often taken as a paradigm case of a theoretical term—be said to designate entities, or must reference to electrons be looked upon as only, for example, the employment of a convenient fiction, instrument, or calculating device? Now, there are, no doubt, real difficulties in the way of understanding precisely what is involved in such assertions as "Electrons exist"; but the perplexity about the ontological status of theoretical entities must be attributed at least in part to this problem's being formulated against the background of the theoretical-observational distinction, rather than entirely to some intrinsic opaqueness of the concept of existence. For, once we make a sharp distinction between "theoretical" and "observational" terms, and lay it down that the former are to be interpreted via the latter, we are easily led to be puzzled, not only about the interpretation of the former, but also about whether they have to do, literally and explicitly, with anything that exists. Observation terms are clearly meaningful and clearly they refer to entities that exist (this is particularly obvious if we "adopt a thing language" to provide our "observational vocabulary") or properties that are real. The problem, with regard to the ontological status, as well as the interpretation, of a class of terms distinct from these is apparently merely in the sharpness of the distinction. (The problem of interpreting theoretical terms was understood to involve, as one aspect, the question of what if anything such terms were "about.") It was all too easy to view the distinction between observational and theoretical as paralleling a distinction between existent and non-existent.

The evolution of the theoretical-observational distinction only made matters successively worse with regard to the problem of ontological status. If theoretical terms could, as early positivism maintained, be exhaustively defined via observational ones, then it was unnecessary to assume the existence of theoretical entities: theoretical terms could be treated as a convenient shorthand, and theoretical entities as convenient fictions. With the view that theoretical terms could be only partially so defined, the puzzle arose again: did the "extra meaning" necessitate reference to entities, even unobservable ones? To avoid this inference (an unpleasant one for positivists) it was proposed that the "extra meaning" was contributed by the place of the term in the "system." Thus again, if theoretical entities can be said to exist in any sense at all, they do not exist in the same sense that tables and chairs exist. There were many who could not but feel uncomfortable at this evident taking away with the left hand what had been given with the right. Finally, the abandonment of the sharp distinction by many, and its replacement by, for example, the "continuum" view—a view whose major purpose is to deal not with the problem of ontological status but with the difficulties of the theoretical-observational distinction itself—seem to leave the problem high and dry. As Grover Maxwell, wrestling with the problem, expresses his discomfort, "Although there certainly *is* a continuous transition from observability to unobservability, any talk of such a continuity from full-blown existence to non-existence is, clearly, nonsense." [13] Where on the alleged continuum is one to draw the line? Similar uneasiness is engendered by the view that certain terms are used "theoretically" in some contexts and "observationally" in others.

Further prominent doctrines of one phase or another of the positivistic tradition co-operated to generate the problem of ontological status—although those further doctrines, also relied to some extent for their precise formulation on the theoretical-observational distinction. Among these other doctrines was the verifiability theory of meaning. According to this view—expressed here in simplified form, of course—a statement is meaningful if and only if there exists a method of verifying it. Theoretical terms, in contrast to observation terms, were alleged to refer to what is unobservable; hence, if methods of verification of existence statements are restricted to observation, statements such as

"Electrons exist" are unverifiable if taken literally and thus are meaningless unless interpreted non-literally. The failure of the verifiability theory, and serious problems regarding the sense in which theoretical entities are "unobservable," did not remove the deep suspicions that, being unobservable, theoretical entities cannot be talked about meaningfully, at least not in any literal sense. And it followed that theoretical terms must be given some interpretation which does not involve reference to such entities.

A second positivistic doctrine which aided in making the problem of ontological status appear unduly serious was the view that scientific theories are "interpreted axiomatic systems." Modeling so much of their approach to the philosophy of science, as they did, on mathematical logic, the positivists talked about the linkages correlating theoretical with observational terms ("co-ordinating definitions," "rules of interpretation," "correspondence rules," etc.) as being analogous to the interpretation of a formal system in mathematics and logic. Such linkages in logic are clearly unlike the kinds of connections which can be asserted to hold between existent things (e.g., causal interaction); and so, the use of this analogy automatically leads one to suppose that, inasmuch as observables exist, theoretical terms also are linked to observational ones in ways different from the ways in which existent entities can be related to one another. Once again we seem compelled *by the very approach employed*—in this case, by the employment of the logical model (although even here the theoretical-observational distinction is relied upon)—to say at the very least that we are puzzled by statements about the "existence" of theoretical entities. It appears that a great deal—not necessarily all—of the gravity of the problem of the ontological status of theoretical entities is the result of adopting a certain approach or set of approaches to the philosophy of science. The problem—or at least part of the problem—arises simply *because* we employ a certain technical distinction (theoretical-observational) as well as certain other doctrines (verifiability theory of meaning) and analogies (logical models of scientific theories and their interpretations) which in turn rely heavily on that distinction. None of these views has proved successful; and so one must suspect that a problem which arises at least in good measure against the background of those views may, to that extent at least, not be the problem it has been made out to be. On the contrary: to some extent, at least, it may well be a pseudoproblem.

### 7. *Need for Reconsideration of the "Ontological Status" Problem*

Thus, many of the puzzling features of such statements as "Electrons exist" can be attributed merely to demanding an analysis of those statements within the context of the theoretical-observational distinction and associated doctrines, to utilizing that distinction as an analytical tool for the framing of problems and solutions in the philosophy of science. But it is by no means necessary that we approach such statements in this way; and in view of the protracted failure of the technical distinction between theory and observation and because of the way in which it (together with associated doctrines) led to the creation, or at least the exaggeration, of the problem of ontological status it may not even be the best way. Indeed, the multitude of weaknesses that have been exposed in the positivistic *grounds* for worrying about the legitimacy of existence-claims for "theoretical entities" may even encourage us to take a new look at that problem. Such a fresh examination would have to be far more radical than the approach of Feyerabend and Kuhn. Because, for all the divergence of their position from the positivistic view that theories are comparable by virtue of their dealing with a common core of "observation terms," they nevertheless approach their problems with that distinction in mind. The "revolution" of Feyerabend and Kuhn does not consist in denying the utility, the centrality, of the notions of theory and observation in stating and dealing with the problems of the philosophy of science, but rather, simply in reversing their respective roles: it is now theory ("high-level background theory" or "paradigm") that determines the meaning and acceptability of observation, rather than vice versa.

In the course of such a radical re-examination of the problem of ontological status further limitations of the theoretical-observational approach may well emerge. After all, designed as it was to deal primarily with the specific problems of meaning, acceptability, and comparability, it is only natural that that distinction might not do other jobs equally well. Indeed, we might expect that, while some problems—like that of the ontological status of electrons, waves, space-time, or fields—might be exaggerated or distorted by formulating them in terms of that distinction, still other problems, and other interesting and important features of science, may well have been pushed into the background and ignored because of the central place accorded to the theoretical-observational distinction and the problems it was introduced to handle. And it may be that those other problems and features have not just *failed* to be noticed because philosophers were too busy with theoretical-observational work; it may be that that distinction actually *obscured* the existence of those problems and features, that the limitations of that distinction and the complex of problems it was designed to deal with actually drew attention away from those other problems and features.

But it is not only the difficulties and limitations of the theoretical-observational approach that make a re-examination of existence-claims in science an attractive venture; there are more positive grounds for the undertaking. For there are a great many clear cases of scientific terms usually classed as "theoretical" which we are strongly inclined (or, as some would have us believe, tempted) to use in statements predicating existence or non-existence of them. The number of such clear cases is surely comparable to the number of clear cases of terms that we feel strongly inclined to classify as either "theoretical" or "observational." We need not fear, therefore, that we will be worse off initially in reopening the problem of ontological status than those who began to approach the interpretation of science by examining clear cases of "observational" and "theoretical" terms and statements. But more: the very multitude of cases in which existence assertions seem so natural in science is itself a fact that cries out for explanation. It suggests that we try to determine what underlies this inclination, this feeling of naturalness, and to see whether what underlies it may not be sounder reasons than have been noticed by the positivistic tradition. Certainly we should at least look into this subject very carefully before agreeing to dismiss this inclination as illegitimate—as a philosophical or metaphysical overlay, imposed on a science which is indifferent to questions of existence, or to which such questions are irrelevant.

Indeed, we will find that there is a sound basis in science for existence-claims of the kinds mentioned above. But the approach that brings this out will also, even if only in a preliminary way in the present paper, enable us to see a number of other classical problems of interest to the philosopher of science in a new light (including problems discussed earlier in this paper); and a number of further problems, hitherto ignored or at least treated lightly by philosophers of science, will be exposed and brought into prominence by the approach used.

## NOTES

1. In the empiricist-positivist tradition, discussion has also centered on the meaning (or meaningfulness) of statements as well as—and sometimes instead of—concepts. But, despite the fact that the history of science contains a multitude of cases in which the introduction of new concepts, or the abandonment of old ones, played a crucial role in developments, little attention has been devoted to the problem of the acceptability of concepts. Presumably the reason for this was a belief, often tacit, that that problem would be solved automatically with the solution of the problem of the acceptability of statements. This belief, however, is highly questionable, and a close examination of reasons for introducing new concepts and abandoning old ones is needed in science.
2. P. Feyerabend, "Problems of Empiricism," in R. Colodny, ed., *Beyond the Edge of Certainty* (Englewood Cliffs, N.J.: Prentice-Hall, 1965), p. 180.
3. *Ibid.*, p. 213.
4. T. S. Kuhn, *The Structure of Scientific Revolutions* (Chicago: University of Chicago Press, 1962).
5. *Ibid.*, p. 102.
6. *Ibid.*, p. 105.
7. *Ibid.*, p. 140.
8. Feyerabend, "Problems of Empiricism," p. 214.
9. D. Shapere, "The Structure of Scientific Revolutions," *Philosophical Review*, 73 (1964): 383–94, and *idem*, "Meaning and Scientific Change," in

R. Colodny, ed., *Mind and Cosmos* (Pittsburgh: University of Pittsburgh Press, 1966), pp. 41–85. See also Chapter 3 and 5 in this volume.

10. Feyerabend, "Problems of Empiricism," p. 219, n. 3.

11. Kuhn, *Scientific Revolutions,* p. 42.

12. *Ibid.,* p. 41.

13. G. Maxwell, "The Ontological Status of Theoretical Entities," in H. Feigl and G. Maxwell, eds., *Minnesota Studies in the Philosophy of Science,* vol. 3: *Scientific Explanation, Space, and Time* (Minneapolis: University of Minnesota Press, 1962), p. 9.

# PART II

# ANALYSES OF ISSUES

CHAPTER 7

# SPACE, TIME, AND LANGUAGE

*An Examination of Some Problems and Methods of the Philosophy of Science*

Physical science in the Twentieth Century has raised two questions of the most profound importance regarding the scientific role of the concepts of space and time. First: Do we, in a physical theory, have to deal with space and time in the terms in which they have almost universally been dealt with or at least implicitly thought of? and second: Is it necessary, in a physical theory, to employ the concepts of space and time at all?

To the first of these two questions, nearly everyone today would give a negative answer. This fact stems from two historical sources, one mathematical, the other physical. Developments in mathematics during the nineteenth and twentieth centuries have led to the construction of consistent systems of geometry in which space is considered to have properties which traditionally had been thought of (when they were even thought of) as self-contradictory. The most famous of these developments was the creation of non-Euclidean geometries; but of almost equal importance was the independent creation of geometries with dimensions greater than three. And perhaps of still greater potential importance for physics is the highly general subject of topology, which goes far beyond classical geometry by dealing with properties of space other than those connected specifically with its measurement.

Yet the mere development of such formal systems of mathematics was not enough, of itself, to convince very many people that physical space was or even could be non-Euclidean, other than three-dimensional and simply connected, and so forth; it was still believed that such formal systems either could not, or at least did not, have any application in physical science—that they were mere amusing games played by mathematicians. But the adoption in the theory of relativity of a non-Euclidean geometry, and the treatment of time as a fourth dimension of this geometry, led to an acceptance of the view that treatments of physical space and time can employ any developed mathematical schemes (see note 1), and theories employing various non-Euclidean geometries have become the rule rather than the exception. More recently, the great advances of topology have been utilized for physical purposes—for example, in Wheeler's "geometrodynamic" view of the universe, which attributes to physical space certain properties which would have been inconceivable to all but a very few farsighted men even half a century ago. (A nontechnical discussion is given by Wheeler, ref. 3.)

Of course, such radical developments have had their conceptual price: for the introduction of such new geometrical and topological notions has raised serious questions for those who wish to understand what science is doing. What, for example, are the grounds for attributing such properties to space and time—that is, what are the reasons for employing, or even for considering, a certain geometry or topology in physical science? Is the choice (or the candidacy) determined by purely factual, empirical considerations, or is it "purely conventional" which sort of geometry or topology we select? Or, for that matter, are the grounds the same for the selection or consideration of topological as for metrical properties, or even for different topological properties? And further, in employing such notions, have scientists departed so far from ordinary conceptions of space and time that they are no longer even talking about space and time at all? Are scientists, with their outlandish geometries and topologies, even concerned with the description of space and time, or with explanation of experience, or are they merely using these formal mathematical schemes as convenient tools of calculation which have no relation to any supposed structure of "reality"?

These last questions, of course, already verge on the second of the two fundamental questions raised earlier, and we see that those two questions are not as distinct as they at first seemed to be. For to wonder if, in employing such hitherto strange notions of space, science has not gotten so far from ordinary conceptions that it *is* no longer talking about space at all (despite its continued use of that term or formal analogs thereof), is already to wonder whether science *need* talk about space at all.

Such views are a far cry from those proposed by Kant, who held not only that science must utilize the concepts of space and time as fundamental categories, experience being impossible except in spatio-temporal terms, but also (at least on the most usual interpretations) that space must be Euclidean and three-dimensional. And they are a far cry, too, even from the view of Samuel Alexander, who declared in 1916 that "It is not, I believe, too much to say that all the vital problems of philosophy depend for their solution on the solution of the problem of what Space and Time are and more particularly how they are related to each other." [4] Certainly, Alexander was right to this extent: that the concepts of space and time *have* figured crucially in philosophy and science, at least since Leibniz and Newton, or perhaps since the Fourteenth Century. But if he meant that they will continue to do so, that the concepts of space and time are even now playing a fundamental role in human knowledge (in spite of the continued presence of the *words*, "space" and "time" or formal analogs thereof), then his point is far from obvious.

In the present paper, I want to depart from a consideration of this question of the extent to which physical theories are, or need be, concerned with space and time. I do not intend, here, to give an answer to this question; rather,

I want to examine the question itself, to see what exactly it is asking, what sort of answer we can reasonably be expected to give, and how we are to go about finding such an answer. Thus, in a very literal sense, this paper will *depart* from questions about space and time, in that it will pass into a consideration of some central tasks which face the philosopher of science today, and of some methods which seem appropriate and promising for dealing with those tasks. The reverse side of this discussion will be a diagnosis of the reasons why the most prevalent approaches to the philosophy of science in this century have so largely failed to do justice to this sort of problem.

In order to prevent any misunderstanding, let me make some qualifying remarks before proceeding. In the first place, I will not be dealing with *all* aspects of the question of the role of space and time in physical theory. For example, the question of the grounds for selecting or considering a geometry or topology for use in a physical theory, though far from settled yet, will be omitted from direct consideration here. And even with regard to the question which is here at issue, some of the reasons suggested by quantum mechanics for doubting the microphysical adequacy of the concepts of space and time (but at the same time arguing for their inescapability) go far beyond the considerations of this paper (see note 2). Furthermore, as the sorts of problems which are raised in the present essay are far from being the only kinds of problems faced by the philosopher of science, or even the only problems regarding space and time, so also the methods which I will propose for dealing with the sorts of questions I do raise are not the only methods which the philosopher of science can properly utilize to answer his questions. They are not appropriate for answering *all* the questions which may confront the philosopher of science, and they are not even necessarily the only methods with which he might well approach the problems I do raise. For although I think that the methods that philosophers of science *have* for the most part been employing have been inappropriate for dealing with these sorts of problems (and therefore these sorts of problems have been largely passed over by philosophers of science), it is not impossible that some other procedures might also cast light on these questions. I only maintain that the methods which I will propose do seem appropriate and promising and clarifying of some tasks of the philosopher of science, and that, for these particular sorts of problems (which are important ones at the present stage of science), the methods which have long been in vogue are not appropriate.

## I

Statements made by competent scientists in the name of science about space and time often seem odd or even self-contradictory to the critical non-scientist. We are told, for instance, that the universe is (or at least may be) expanding, and this, we find on further inquiry, means not that the galaxies in space are moving away from one another into still deeper space, but rather that space itself is expanding. Yet ordinarily, space is thought of as something *in* which motion can take place, while it itself *cannot* (logically) move or change size; and even if it could expand, what would there be for it to expand into, if not further space? We are told that space has a curvature, and that it has (or at least may have) different properties in different places, or even different properties in different directions in the same place, and, furthermore, that these properties may change with time. We are told that there is no such thing as absolute simultaneity of events; and elementary particles may move backward in time.

Such apparent contradictions are not limited to scientific assertions about space and time; they arise also in connection with a multitude of other terms used in science: for example, we are told that particles may jump across space, or pass through two holes at once; and the oddness of such assertions seems to relate as much to the term "particle" as to the term "space." Nor do such puzzles arise only in connection with physics and its cousins: the introduction of the concept of "the unconscious" in psychology makes possible statements about human motivations which, from the viewpoint of common talk about mental activity, are self-contradictory. Thus the problems raised here regarding space and time are instances of a fairly widespread type of problem regarding scientific concepts.

What bothers us about such statements is not so much that they appear to be *false*—as though, while they might reasonably be considered, they happen not to describe the way things are; it is rather that they seem to assert something that *could* not be true of space and time and particles and motives; and this is to say that they seem to be incompatible with the ways in which the terms "space" and "time" and the rest are used in ordinary contexts.

This distinction between statements which bother us because we think they are false, and those which bother us because they seem self-contradictory, needs to be developed further. Let us say of any belief that can be stated in a language, and whose contradictory also can be stated (makes sense) in that language, that it is *expressible* in that language. Beliefs expressible in a language which are accepted, implicitly or explicitly, by the majority of people who speak the language include what are often called *common-sense beliefs*. Now, philosophers certainly can be concerned with bringing such beliefs to light and subjecting them to criticism; but these are not the sorts of beliefs with which I am concerned in this paper. I am concerned rather with beliefs which can be stated in a language, but whose denials make no sense, i.e., are self-contradictory (at least when interpreted in any ordinary way). Such beliefs can be said to be *incorporated in* (or *embodied in*) the language. (In these terms, we can see why such frequently heard slogans as, "Science contradicts common sense," or, "Science is the long arm of common sense," are misleadingly vague.)

Although I know of no more appropriate word, the use of the term "belief" in the context, "belief that is incorporated in a language," is certainly an odd one. What would we make, in any usual circumstance, of the statement, "I believe that all bachelors are unmarried males"? (What would we be *dis*believing if we said that?) And insofar as it is a queer use of the word, it is also queer to say, with Professor Austin (who unfortunately did not explain his use of the expression "incorporated in"), that ordinary language can incorporate "superstition, error, and fantasy." [13] But notice: to the extent that it is odd to say that a language incorporates errors, it is also odd to say that scientific usage has *corrected the errors* of ordinary language. This consideration will shortly make us hesitant about this view of the relation between scientific and ordinary usage.

It is important to note that not all ordinary uses of the terms "space" and "time" are relevant to our problem. This is true not only of such expressions as "time on my hands," "the time of your life," and "a new dimension of color," which are readily interpreted as metaphorical; it is also true of some statements which do not, superficially, conflict with the scientific usages noted earlier. For instance: "We are doubling our (office) space," and "Time is passing slowly during this lecture," which are superficially similar to statements about space expansion or time passage in certain areas of science. Whether such usages are to be interpreted as "metaphorical" or "elliptical" or "secondary" in some sense, they are not the sorts of usages we have in mind in noting the oddness or paradoxical character of the scientific statements. On the other hand, not all scientific usages of the terms "space" and "time" are relevant here either (at least in any obvious way): For example, the expressions "sample space" and "configuration space," while their precise interpretation raises extremely difficult questions, do not suggest application of the term "space" to physical space, but rather utilizations of certain mathematical tools for other purposes. (See note 3.)

Again, it might be suggested that the sense in which such statements as "Space is expanding" seem odd is simply the sense of Newtonian physics: it might be suggested, that is, that ordinary usage has become infected with Newtonian concepts—that such concepts have become incorporated into our ordinary usage—so that the contradictions are merely between Newtonian and Twentieth Century science. This is possible, though it presupposes a clarification of the sense in which such beliefs can become "incorporated" into a language, and that linguistic changes of these sorts take place. Such questions are interesting and important in their own right; but for present purposes, we need not consider them, since we are concerned with conflicts between present usage, or certain aspects thereof, and scientific usage; whether or not the former is "Newtonian" is irrelevant to our present problem.

## II

Generally, those who have reflected on the paradoxical character of the kinds of scientific assertions noted above have tended to draw one of four sorts of conclusions from them:

*First:* that science is self-contradictory, and either must be rejected out of hand as a source of knowledge, or must be seen as giving only a part or an aspect of the truth, its self-contradictoriness being, in some mysterious way, a consequence of its incompleteness.

*Second:* that ordinary concepts are somehow unsatisfactory—either as being vague, or unworkable, or self-contradictory—and that these deficiencies have been corrected by science. [The terms "concept" and "meaning" as used in this paper should not be taken as referring to Platonic or mental entities, but rather as being elliptical for (very roughly) the following: the roles played in a language and its employment by a family of terms whose roles are very similar.]

*Third:* that the scientific usages of the terms in question have nothing whatever (at least in any interesting or enlightening way) to do with their ordinary usages, and so, if the scientist during his working hours uses the terms "space" and "time" in ways which seem incompatible with their ordinary uses, this should occasion no more surprise than the fact that, in ordinary language, the term "pen" is used sometimes when we are talking about writing instruments and at other times when we are talking about the homes of pigs; or, in a more charitable version of this view, it should worry us no more than the fact that the word "star" is used to refer to certain large hot objects and also to movie personalities. In other words, at worst there is no relation at all between two different uses of the same sort of sound or inscription (scientific "space" and ordinary "space" are no more related than Parker pens and pig pens), or at best the relation is "merely metaphorical" (as in the use of the word "star" to refer to Kim Novak and Alpha Centauri). And in neither case (so the argument runs) can there be a contradiction between the uses.

*Fourth:* that the ordinary and scientific meanings of such terms as "space" and "time" have a literal and not merely metaphorical relationship to one another. This view, too, comes in two main varieties. Version A claims that ordinary concepts, while they are not unworkable or wrong or self-contradictory or even vague in any disastrous sense, are useful only within a very restricted domain, for very limited purposes, and that, when the ordinary and scientific senses of these terms are properly understood, it will be seen that the ordinary concepts are "special cases" of the scientific ones, derivable somehow from the latter with "suitable restrictions." Version B, agreeing that there is nothing wrong with ordinary language that a little understanding won't cure, holds that scientific language does retain certain aspects of ordinary usage but rejects others as irrelevant to its purposes or unusable by its methods; the paradoxes, according to this view, arise because of the abandonment of those aspects of ordinary usage.

Of these four types of view, the first—the view that dismisses science as self-contradictory—was, in one form or another, held by many Idealists of the nineteenth and early twentieth centuries. (The Idealists, of course, found contradictions everywhere, in ordinary language as well as in science.) But this view, far from having been established by its adherents, is now almost universally rejected by thinkers. The reason for this is not just the unclarity of many of the fundamental notions employed by such critics of science, or the fact that many of the specific arguments by which they tried to establish the inconsistency of science have proved to be fallacious; the general dissatisfaction with this attitude toward science stems even more from the rather healthy feeling that, when a subject is as well-developed and successful as science is, it might be wiser to suspect the arguments which purport to show the inadequacy of the subject than to impugn the subject itself.

The second position—that the ordinary meanings of these terms are self-contradictory or unworkable or vague in some viciously self-defeating sense—is one that was popular among British, American, and Viennese philosophers for three or four decades of the present century, beginning about 1914. It still finds scattered support among philosophers, and can be found asserted *ex cathedra* in the introductory pages of many books by scientists. Here, too, however, the main line of philosophy has, under pressure of criticism, been forced to retreat—and, ironically, for reasons very like those which led to the abandonment of Idealism. For in the face of half a century of analysis, philosophy cannot avoid confessing that ordinary language is, after all, workable (it works); that the supposed contradictions that philosophers have held make it unworkable themselves rest on confusions; that ordinary language is not vague in the calamitous way some philosophers have supposed it to be; and that, finally, if it is in error, then (as we saw earlier) the kind of error involved must be of a very special and peculiar sort, whose nature must itself be explained in detail. For unlike ordinary errors, which are errors because their denials happen to be true, the denials of the sorts of "errors" that ordinary language is alleged to commit do not even make sense.

Thus the first two views—one holding that there is something wrong with science, the other that there is something wrong with ordinary language—have had to be abandoned. This is not to say that, after further investigation, we might not find that there was *some* sort of point (though obscurely and misleadingly stated) behind them—that, for example, we might not find *some* useful sense in saying that ordinary language is "in error." But as matters stand, the case is very much against those views, and at the very least they have not shown us the way out of the apparent conflicts between scientific and ordinary usage which occasioned our inquiry. So we are left with the last two views: either there is no relation between ordinary and scientific usage, at least not the kind that can engender contradictions, or there is a relation between them, and of such a kind that the apparent contradictions between them will be resolved.

The extreme version of the third view—that there is no relation whatever between ordinary and scientific usage of terms like "space" and "time" (or their formal analogs)—has been advanced in several forms: for instance, in the view that ordinary language and science have to do with different "realms" or "kinds of facts," or that they have to do with different "purposes," and that these differences imply that the concepts they use in dealing with those facts or in achieving those purposes are wholly dissimilar. It is hard to avoid thinking that this sort of view is an overhasty reaction to the paradoxical relationships between scientific and ordinary usage. It seems much more likely that, for *some* uses, at least, of such terms as "space" and "time" (or their formal analogs) in *some* scientific theories, there are *some* sorts of relationships to at least *some* facets of the ordinary usage of those and related terms. After all, science has taken, and still does take, much of its impetus from practical considerations—considerations which are relevant or potentially relevant to everyday life, that same everyday life in which ordinary language plays a part. And new scientific theories, in turn, are not developed in a vacuum, but against the background of other theories and their shortcomings. Part of Galileo's interest was in the purely practical matter of giving an account of ballistic trajectories; and the laws of motion, of change of spatio-temporal position, which he formulated were incorporated into Newton's theory—the same theory whose deficiencies are taken care of by the theory of relativity, which in turn serves as the basis for much of current theorizing about the structure of space and time. It seems likely, therefore, that at least in the early stages of the development of a branch of science, the terms are utilized in ways fairly closely related to their ordinary uses, and that later theories, for various reasons, depart from the uses in previous theories, so that successive developments of that branch of science lead finally to uses of those terms in ways which are very unlike, and even in certain respects contradictory to, their ordinary uses. And it is important for understanding science —for being able to see what it says and does—to trace the paths of those successive changes, and to become clear about the reasons (the *good* reasons) why the changes had to be made.

What, then, is the character of those departures? Are they adequately described, in all cases, by saying (with the "liberal" version of the third view described above) that successive shifts of usage in science are related to their predecessors "metaphorically"? Like the preceding sort of view, such interpretations of scientific usage all too often have the effect of interpreting science away. Further, considering the lack in philosophy of any thorough analysis of sorts and functions of metaphors, any such assertion about scientific usage cannot be viewed with much confidence. (Is it really true, for instance—as this view must suppose if it is to remove the paradoxes—that metaphorical uses of words cannot contradict more "literal" uses?) And finally, it is hard to believe that a subject which tries to be so literal and claims to be descriptive and explanatory, and which abandons theories and accepts new ones for such varied and complex reasons, can be easily brought under this single, simple analysis without doing violence either to science or to the word "metaphor."

What, then, of version A of the fourth view—that ordinary language is related to scientific language in something like the way in which Newtonian physics is often said to be related to relativity physics—as "what the general case becomes under suitable restrictions," or as "special case to general case," the one being "reducible to" the other? But this view, too, is of little help to us; for if it is true, as many recent investigators have argued cogently, that the meanings of certain terms, like "space," are not constant from one theory to another, then, as Professor Feyerabend has reminded us,[16] the relation of "reducibility" which is asserted to hold cannot be interpreted (as it usually is) as one of "deducibility." Therefore, that this is an adequate account of the relation between Newtonian and Einsteinian physics is itself questionable (see note 4); and if the sense of "reducibility" is not that of "deducibility," then in the absence of any account of what the relation *is*, we must confess that this view leaves the important work of analysis still to be done. This difficulty is only compounded when the view is transferred to the relation between scientific and ordinary usage. The specific way in which ordinary usage is supposed to be "reducible" to scientific usage (or *vice versa*), and in such a way that contradictions between the usages are possible, has never been clarified. (In the case of the relations between scientific theories and ordinary language, surely the notion of a "boundary condition" loses its literal meaning.)

Finally, then, we are left with version B of the fourth type of view—that scientific usage retains some aspects of ordinary usage, but abandons others as irrelevant or unusable. Such a view has been presented recently by Max Black in a review of G. J. Whitrow's *The Natural Philosophy of Time.*[18] His account, being presented in a review written for non-philosophers, is necessarily sketchy; but it is limited in more important ways also. His analysis is restricted to the problem of time, and in particular he applies his view of the relation between ordinary and scientific usage to one problem only, that of "the direction of time." He suggests that there are no similar paradoxes in the case of space; for in passing from the ordinary to the scientific concept of space, he declares, "the loss of conceptual content entailed by such idealization is relatively unimportant" [ref. 18, p. 183]. Finally, by talking of "the scientific concept," he suggests that there is only one scientific concept of time.

Black is aware that there is a problem regarding the relations between the ordinary and scientific concepts of time:

> . . . it is necessary to trace the logical connections between the idealized scientific concept and the layman's rough concept from which it is derived. Such connections must be present between what the man in the street calls "time" and what the theoretical physicist calls "*t*," however hard the concepts may be to disentangle and analyze. Otherwise to speak of *t* as time, as Whitrow, like all of us, constantly does, would be to indulge in an outrageous pun. [Ref. 18, p. 180.]

After criticizing Whitrow, Black proceeds to outline his own view of the relation between scientific time and "the common-sense notion" (by which he means the use of the term in ordinary language).

> So wide is the gap between the common-sense notion of time and the physicist's *t* that clarity would be fostered if physicists were to imitate the psychologist's practice of talking about *g* rather than about intelligence by referring to their own concept as *t* rather than as "time." It is not shocking to be told that *t* may have a unique origin like the absolute zero of the temperature scale, or may have several dimensions, or even that it may "run backward"; it is only when such aphorisms are transformed into the corresponding statements about time that paradox emerges and philosophical hackles rise.
>
> The basic connection between *t* and time is not hard to discern: clocks measure only lapses of time, answer only questions about how much later one point-event (a tick, a flash, a beep) occurs than another point-event. All the other logical features that enter the constitution of the common-sense concept of time are deliberately excised. [Ref. 18, pp. 182–183.]

In terms of this account, Black offers a tentative diagnosis of the physicist's assertion that time (or rather *t*) can "run backward":

> The reason that physics can rove backward in time as freely as forward may be that the subject has limited itself to the investigation of relations that are essentially symmetrical. [Ref. 18, p. 184.]

Thus, apparently, the asymmetrical relation of beforeness-and-afterness is supposed to be "expurgated" or "excised," along with "all the other logical features that enter the constitution of the common-sense concept of time," leaving only symmetry. (See note 5.) Instead of saying, "*A* occurs 60 seconds before *B*," physics says something like, "*A* occurs 60 seconds from *B*." This may well be an important aspect of the difference between ordinary and physical time—though if it is, if the only feature of time that physics can talk about is time difference (see note 6), then Black has stated the problem of the "direction of time" misleadingly. For then the problem would not be that "physics can rove backward in time as freely as forward," since physics would not even have the conceptual facilities for talking about such roving; physics would neither rove forward nor backward in time, and *t* could not "run backward." (See note 7.) Presumably it would only seem to us, interpreting *t* as our ordinary time, to do so. How exactly such an interpretation creates this appearance of talk about backward time flow would then have to be worked out. (See note 8.)

Even if it is true that some paradoxes arise because of the deletion of features from ordinary concepts, others may be better described as arising because of features that are added. Indeed, without a more precise examination of the "logical features" of ordinary and scientific time, Black's suggestion that "all the other logical features" of "the common-sense concept of time" are deleted is gratuitous. But Black's way of describing the relationship between ordinary and scientific concepts raises yet more disturbing doubts; for this "deletion-and-addition" way of talking about conceptual changes is itself highly suspicious. Surely a discussion of the "logical" differences between Newtonian and Einsteinian time would not employ as tools of analysis simply the notions of "addition" and "subtraction" of "features"; the alterations which Einstein imposed on the concept of time are too complex and far-reaching to be described adequately in terms of a hacking away here, a tacking on there. For questions can be raised regarding the "space" and "time" of relativity (particularly the general theory) that *could* not be raised regarding Newton's. Yet deletion of features (unless this expression is to mean something stranger even than it sounds) could only negate the possibility of asking certain old questions; while, although addition of features might make some new questions possible, it could not make possible questions which were *logically* excluded by the previous features (as the new Einsteinian questions were excluded by the Newtonian concept of space). (See ref. 20.) And surely similar considerations must hold for the relations between ordinary and Einsteinian concepts of space. Black's analysis, like that of Whitrow and his predecessors, seems (in Black's own words) "dominated by a simplified conception of the logical relations" (ref. 18, p. 180) between ordinary and scientific concepts.

To summarize our results so far, then: The views which deny the adequacy either of science or of ordinary language, and the views which deny the existence of any relation between scientific and ordinary concepts, have been rejected; we must suppose, at least as a working hypothesis, that scientific concepts are descended from ordinary ones. And the existence of contradictions between the ways in which the ordinary and the scientific concepts of "space" and "time" (and other concepts) behave is evidence that those scientific usages (and perhaps also the ordinary usages) have shifted, that mutations have occurred, in the course of the development of science (and perhaps also of ordinary language). (Differences in usage of terms in different scientific theories guarantees that the changes have not been confined to ordinary usage.) But all efforts so far to characterize the relations and the shifts have been inadequate or at best incomplete. If the sorts of paradoxes noted at the beginning of this paper are to be removed or understood, then philosophers of science must undertake to trace the relationships between ordinary and scientific usages of the terms involved in the paradoxes, and to examine the reasons why the departures of usage have been made. But the removal of paradoxes will not be the only fruit of such an investigation: for it, if anything, can be expected to reveal for us the rationale behind scientific advance, and so make clear to us the sense in which science does advance.

It is now possible to understand the hard core, or at least a large part of the hard core, of the question, To what extent does, and need, a physical theory deal with space and time? For one of the main reasons for asking that question is that many physical theories seem to be talking in such an odd, paradoxical way when they use the terms "space" and "time" or their formal analogs; and this is to say that their uses of those terms are odd when contrasted with ours. To this extent (which is the extent to which we have here considered it), the question can now be put in a new and more enlightening way: What are the relations (supposing there to be some) between the uses of the terms "space" and "time" (or their formal analogs) in a scientific theory and certain ordinary uses of those terms? And a full answer to this question, with regard to any particular scientific theory, would consist in a tracing out of the successive scientific departures from ordinary uses, and the reasons for those departures, which eventuated in the use of those terms in the theory in question. (See note 9.)

Questions like, Does general relativity really deal with space and time? Does quantum mechanics deal with particles? Does science talk about causes? Does it give explanations? cannot be answered by a simple "Yes" or "No." For different theories, especially highly sophisticated modern ones, use words like "space," "time," and "particle" in ways related, though distantly, to our ordinary corresponding terms, and give "explanations" which are related, though often distantly, to the kinds of things we ordinarily call explanations. Whether we decide to say that such theories deal with space, time, or particles, or give explanations or insight into causes, will be an indication of the distance of that relation; but beyond that, such choice of words will throw little light on the logic of the theory. The question is not *whether* science deals with space and time, but rather *what* it is doing when it uses the term "space" or "time" or a formal analog thereof; and also why it uses the term in those ways (for what good reasons it has come to use it in those ways) rather than in the ways in which we ordinarily use it.

## III

In approaching this problem of tracing the changes in the concepts of space and time from ordinary language through their evolution in scientific theories, what help can we hope to obtain from results already achieved by philosophical analysis? Do we at least have at our disposal analyses of the ordinary meanings of "space," "time," and their near relatives, and of the uses of corresponding terms in scientific theories, so that it will remain for us only to put these analyses side by side, compare them, extract the similarities and differences between the uses, and locate the rationale behind the differences?

Unfortunately, we get very little assistance in these directions from work accomplished so far by those interested in the analysis of ordinary language. For those philosophers have not yet applied their techniques of analysis to the terms "space" and "time" and their cognates with sufficient thoroughness. It is a pity, too, for despite Professor Austin's warning about avoiding any fields which (like Time) are not "rich and subtle," or which have been "trodden into bogs" by philosophical disputes,[13] this is a fertile and important area for cultivation. (See note 10.)

In our attempt to understand the role of the concepts of space and time in scientific theories, questions like the following will arise: What are the reasons why shifts of meaning are made from one theory to another? Are the shifts made because of "facts" which, while they are independent of theoretical interpretation, are accounted for by one theory but not by the other? Are there other considerations which, either in conjunction with theory-independent facts or, if there are no such, then by themselves, force or make advisable the shifts? And what is the character of these "further considerations"? In their studies of the different sorts of statements and concepts employed by science, and of the reasons for considering those statements and concepts acceptable or not acceptable, those philosophers of science of the past four decades who have allied themselves with the "Logical Empiricist" trend of thought have, at the very least, in their failures, brought to light some kinds of positions which cannot be accepted without drastic modifications. There are three main reasons why much of their work, and more generally their way of approaching problems, has failed to do justice to the questions and tasks outlined here.

First is the view, widespread for so long among Logical Empiricists, that there is something "wrong" with ordinary language, and that, therefore, if ordinary usage of the word "space" is contradicted by scientific usage, so much the worse for ordinary language. But as we have seen, the sense in which a concept occurring in a language can be "wrong" is itself in need of clarification; so even if there were something to the view they expressed, their way of putting it does not give much enlightenment about the paradoxes we have noted.

In the second place, Logical Empiricist approaches to the analysis of the internal structure of scientific theories tended to give a distorted picture of the roles of concepts within those theories. For, holding as it did that scientific theories are like interpreted axiomatic systems, Logical Empiricism suggested that a complete account of the concepts of a particular scientific theory is achieved when it is shown how those concepts can be defined in terms of a few primitive concepts (also occurring in the theory) which are related by "correspondence rules" to experience. But such a view of conceptual analysis ignores most of the work done by concepts within a scientific theory. It treats a scientific theory, in Wittgenstein's words, "like an engine idling." To illustrate this, let us note that the axiomatic view assumes that concepts in an axiomatic system are related to logically more fundamental ones (though perhaps not to epistemologically more fundamental ones) by definitional reducibility or not at all (see note 11). But is there not another possibility? Might it not be that the concept of space, for example, is related to other concepts in a scientific theory in a more intricate and interdependent way? In another paper,[20] I have argued that this is indeed the case: the geometrical concepts within a theory function partly to determine what can count, within that theory, as a fact, what can count as an entity, what can count as behavior of an entity, and what can count as an explanation of the behavior of an entity; and shifts in the spatial concepts of the theory entail corresponding shifts in the concepts of "fact," "entity," "behavior of entities," and "causal explanation" within the theory. If this is so, then reducing one such concept to another will lose much of the richness of the connections that exist between them within the theory; it will even falsify the way those concepts work together to form what we call a "theory."

Professor Feyerabend, in his contribution to this seminar, has called attention to a classical assumption about scientific theories, which he has called "The Condition (or Postulate) of Meaning Invariance," i.e., the assumption that the meanings of terms are constant from one theory to its successors (see note 12). Professor Kuhn has also voiced dissatisfaction with that supposition.[17] In this paper, I have shared that dissatisfaction. The point which I have just been discussing, however, may be said to be another unquestioned assumption of classical philosophy of science; it can be called "The Postulate of Meaning Independence," i.e., the view that, in an axiomatization of a scientific theory, there are only two sorts of terms: primitive terms, which are logically independent of one another, and defined terms, which are related to primitive terms by definitional reducibility; and further that, beyond these relations (and the relations of some terms to others which are correlated directly with experience), there are no further relations between the terms which are of importance for understanding the theory.

Finally, the third reason why the Logical Empiricist approach to the analysis of science has not dealt adequately with the questions asked here is that it tended to neglect important relations between scientific theories. For although it did raise questions about the comparative adequacy of different theories—their consistency, comprehensiveness, simplicity, and degree of confirmation or evidential support—nevertheless, in its concentration on axiomatization of single theories, Logical Empiricism was concerned, and believed it need be concerned, only with "complete" or "fully-developed" theories. As far as we learn from Logical Empiricism, a theory (like the theory of relativity or quantum mechanics) might just as well have been presented at any time, without any special problems to answer; the question of its acceptability or unacceptability need not depend on any consideration of any other scientific theory that has ever been presented. Hence the good reasons why, in the light of the shortcomings of preceding theories, particular sorts of changes of meanings and doctrines were introduced, were passed over by the Logical Empiricists; their view held no place for any "logic of discovery." Theirs was essentially a static picture of science, and therefore the dynamic questions, raised in this paper, of the logic of scientific change, of scientific advance, tended to be precluded from adequate treatment (to the extent that they were even admitted to be relevant) by the very approach itself.

## IV

In recent years, the growing dissatisfaction with the Logical Empiricist approach to the philosophy of science has led not only to increased concern with problems relevant to ours which that approach tended to slight, but also to efforts to formulate new approaches to the philosophical analysis of science. In connection with such efforts, one subject that has been much discussed lately is that of the relevance of the history of science to the philosophy of science; for it follows from the Logical Empiricist's static approach that the history of science is irrelevant to the philosophy of science. "The Mutual Relevance of the History and the Philosophy of Science" was the subject of a recent symposium [21] with Professors Adolf Grünbaum and Norwood Hanson as participants.

Neither Grünbaum nor Hanson attempts, in this symposium, to give us a clear picture of their conception of the aims and methods of the philosophy of science, and in the absence of any such account it is difficult to see how they could expect to give any coherent argument as to how the history of science is or is not relevant to those aims and methods. In any case, Grünbaum, who is generally more concerned in his paper with the relation of the philosophy of science to the history of science than with the converse, makes only the rather disappointing admission that "historical knowledge may indeed contribute to carrying out philosophical analysis of a theory. Thus, in the case of the special theory of relativity . . . historical inquiry might disclose, for example, the vicissitudes in Einstein's own philosophical orientation, thereby explaining . . . why advocates of contending schools of philosophic thought can each 'find some part of Einstein's work to nail to his mast as a battle flag against the others.'" However, according to Grünbaum, "what is known reliably so far about the *history* of the [special theory of relativity] has failed signally . . . to contribute to the clarification of the more subtle questions concerning its *epistemological* basis."

Hanson's view gives the history of science a slightly more positive role to play: according to him, history and philosophy of science have "a common concern in the structure and function of scientific *arguments*." From this he somehow concludes that therefore philosophers of science should be interested in past scientific arguments. Whatever this means, and however Hanson arrives at it, he believes that history of science has "*no* logical relevance whatever" to philosophy of science. The philosopher of science should, according to him, cite historical facts; but once this is done, those "facts are not germane to the sophisticated professional appraisal of the intellectual flight and logical maneuvers demonstrated thereafter." That is, as Hanson says, the historical facts are not relevant to the question of whether the argument is sound or unsound; they are only brought in to show the argument to be relevant to science. But what a strange contention this is! Surely the fact that the "soundness or unsoundness" of the argument must, in one sense, be judged according to the canons of logic (and in this sense, the "subject-matter" or "content" of the argument *is* irrelevant) does *not* justify Hanson's conclusion that therefore those historical facts and their interpretation have "no logical relevance whatever" in the sense that those facts "are not germane to the sophisticated professional appraisal" of the argument *as an argument about science.* For once the argument has been judged logically sound, it is exactly those adduced facts and their interpretation which will serve as the basis of sophisticated professional appraisal.

In the light of the arguments of the present paper, the function of the history of science *vis-a-vis* the philosophy of science is not the merely negative one of preventing philosophical error or confusion based on misunderstanding (although that is important). Nor does it merely furnish illustrative examples for contentions whose validity is to be judged independently of those examples. It furnishes at least part of the subject-matter for studying and analyzing those shifts of usage, and the reasons why such shifts were made, that lie behind the contradictions between ordinary and scientific language; and in this sense the history of science is logically related to the philosophy of science—at least as conceived in this paper. This concern with the history of science does not mean that the philosopher of science must commit the "genetic fallacy" of confusing the reasons for scientific change with its causes; for our concern is with those reasons which are generally taken, by competent scientists, to be *good* reasons for the changes (and, where there is widespread disagreement as to which reasons are good, with the reasons for the disagreement). Nor does the concern with the history of science mean that the latter (along with the relevant portions of ordinary language) constitutes *all* the subject-matter for the philosophy of science: for naturally, it is current science which, above all, furnishes the *live* problems, the paradoxes that bother us now.

Philosophy of science, then, at least insofar as it is concerned with problems such as have been outlined here, must examine the comparative anatomy and physiology of scientific theories and ordinary language. This is not to suggest that philosophy of science must limit itself to description and comparison of the structure and function of scientific concepts and theories, and of corresponding portions of ordinary language. For just as the study of comparative anatomy and physiology can disclose that certain organs in certain species are useless, and why, philosophy of science can do the same for science. It may even come across some new possibilities.

NOTES

1. Even this view, however, fails to do justice to the actual role of mathematics in physics. For the history of science reveals that in fact physics has managed successfully with mathematical techniques, the basic concepts of which were far from clearly developed. The calculus was not provided with a rigorous foundation until the work of Cauchy, Weierstrass, and others concerning the concepts of limit and continuity removed the vagaries

(criticized as early as Berkeley) of the intuitive notion of an "infinitesimal." [1] Likewise, a precise definition of "dimensionality" was not achieved until Brouwer's unnoticed work of 1913, and was not brought to the general attention of mathematicians until the work (independent of Brouwer and of each other) of Menger and Urysohn in 1922—long after geometries of various dimensions began to be used self-consciously by physicists.[2] The picture which philosophers often give of science as always taking over from mathematics fully precise and clear concepts is thus not correct.

2. Chief among these reasons suggested by quantum mechanics are arguments based on the uncertainty relations, according to which spatial position and momentum of a particle are not simultaneously accurately measurable; the same is true of energy and time. Thus, it is argued, these pairs of concepts are "complementary," and the concepts of space and time are not accurately applicable to microphysical events; the only reason that they seem to be accurately applicable on a macrophysical scale is the relative smallness of Planck's constant. (Sometimes the complementarity is expressed as holding between the concepts of space and time on the one hand and causality on the other: cf. references 5 and 6.) Moreover, according to the "Copenhagen Interpretation" of quantum theory, it is either unlikely or (in some statements of the view) impossible that we will be able to avoid classical descriptions of quantum phenomena; hence, though the concepts of space and time are, strictly speaking, inapplicable to such phenomena, we have no alternative but to use them. This orthodox view was well expressed by De Broglie in one of his writings: "The data of our perceptions lead us to construct a framework of space and time where all our observations can be located. But the progress of quantum physics leads us to believe that our framework of space and time is not adequate to the true description of reality on a microscopic level. However, we cannot think otherwise than in terms of space and time, and all the images that we can evoke are connected with them. Furthermore, all the results of our observations, even those which bring us the reflection of realities from the microphysical realm, are necessarily expressed in the framework of space and time. That is why we seek, for better or for worse, to represent to ourselves microphysical realities (corpuscles or systems of corpuscles) in this framework which is not adapted to them." [7] Elsewhere, De Broglie's view is somewhat more temperate: "In truth, the no-

tions of space and time drawn from our daily experience are valid only for large-scale phenomena. It would be necessary to substitute for them, as fundamental notions valid in microphysics, other conceptions which should lead to our finding asymptotically, when we go from the elementary phenomena to observable phenomena on the ordinary scale, the customary notions of space and time. Need we say that this is a difficult task? We might even wonder if it is possible that we could ever succeed in eliminating from this matter something which constitutes the very framework of our everyday life. But the history of science shows the extreme productivity of the human mind and we must not give up hope. However, as long as we have not yet succeeded in expanding our concepts in the indicated direction, we shall have to strive to make microscopic phenomena enter, more or less awkwardly, into the framework of space and time and we shall have the painful feeling of trying to enclose a jewel in a setting which was not made for it." [8] And still later, in describing his earlier views, to which, in the light of Bohm's anti-Copenhagen work, he has now returned, he says that "All real phenomena must be describable within the framework of space and time, and hence I felt that Schrödinger was wrong to think that the problem of *n* particles in interaction could be solved only by considering the propagation of a fictitious wave in generalized space. To my mind, the problem had to be posed and resolved by considering the interactions of *n* waves with regions of singularity in three-dimensional, physical space." [9]

Although the Copenhagen Interpretation, or some variation thereof, has become generally accepted, it has come under severe attack from many quarters in recent years, and a number of different viewpoints with regard to the role of space and time in quantum physics are currently distinguishable. Feyerabend, in criticism of the view that we cannot do without classical concepts, holds that "Although in reporting our experiences we make use, and must make use, of certain theoretical terms, it does not follow that different terms will not do the job equally well, or perhaps even better, because more coherently. . . . Any restrictive demand with respect to the form and the properties of future theories, any such demand can be justified only if an assertion is made to the effect that certain parts of the knowledge that we possess are absolute and irrevocable. Dogmatism, however, should be alien to the spirit of scientific research." [10] Nevertheless, detailed proposals of such alternative sets of concepts have so far not proved very successful; and Bohr has argued that "it would be a misconception to believe that the difficulties of the atomic theory may be evaded by eventually replacing the concepts of classical physics by new conceptual forms." [11] [See also ref. 12.]

The questions with which we will deal here will be, however, independent of such quantum-theoretical considerations and the problems to which they give rise. They will even, in certain respects, be logically prior to those problems, for we will be asking about the meanings of the terms "space" and "time" in the classical physics which the defenders of the Copenhagen Interpretation claim to be necessary for dealing with quantum phenomena.

3. Sometimes configuration space is taken to be "real" and "physical," but usually it is opposed (none too clearly) to "physical space." Thus one well-known text, in the context of an illustration of Hamilton's principle, states that "A merely *represents* for convenience the state of a system of several particles and is a point in what may be called *configuration* space to distinguish it from ordinary three-dimensional physical space. Hence the paths we are speaking of are not in general paths in physical space." [14] But this statement only tells us that talk about "configuration space" is *not* to be understood as talk about "physical space"; the role of the term "space" in the former connection, the rationale for using the terms "path" and "space" in that connection, the sense in which the subject of discussion "merely *represents* for convenience" (as opposed to exactly what?), all still require clarification.

Heisenberg relies on the peculiarities of the term "configuration space" for one of his criticisms of David Bohm: "Bohm considers the particles as 'objectively real' structures, like the point masses in Newtonian mechanics. The waves in configuration space are in his interpretation 'objectively real' too, like electric fields. Configuration space is a space of many dimensions referring to the different coordinates of all the particles belonging to the system. Here we meet a first difficulty: What does it mean to call waves in configuration space 'real'? This space is a very abstract space. The word 'real' goes back to the Latin word 'res,' which means 'thing'; but things are in the ordinary three-dimensional space, not in an abstract configuration space. One may call the waves in configuration space 'objective' when one wants to say that these waves do not depend on any observer;

but one can scarcely call them 'real' unless one is willing to change the meaning of the word." [W. Heisenberg, ref. 15; the same criticism is made in his "The Development of the Interpretation of the Quantum Theory" (ref. 12, p. 17).] To characterize "configuration space" as "a very abstract space," however, does not go very far toward clarifying the use of the expression. In general, the mathematical usages of the term "space," and the physical employment of those usages, have not been considered carefully enough by philosophers.

4. That this is not an adequate account of the relation between Newtonian and Einsteinian physics has been argued recently by T. S. Kuhn; [17] but see the criticism and amendment of Kuhn's argument in my review of his book (forthcoming in *Philosophical Review*).
5. This certainly goes too far in the case of the conceptions of time in most scientific theories: for one thing, not all the topological features are deleted or changed.
6. That physical talk about metric properties of time is merely talk about time differences is itself questionable: even the measurements of those differences presuppose specification of a time metric. But certainly the topological characteristics of time bring in features which go beyond talk about clock-measured differences.
7. This is, indeed, the way the difficulty about "the direction of time" makes itself felt when one tries to understand science: well-established physical laws (the Newtonian and Einsteinian laws of gravitation and dynamics, the laws of electrodynamics, those of quantum theory, etc.) do not provide any means of expressing our intuitive notion that time has a "direction"; the fact that events take place in the temporal order they do, rather than ever occurring in the opposite temporal order, thus appears from the viewpoint of physical theory to be accidental. But this conclusion is not very palatable: "Could we argue that all this is accidental and that we will discover some other physical law which clearly specifies the sense of time and which is responsible for giving us our ideas on the subject? This, I think, is not a plausible explanation, since systems we understand in detail seem to show time's arrow." [19] Then is this a shortcoming of current (or perhaps of any) physical theory? Or is our intuitive notion of time's "direction" a "subjective" notion which is inapplicable to physical reality?

It must be noted that Whitrow's discussion of the problem in

*The Natural Philosophy of Time* is extremely confused, conceiving the problem in terms of physical facts (observed initial and final conditions) rather than in terms of physical laws.

8. The misleadingness of Black's account is further complicated by his unusual conception of the distinction between symmetrical and asymmetrical relations. For a relation to be not asymmetrical does not, for logicians, imply that it is symmetrical ("loves," tragically, is neither), so that to say that the asymmetry of ordinary time-relations is "expurgated" does not, by itself, leave the remainder symmetrical. This confusion reflects, of course, on Black's view that the shift that has taken place is due to mere deletion.
9. An examination of current controversies about the interpretation of science might give one the impression that this task is only an incidental one where modern physics is concerned. For it would appear that developments in twentieth century physics, especially relativity and quantum theory, have produced experimental results and employed mathematical techniques for which no adequate language existed; and that the chief task of the philosophy of science is to contribute to the development of such a language. (Behind this view may lurk the idea that such a language is incidental to the theory itself; but I will not discuss this point here.) On this view, we should not await the development of such a language, in order to describe its relations to its predecessors–for then the issue will be dead; rather, we should enter the going battle, and help state criteria for the development of the new language. Thus Heisenberg speaks of the controversies that "developed around the problems of space and time raised by the theory of relativity. How should one speak of the new situation? Should one consider the Lorentz contraction of moving bodies as a real contraction or only as an apparent contraction? Should one say that the structure of space and time was really different from what it had been assumed to be or should one only say that the experimental results could be connected mathematically in a way corresponding to this new structure, while space and time, being the universal and necessary mode in which things appear to us, remain what they had always been? The real problem behind these many controversies was the fact that no language existed in which one could speak consistently about the new situation" (ref. 15, p. 174). The same fundamental problem, according to Heisenberg, lies

behind the controversies about the foundations of quantum theory.

Certainly such questions are of extremely great importance; and philosophy of science should be concerned with the live issues connected with understanding science. But a realization of this fact only increases the urgency of the tasks outlined in this paper; for what sorts of arguments might be adduced in recommending a new way of talking in science, or in setting up criteria for the adequacy of such a language? It may very well be that a critical survey of *past* reasons for making such changes will not provide *sufficient* grounds for endorsing a certain sort of change under new circumstances; for it may be that the new circumstances demand even fresh criteria of scientific adequacy. (Or it may be that ultimately a new scientific language is not adopted for any sort of rational reasons at all, but merely on the basis of "the climate of the times," or "metaphysical bias," or something of the sort—but in that case there is no question of argument anyway.) But can anyone deny that a clear picture of the ways in which science has found itself able to change, and of the reasons behind those changes, is of the utmost relevance to the discussion? Even where entirely new criteria seem called for, we may have much to learn about new options from a close examination of what has happened in previous similar circumstances.

10. Not only is such analysis needed for the tasks outlined here: an analysis of the ordinary usage of the terms "space" and "time" and their cognates also has important bearing on the analysis of other terms of ordinary language. For example, the concept of an individual, and the related concept of personal identity, are often analyzed in ways that utilize the terms "space" and "time"; but these latter terms are often taken for granted in such treatments, as though there were no problems concerning them. Yet one traditional version of the so-called "relational theory of space and time," to which much lip-service is still paid, attempts to remove certain problems concerning space and time by analyzing them in terms of relations between individuals. Surely one wants to know what is behind the terms "space" and "time" which are ingredients in these accounts of "individuals": for if space and time require for their analysis the very notion of an individual which they are supposed to clarify, then we are entitled to ask what sort of clarification has been achieved.

11. The parenthetical qualification is intended to bring out the irrelevance here of the view that theoretical terms are not definitionally reducible to observation terms. The point here is not of "epistemological" definability in experiential terms, but rather "logical" definability in terms of primitives of an axiomatic system.
12. Feyerabend, "How to be a Good Empiricist," this volume, p. 3. My disagreements with Feyerabend's presentation of his "Postulate of Meaning Invariance" stem largely from the fact that, in his anxiety to argue *against* a certain view (classical empiricism), he concentrates so heavily on the differences between the uses of terms in different theories that he fails to appreciate the positive implications of meaning variance in defining tasks for the philosophy of science.

## References


1. E. T. Bell, *The Development of Mathematics*, McGraw-Hill, New York, 1940, Ch. 13.
2. W. Hurewicz and H. Wallman, *Dimension Theory*, Princeton University Press, Princeton, New Jersey, 1941, Ch. 1.
3. J. A. Wheeler, "Curved Empty Space-Time as the Building Material of the Physical World: An Assessment," in E. Nagel, P. Suppes, and A. Tarski, eds., *Logic, Methodology and Philosophy of Science*, Stanford University Press, Stanford, California, 1962, pp. 361–374.
4. S. Alexander, *Space, Time, and Deity*, MacMillan, London, 1920, p. 35.
5. W. Heisenberg, *The Physical Principles of the Quantum Theory*, Dover Publications, New York, 1930, pp. 63–65.
6. A. D'Abro, *The Rise of the New Physics*, Vol. II, Dover Publications, New York, 1951, pp. 950–955.
7. L. De Broglie, "L'espace et le temps dans la physique quantique," *Proceedings of the Tenth International Congress of Philosophy*, Vol. I, North Holland Publishing Company, Amsterdam, 1949, p. 814. (Quoted in M. Jammer, *Concepts of Space*, Harvard University Press, Cambridge, 1957, p. 187.)
8. L. De Broglie, *The Revolution in Physics*, Noonday Press, New York, 1953, p. 220.
9. L. De Broglie, *New Perspectives in Physics*, Basic Books, New York, 1962, pp. 164–165.
10. P. Feyerabend, "Problems of Microphysics," in R. Colodny, ed.,

*Frontiers of Science and Philosophy,* University of Pittsburgh Press, Pittsburgh, Penna., 1962, pp. 230–231.

11. N. Bohr, *Atomic Theory and the Description of Nature,* University Press, Cambridge, England.
12. W. Heisenberg, "The Development of the Interpretation of the Quantum Theory," in W. Pauli, ed., *Neils Bohr and the Development of Physics,* McGraw-Hill, New York, 1955.
13. J. L. Austin, "A Plea for Excuses," *Proceedings of the Aristotelian Society,* **LVII,** 1–30 (1956–57).
14. R. B. Lindsay and H. Margenau, *Foundations of Physics,* Dover Publications, New York, 1957, p. 129.
15. W. Heisenberg, *Physics and Philosophy,* Harper, New York, 1958, p. 130.
16. P. K. Feyerabend, "Explanation, Reduction, and Empiricism," in H. Feigl and G. Maxwell, eds., *Minnesota Studies in the Philosophy of Science,* Vol. III, University of Minnesota Press, Minneapolis, 1962, pp. 28–97.
17. T. S. Kuhn, *The Structure of Scientific Revolutions,* University of Chicago Press, Chicago, 1962, pp. 100–101.
18. M. Black, Review of G. J. Whitrow, *The Natural Philosophy of Time,* in *Scientific American,* April, 1962, pp. 179–185.
19. T. Gold, "The Arrow of Time," in *Recent Developments in General Relativity,* Pergamon Press, New York, 1962, pp. 225–226.
20. D. Shapere, "The Causal Efficacy of Space," *Philosophy of Science* (to be published).
21. Meeting of the Eastern Division of the American Philosophical Association, Christmas, 1962; published in *The Journal of Philosophy,* Oct. 11, 1962.

CHAPTER 8

# INTERPRETATIONS OF SCIENCE IN AMERICA*

In the attempt to understand the nature of science – its aims and its methods – the dominant approach during the 1930's and 1940's was logical positivism, or, as it later preferred to call itself, logical empiricism. It is important to recognize that logical positivism was a movement rather than a unified doctrine, a movement within which there was much room for disagreement in matters of detail, in matters of emphasis, and in matters of principle. More and more, however, its adherents came to agree on three basic points.

(1) *The conception of what it means to "try to understand science."* The members of the movement came to agree that "to understand science" would be to understand concepts (or terms) used in *talking about* science. Thus, for example, one might say that "'$F = ma$' is a law," talking *about* the scientific statement $F = ma$ by assigning a predicate to it; and the problem of "understanding" is, What are we saying about $F = ma$ in calling it a "law"? Such terms as "law," and also "theory," "explanation," "confirmation," "evidence," "observation," they called *metascientific* terms (or concepts), in contrast to properly *scientific* terms like "force," "mass," "acceleration," "catalyst," "gene," or "superego." And to understand these metascientific terms – to understand, for example, what is involved in something's being an explanation (to understand the criteria which something would have to satisfy in order to count as an explanation) – would be to understand the "nature" of science.[1]

(2) *The central role, in such analysis, of formal logic*. To obtain such understanding of metascientific concepts, one was to concern oneself not with particular scientific ideas, which come and go, are accepted and rejected, in history, but rather with the general characteristics of, for example, *any possible* scientific theory, law, or explanation. One might say that they thought of these concepts as having a universal, or "eternal," meaning: to decide whether any particular scientific view, at any given time in history, was or was not a (possible) explanation of something, it would have to satisfy general, essential criteria. And the subject that dealt with "the limits [or range] of all possibility," as Wittgenstein had put it in his influential *Tractatus Logico-Philosophicus,*[2] was the newly-invented mathematical logic. The analysis of metascientific concepts was to be undertaken, and

formulated, then, in logical terms. Philosophy of science, the attempt to understand science, was thus, in Carnap's phrase, "the logic of science."[3]

(3) *The distinction between "theory" and "observation," or, more specifically and importantly, the independence of the latter from the former (and the dependence of the former on the latter).* As the preceding point, about the role of logic, accounts for the "logical" aspect of "logical empiricism," so the present one ties the movement to traditional empiricism. From the very beginnings of modern empiricism in the writings of Bacon, Locke, and Hume, one of the major motivations for making this distinction or its traditional analogues was to lay the basis for a defense of the claim that all ideas ("meanings," "concepts") and beliefs ("claims," "propositions," "theories") about the world of nature must be founded on the observable facts of that world. Observation, it was alleged more generally, is the basis of the interpretation, assessment, and comparison of theoretical ideas (concepts, propositions, *etc.*). For the moment let us focus only on the question of the assessment of theories. The argument then continued by saying that, if the assessment of individual theories is to be "objective" (as science, pre-eminently among human activities, was held to be), then that against which their adequacy is to be assessed, in terms of which their acceptability is to be tested, must be *independent* of those theories. Just as a person cannot be expected to be a fair and impartial judge of someone else if he is dependent on the person he is expected to judge, so also, in general, if judgment is to be unbiassed, unprejudiced, objective, then the grounds on which the judgment is to be based must be impartial; and, to the logical empiricist as to the traditional empiricist, this impartiality or objectivity required the absence of any "theory" permeating and slanting the observational evidence.

There were two important corollaries of these three theses. First: the object of study, in the attempt to understand science, was not the actual content of ongoing science, but rather the "logical form" of metascientific concepts. The conclusions of the analysis would be *applicable* to actual science (though it was never very clear whether those conclusions were to be taken as describing how science actually is, or as prescribing how it ought to be); nevertheless, those conclusions were *derived from* logical analysis, rather than from a study of actual scientific cases.[4]

Secondly: the attempt to understand science, concerned as it was with static, eternal, and unchanging conditions for being a theory (for example), was not to concern itself with the *processes* of development of science, but with the finished *product*. Indeed, as far as assessment of theories was concerned, only *justification after* formulation was of interest; there was no "logic of discovery."

But positivism was not destined to hold unchallenged sway as an approach to the interpretation of science. On nearly every point in its program – with regard to its specific attempts to analyze, for instance, the metascientific concepts of "law," "explanation," and "confirmation" – logical empiricism faced increasingly numerous and serious difficulties. But the focus of trouble, the weak pillar whose collapse was destined to bring the whole temple down, was the failure of the theory-observation distinction. Here again there were specific problems. It proved difficult to point to any clear, pure examples of either "theoretical" or "observational" terms, much less to formulate general criteria which would serve in any given case to categorize a given term or concept unambiguously as one or the other. But the ultimate reason for the reaction against the approach which began in the 1950's and reached a peak in the following decade, was not these specific failures: after all, the theory-observation distinction *was* based on an important motivation, namely, to explain how science managed to be "objective." And as long as that motivation appeared well-founded, the mere failure to have it satisfied *so far* was not sufficient to cause abandonment of the attempt. The reaction was based on something much deeper. For over the third through the seventh decades of the present century, it became increasingly clear that logical empiricism, and indeed empiricism in general, at least with regard to the above-mentioned motivation, involved a fundamental paradox: that the more theory-*independent*, theory-*neutral*, the observation-language was required to be, the more *irrelevant* to science it had to become. If, in order to guarantee the objectivity of scientific reasoning, the "observation-language" had to be entirely free of, independent of, theory, then how could it be relevant enough to serve as a *test* of theory? There thus appeared to be a tension between two necessary but apparently conflicting requirements of any analysis of science: on the one hand, the *condition of testability* of a theory in terms of observational evidence (or, alternatively, the *condition of relevance* of the observational evidence to the theory to be tested), which seemed to require that the latter (observation) have something to do with the former (theory); and, on the other hand, the *condition of objectivity*, which seemed to require that observation (or "the facts," "the evidence") be free of, independent of, theory.

A massive literature was devoted to trying to establish connections between theory and observation (as regards both meaning and acceptability of the former). It was alleged, for example, that theories, or theoretical statements, had as logical consequences observational predictions in terms of which they could be tested; but it was not easy to see how theoretical premisses could lead to conclusions in which an entirely different sort of term occurred; and even if observational premisses (*e.g.*, statements of "initial conditions") were added, the lack of a common term in the theoretical and observational premisses left the same difficulty of how the transition to the conclusion could be achieved. Such difficulties might have been alleviated or removed had it been possible to show that theoretical *terms* could be defined, at least partially, through a pure observation language; for such definitional postulates, connecting the theoretical and observational terms in the other premisses and in the conclusion, could have been added to the list of premisses. Unfortunately, all such attempts failed. Retreats from insisting that the observation-language be "phenomenalistic" to less stringent requirements (for example, that it be a "physical-object" language) accomplished little or nothing, or less, in the end. For the above problems still remained, or were even aggravated. If "electron" could not be defined in terms of a cluster of "pure observation terms," presumably something like "red-here-now" or "pointer reading at $x$," then surely it could not be defined more easily *via* a "physical object language" made up of words like "table" and "rock." But in the end what was far worse was that such retreats seemed to critics to amount to admitting theoretical interpretation into the observation language: isn't calling that brown patch a table an interpretation, implying the presence of a solid three-dimensional object? But such an admission would be tantamount to abandoning the view that our concepts and beliefs are ultimately grounded in observation. And the abandonment of the effort to set up a one-way dependence, making theory dependent on observation but not vice-versa, appeared to sacrifice the foundational motive of the entire empiricist tradition, the condition of objectivity. Whatever other motivations there were (particularly the condition of testability or relevance) that led to attempts to relate theory and observation through meaning or deduction (and even apart from the universal failure to specify an unambiguous "observation language"), any such chain of connection became either so short as to be open to the threat that theory had been surreptitiously introduced into the alleged observation language, as with Carnap's admission of a "physicalist" vocabulary as the observation language, or else, in the effort to satisfy the condition of objectivity, so long, tortuous, and tenuous as to be invisible.

It would not be completely accurate to say that the difficulties were understood in just the way I have outlined them during the critical period of the 1950's and 1960's; indeed, it was not until a strong reaction had set in, and not even until the deficiencies of that reaction had themselves been fully grasped, that, in retrospect, this tension between the conditions of testability and objectivity could be perceived as in fact lying at the root of the issue. But since it can now be seen that it did, I will interpret the course of events, and the problems at issue, in terms of this tension.

Reaction to positivism began to arise in the early 1950's. N. R. Hanson, responding partly to the difficulties of making the theory-observation distinction in a sharp, clear way, maintained that what we consider to be observational is, and must be, permeated with theory – that is, in his graphic and revealing phrase, observation is "theory-laden."[5] (I will return later to why this phrase is revealing.)

Far more extreme views began to appear in the late 1950's and early 1960's. Paul Feyerabend suggested that experience, observation, requires interpretation in order to turn it into evidence relevant to theory.[6] And by the early 1960's he was maintaining that even what we count as an observational fact is determined by a prior and fundamental theoretical perspective, a background of presuppositions, and that different such perspectives uncover – or perhaps even create – different sets of observations.

Thomas Kuhn joined the attack in his 1962 book, *The Structure of Scientific Revolutions*, without question the most influential book on the interpretation of science in the past quarter century. Despite their differences, Feyerabend and Kuhn agreed on three basic points.

(1) That what counts as an observation – and also what counts as a scientific problem, a scientific explanation, law, *etc.* – depends crucially on some background of presuppositions (Feyerabend: "high-level background theory"; Kuhn: "paradigm").

(2) That different high-level background theories or paradigms determine different observations, problems, explanations, *etc.*, or criteria thereof.

(3) That such changes of fundamental perspective have taken place (and perhaps should, and perhaps must, take place) in the history of science.

From the standpoint of the Feyerabend-Kuhn view, the criticisms of logical empiricism, though usually implied rather than explicit, are profound. If what counts an observation, for example, or an explanation, law, *etc.*, varies from paradigm to paradigm, then the metascience-science distinction, *at least in its positivistic version,*[7] must also fail. For that distinction was based on the idea that the criteria for being an observation or an explanation had to do with *all possible* observations or explanations. And with the demise of that idea, the central role in the analysis of science of formal logic, conceived as the study of the total range of possibilities, also had to be abandoned. And, finally, the two corollaries of the three propositions with with we began also had to be rejected. If what counts as an explanation is dependent on high-level presuppositions which are themselves inseparable parts of the changing content of science, then to try to understand science we *must* return to an examination of actual science and its historical development, its alterations of fundamental perspective or paradigm, its shifts of criteria for being an explanation. Philosophy of science is then no longer a branch of logic, loftily disdaining, except for purposes of occasional illustration, the ephemeral fates of particular scientific ideas. *Even the conception of what it means to try to understand science has undergone a radical and fundamental change.*

However, the price of reaction – as is so often the case, perhaps necessarily, with reactions – was high. It was not simply that concepts like Kuhn's "paradigm" or Feyerabend's "high-level background theory" were distressingly vague and ambiguous – so much so that many of their claims appeared supported by twisting the history of science into their mould, rather than taking their inspiration from it. Nor did the dissatisfaction with their views result simply from the others of the multitude of specific objections that have been levelled against them.[8] As in the case of the rejection of logical empiricism, it was not so much the specific objections, however impressive they became in their cumulative effect, that has led to the search for an alternative approach. Rather – again as with logical empiricism – it was ultimately a matter of more fundamental principle. For in tying observation, or evidence, or, more generally, the reasons for accepting or rejecting a theory (at least a "fundamental" theory or "paradigm") to that theory, the role of that "evidence" *as evidence* for or against the theory, was destroyed. If "observation" is determined by the (high-level) theory (paradigm), then how can it serve its traditional purpose, as the ground for assessing that theory? In thus relating observation so tightly to theory, Feyerabend and Kuhn appeared to have gone too far in reaction against positivism or empiricism, and to have made observation – the very thing that should serve as independent test of theory – completely determined by it. In assuring relevance, they had destroyed objectivity – as logical empiricism had done the exact opposite, destroying relevance in the process of trying to assure objectivity.

The point is so important that it deserves re-emphasis. The logical empiricists, motivated by the desire to analyze and ensure the objectivity of science, focussed on the "independence" aspect; but in doing so, they found themselves confronted not only with the difficulty of specifying any such completely independent element, but also by the fact that, if there were anything as independent as they tended to demand, it would be *too* independent to do the very job they wanted it to do; it would, that is, be irrelevant. The critics, impressed as they were – partly through the history of science, and partly through purely *a priori* considerations – by the influence of presupposition on the interpretation of evidence, and by the need for such interpretation in order to ensure relevance, focussed on that aspect to such an extent that the "independence" aspect was lost, and with it the very objectivity that had been the empiricists' pride in science.

Thus, in their very attempt to reject or revise the "theory-observation" distinction, the critics of logical empiricism have left unsatisfied the original motivation – which seems a legitimate one – of the distinction, to explain the sense in which science can be "objective" and "rational." The result, for Feyerabend and Kuhn, has been a complete relativism with regard to science and scientific change, or at least major scientific change: each paradigm, tradition, community determines what counts as a good or bad reason; different paradigms, different traditions, different communities count different things as good and bad reasons, and there is nothing left in terms of which to assess the relative merits of different paradigms. The logical conclusion is clear in Kuhn's words, however much he attempted to qualify them in other places or later writings. A paradigm is, according to Kuhn, "the source of the methods, problem-field, and standards of solution accepted by any mature scientific community at any given time";[9] after a scientific revolution "the whole network of fact and theory . . . has shifted."[10] Hence "the reception of a new paradigm often necessitates a redefinition of the corresponding science. ... The normal-scientific tradition that emerges from a scientific revolution is not only incompatible but often actually incommensurable with that which has gone before."[11] It follows that "the competition between paradigms is not the sort of battle that can be resolved by proofs,"[12] but is more like a "conversion experience."[13] "What occurred was neither a decline nor a raising of standards, but simply a change demanded by the adoption of a new paradigm."[14] "In these matters neither truth nor error is at issue."[15] Similar remarks could be cited from Feyerabend.[16]

It seems probable to me that Kuhn originally wanted to show that science, far from being a routine mechanical cranking out of results according to a prescribed method, without interesting intellectual content, was really creative, like art. But the implication was really profoundly *anti-scientific*; for it was, rather, that what one accepts in science is only a matter of taste, like art. (We may for the present leave aside the question of whether all his may be unfair to art.)

Whatever the original intent, the relativism consequent on the views of Kuhn and Feyerabend has left the philosopher of science with a central problem today: whether to accept that relativism, or to attempt to escape from it by a closer analysis of scientific reasoning. Are there any indications in the present situation that suggest that the latter course may indeed be possible without returning to the sterility of the older positivistic empiricism? A resurvey of both logical empiricism and the views of Kuhn and Feyerabend will indicate that it is far from necessary to accept the more disturbing conclusions of either, and will further suggest some directions in which a new approach might hope to proceed with success.

The logical empiricist argument which we have surveyed may be summarized as follows:

(1) Observation must be independent of, neutral with respect to, the theory to be assessed. (This is the condition of objectivity.)
(2) "Interpretation" is always in terms of "theory."
(3) Observation must be neutral with respect to *all* theory.
(4) The observation language must be the *same* for all theories.

There are some enormous logical gaps in this sequence of claims. That observation must be "neutral" with respect to the theory to be assessed does not imply that observation cannot be interpreted by *some* theory. The condition of objectivity requires merely that the particular observations which will be used in testing a particular theory must not be "interpreted" (at least in any dangerous way) by *that* theory; some other theory may, however, be utilized in the interpretation, as long as there is justification for using that theory. Thus "interpretation" might be in terms of prior theory – or, more generally, some prior information available to the tester. (I will not discuss 2, which the two views hold in common.) Similarly, it does not follow – at least from the condition of objectivity alone – that observations cannot be interpreted differently (expressed in different descriptive terms) under different circumstances, in different theory-testing situations. If they can be, the "observation language" itself might conceivably develop with the development of science.

The Feyerabend-Kuhn view asserts the following propositions.

(1′) Observation, if it is to be relevant, must be interpreted. (Condition of testability or relevance.)
(2′) That in terms of which interpretation is made is always theory.
(3′) The theory that interprets is the theory to be tested.

(4′) The theory to be tested is "the whole of science" (or of a branch thereof).

(5′) This whole forms (some sort of) a unity ("paradigm," "high-level background theory").

(6′) This unified whole not only serves as a basis of interpretation, but also determines ("defines") what counts as an observation, problem, method, solution, *etc.*

The transition from (1′) to (6′) might be called "the rocky road to relativism"; for, as should by now be clear, relativism with regard to so-called scientific "knowledge" is the inevitable consequence of (6′). But is this proposition itself a consequence of the condition of relevance? Indeed, are any of (3′) to (6′) consequences of that condition? It does not seem so: again, why must the theory in terms of which interpretation is made be the theory under test? This certainly does not appear to follow from (1′). And similarly there seems no ground for believing that (4′)–(6′) follow either.

Thus the unacceptable theses of either side of the controversy do not seem to follow from the intuitively plausible insights from which they depart. Of course, there may be some subtle way in which those consequences do follow from some subtle interpretation of (1) and (1′); but, though the possibility of such interpretations must be examined, the conclusions in question do not seem, at least *prima facie*, to follow. Again, (3)–(4) and (3′)–(6′) may be true on grounds *other* than (1) and (1′); or they may follow from (1) and (1′) in conjunction with *other* propositions. This question, too, must be investigated. Finally, our intuitive feeling that (1) and (1′) are adequacy criteria for any philosophy of science might be mistaken; and this possibility also must be examined. However, if the undesirable consequences of the two conditions do not follow from them – as I have argued they do not, at least on any immediately apparent interpretation of them – then there is also no obvious reason why they cannot be compatible with one another. More exactly, the condition of objectivity might be satisfiable even where interpretation is present: for the observations being applied in the test of a theory might be interpreted in terms of other theory (or other information, more generally).

And so it might be possible after all to maintain the legitimate motivations underlying both logical empiricism and the contentions of Kuhn and Feyerabend. Naturally, that must be shown; all I have argued here is that some of the apparent barriers are not there. But this line of attack is now open, and at least initially plausible. Taking such a course would require the development of a revised understanding of the concept of "being objective," one which would recognize that, while ideas are subject to test in terms of independent criteria, those criteria themselves are constructed on the basis of prior ideas; the theory tested, however, must be fundamentally independent, at least in any "vicious" way, of those particular prior ideas.[17] That is, though "observation" may be "theory-laden," it need not be "loaded" by the theory it is to test; and its relevance to the latter theory may be determined on grounds that do not destroy its independence. Whether such a line of attack can escape the relativism into which philosophy of science was plunged in the 1960's remains to be seen.

## NOTES AND REFERENCES

[1] The logical empiricists, of course, did not often speak of themselves as attempting the analysis of the "nature" of science; however, especially in view of the second of the three major points in my summary of the movement, that term seems quite appropriate.

[2] L. Wittgenstein, *Tractatus Logico-Philosophicus*, London, Routledge and Kegan Paul, 1949.

[3] R. Carnap, *The Logical Syntax of Language*, London, Routledge and Kegan Paul, 1937.

[4] Many of the early logical empiricists were scientists themselves, or at least had extensive knowledge of science; and many of them did, in their writings, make detailed studies of scientific cases. However, the general positivistic approach did not encourage such studies, and, more and more in the 1930's and 1940's and after, books and articles representative of that movement, while they were littered with logical symbolism, contained little or no mention of actual science. Even those positivists who did attend to actual science in their writings rarely made reference to the historical development of scientific ideas; indeed, given the quite primitive state of study of the history of science in the 1920's, they were not in a position to do so with any success.

[5] N. R. Hanson, *Patterns of Discovery*, Cambridge, Cambridge University Press, 1958.

[6] See especially P. Feyerabend, "Problems of Empiricism," in R. Colodny, ed., *Beyond the Edge of Certainty*, Englewood Cliffs, N.J., Prentice-Hall, 1965.

[7] The distinction might be retained in various ways. For example, a distinction might be made between the "meaning" of a term and the "criteria of application" of that term (criteria for determining "what counts" as falling under that term). In that case, it could be admitted that the latter change over the history of science, while the former do not; and the philosopher of science could be held to be concerned with the former. (However, this approach cannot be maintained: *cf.*, the discussion of the concept of a "concept" in the Introduction to the present book and in a number of the included papers. [Comment added, 1982.]) It might be objected against the critics of positivism that they must presuppose a transparadigmatic (or transtheoretical) criterion for identifying what counts as, *e.g.*, a criterion of being a possible explanation within a paradigm, a presupposition which would seemingly contradict their contention that such transparadigmatic criteria do not exist.

[8] In addition to the relevant essays included in Part I of this volume, see also I. Lakatos and A. Musgrave, eds., *Criticism and the Growth of Knowledge*, Cambridge, Cambridge University Press, 1970, especially the essay by Margaret Masterman.

[9] T. S. Kuhn, *The Structure of Scientific Revolutions*, Chicago, University of Chicago Press, 1962, p. 102.

[10] *Ibid.*, p. 140.

[11] *Ibid.*, p. 102.

[12] *Ibid.*, p. 147.

[13] *Ibid.*, p. 150.

[14] *Ibid.*, p. 107.

[15] *Ibid.*, p. 150.

[16] See D. Shapere, "Meaning and Scientific Change," in R. Colodny, ed., *Mind and Cosmos: Essays in Contemporary Science and Philosophy*, Pittsburgh, University of Pittsburgh Press, 1966, pp. 41–85.

[17] (Note added in 1982:) This is not, however, the solution I ultimately came to adopt; *cf.* the following remark from "The Concept of Observation in Science and Philosophy" (*Philosophy of Science*, December, 1982). "Some philosophers (most vociferously, Feyerabend) find grounds for accusing science of subjectivity in the fact that the same theory is employed as background information in the theory of the source, the theory of the transmission, and the theory of the receptor of information. . . . How, these philosophers ask, can we expect objectivity in science if the same theory is used in formulating the "hypothesis to be tested" (theory of the source) as in setting up the test of that hypothesis (theory of the receptor)? Indeed, we may add that, given the way we have found the world to be, it is *necessary* that this be the case: for all quantum-based theories imply certain symmetries between particle emission and particle absorption, and that implies, for example, that neutrino emission and capture are both described by the same theory (weak interaction theory). But this fact by no means makes it impossible that weak interaction theory might be questioned, modified, or even rejected as a consequence of the experiment. It is not a logical or necessary truth that it could be so questioned; but *as a matter of fact*, we find that, despite the employment of the same theory in our account of both the source and the receptor, disagreement between prediction and observation results. And that disagreement could eventuate in the alteration or even rejection of weak interaction theory despite its pervasive role in determining the entire observation-situation. Feyerabend's argument supposes that conflict between prediction and observation will not, and perhaps could not, arise in such a case, and this mistaken idea is connected with the view that the world is a mere construction of our theories. What better proof that there is a theory-independent world could we ask for than the occurrence of such a conflict – the fact that there is "input" into the observation-situation which is independent of, and can conflict with, our theories?" Further discussion is, of course, to be found in that article and in the summary included in this volume; still more will be found in the forthcoming book of the same title.

CHAPTER 9

# UNITY AND METHOD IN CONTEMPORARY SCIENCE*

*Listen: there's a hell of a good universe next door; let's go.*
—e.e.cummings

## I

One of the most striking features of the history of science is that, despite numerous setbacks along the way, there has, overall, been a tendency toward the development of more and more comprehensive unification of the various fields of science. Newton fused terrestrial and planetary motions into a unified theory at the end of the seventeenth century. But it was in the second half of the nineteenth century that unification of scientific fields began in earnest. In spite of the differences between electricity and magnetism (and between various types of electricity) which had been noted by investigators beginning with Gilbert, Faraday was able to provide a unified treatment of those types of phenomena. His approach was developed and given a mathematically precise formulation by Maxwell, who was also able to extend it further by incorporating light into the same theory. The study of heat too was assimilated to other areas, partly to the theory of radiation and partly to the theory of the motion of particles. As if in protest against what Einstein called the "profound formal difference" between the nineteenth century's treatment of electromagnetism and light by a continuous (wave) theory on the one hand, and of matter in terms of a discrete particle theory on the other, the twentieth century provided a unification of those domains in

* [Note added, 1982] Some of the scientific ideas described in this paper – particularly in the portions on elementary particle physics – are by now outdated, the paper having been written in 1976. However, the more general points I made here remain valid, embodying as they do features which have been characteristic of the scientific approach throughout recent times and remain so despite the dramatic advances of the past six years. Since those points, and the features of the scientific cases on which they rest, are important for the general viewpoint developed in this book, I have therefore kept the article in its original form. There would have been little point in bringing the scientific discussions up to date, only to find them again superseded in a few years, unless it could be shown that science has altered so drastically in the past six years that those features are no longer found in it. And even then, as always, there might be important lessons to learn even from such ancient science as that of the mid-1970's.

the quantum theory. With that theory came, too, an understanding of the periodic table of chemical elements and of the bonding of those elements into larger complexes, as well as of the spectra of chemical substances.

The years between the development of quantum mechanics in the late 1920s and the Second World War were concerned largely with extension rather than with unification of pre-existing areas: the period saw, among other things, the beginnings of understanding of the electromagnetic force (beginnings of quantum electrodynamics), of radioactive decay, and of the nuclear force (Yukawa exchange-particle theory). Postwar physicists generalized these results in a clear recognition that there are three fundamental types of forces important in elementary particle interactions, the electromagnetic, the weak, and the strong, and in particular developed the theory of the electromagnetic force, quantum electrodynamics, into the most successful scientific theory ever devised. (The fourth recognized force, gravitation, is too weak to play any effective role except where large masses are involved.) But the vast number of particles discovered during the postwar period, and the existence of four apparently independent fundamental forces, again awakened the urge toward unification. As regards the multitude of elementary particles, Gell-Mann and Ne'eman showed that the hadrons (particles interacting via the strong force) fall into well-defined families which are representations of the symmetry group SU(3). A further representation of that group, corresponding to no known particles, constituted a family of three; when appropriate quantum numbers were assigned to the members of this family, it was found that the quantum numbers of all hadrons could be obtained as the result of adding the quantum numbers of this family in pairs (mesons) or triplets (baryons). These "quarks" could therefore be considered to be the constituents of hadrons. More recently still, the addition of further quantum numbers ("charm" and "color") and of a fourth quark, in order to account for the stability of the $\psi$ (or J) particle, has produced the suggestion, at least, of an analogy between hadrons and leptons (particles not participating in the strong interaction): as there are four fundamental leptons, (electron, muon, and their respective neutrinos), so also there are now four fundamental hadrons; like the leptons, the quarks appear to be dimensionless points and therefore true candidates for the status of being "fundamental" (i.e., with no further internal structure); and the leptons and quarks have the same spin (½). The analogy is far from complete: for example, leptons do not combine to form composite structures, as do quarks; further, leptons are observed in experiments, while quarks, if they exist as free particles, seem somehow to evade all attempts to observe them. (The fact that the charges of quarks are fractions of those of leptons may not be a serious disanalogy, as Han and Nambu have shown.) Nevertheless, the analogy is there, and its existence, coupled with increasing experimental success of the four-quark hypothesis, cannot help but suggest some deeper relationship between those particles which interact via the strong force and those which do not.

As regards the four fundamental forces, after enormous obstacles had been overcome, a unification of the electromagnetic and weak interactions is now available (unified gauge theory of Salam and Weinberg), and there is hope of extending that theory to cover the strong interactions. The unified gauge and colored- and charmed-quark theories are bound together, so that the prospect of a giant step in the direction of a unified theory of elementary particles and three of the four fundamental forces now lies before us. (The possibility that there are further forces—and indications are growing that there are at least two others, a "superweak" and a "semistrong" force—could not detract from the significance of such unification as has, hopefully, been achieved, but would only set a challenge for further unification.)

The application of spectroscopy to analysis of chemical composition made it possible—despite Auguste Comte's pronouncement of the impossibility of our ever knowing the composition of the stars—to ascertain that the stars are made of the same substances as are found on earth. The development of an understanding, in terms of quantum theory, of how spectra are produced, together with the theory of elementary particle interactions, has provided an understanding of the processes of stellar evolution, and has, with the further occasional cooperation of the theory of general relativity, led to an understanding of the synthesis, evolution, and relative abundances of the chemical elements. The alliance of general relativity and elementary particle theory, especially when coupled with recent observations, has even made possible reasonably-based theories of the origin of the universe in a "hot big bang."

Thus far I have surveyed some examples of unification in the physical sciences; but the biological sciences cannot be omitted from this picture. Some examples of unification in those areas are the following. Despite the apparent conflict between Darwinian evolutionary theory and Mendelian genetics at the beginning of the twentieth century, those areas were shown to be consistent by Fisher, Haldane, and Wright; further integration at the hands of Dobzhansky, Rensch, Simpson, Mayr and many others, produced a "Synthetic Theory of Evolution" which was at least consistent with, and to a considerable degree explanatory of, evidence from a number of fields which had hitherto seemed incompatible. Chemical understanding of biological inheritance began to be achieved in detail in the mid-twentieth century, and has begun to penetrate the area of organismic development. The work of Oparin and his successors showed that Darwin had been too pessimistic in forecasting that the origin of life could never be an object of scientific investigation, and numerous mechanisms are now known for the production of at least the basic constituents of self-replicating macromolecules. The study of the animal brain, though still rudimentary, has produced associations of various conscious functions with specific regions of the brain, giving reason to expect that improved understanding of, for example, human psychology will ultimately be forthcoming from such investigations. (Except for this remark, I will limit my discussion in this paper to the physical and biological sciences.)

Taken together, the developments I have described provide a broad and coherent picture of the universe and man's place in it. In outline, the picture goes something like this. The universe—at least the one we know—began in (or at least became after a fraction of a second) a hot (of the order of $10^{12}$-$10^{13}$ K) dense (around $10^{14}$ g/cm$^3$) soup of elementary particles. From this state, after a few minutes (for which period the detailed calculations that can be made are nothing short of mind-boggling), emerged a matter-radiation equilibrium in which the matter consisted of roughly 75 percent hydrogen nuclei (protons) and 25 percent helium nuclei (alpha particles). Further cooling enabled electrons, after a few hundred thousand years, to combine with these nuclei, ending the matter-radiation coupling. At some time a few million years later, inhomogeneities developed in the cloud of matter; these inhomogeneities, or at least those of an appropriate size, collapsed gravitationally to form clusters of galaxies, sub-inhomogeneities (or at least those of an appropriate size) collapsing within the larger inhomogeneities to form galaxies. As the galaxies took shape, stars began to form and evolve within them; under the conditions of high temperature and pressure existing in the interiors of those stars, hydrogen and helium underwent nuclear reactions which produced many of the heavier elements; those stars which ultimately died in violent explosions (and which, in the process of exploding, created further heavy elements) spewed those heavier elements into the interstellar medium, where they became available for the birth of later-generation, heavy-element-enriched stars. Some proportion of those later-generation stars can be expected to be born with planetary systems; on some of these planets, under favorable circumstances, the production of complex molecules, and ultimately of self-replicating macromolecules, will eventuate in the evolution of higher forms of life. The transmission of hereditary information from generation to generation of these living creatures, as well as the variations in that information which lead to evolutionary changes, and the development and functioning of individual organisms, can be understood in terms of chemical processes and the physical processes which affect and ultimately explain them.

We thus obtain a coherent view of the evolution of the universe and of life in it as a continuous process understandable in terms of the same ultimate laws. Not all of the parts of this picture are equally well-grounded. I am not thinking here of the fact that the "details" of the picture are far from being worked out, or that the picture has yet to be extended to many areas; although the working out of details often produces surprising problems which ultimately upset the grander scheme, the difficulties I am thinking of are known ones which cannot be dismissed as due merely to lack of detailed knowledge. Nor am I thinking of the fact that parts of the picture may ultimately come to be looked on as mere "limiting cases" of some larger theory. Let me explain the sorts of problems I have in mind by beginning with an example of a difficulty that *can* be considered to be the result of a need to "fill in details." This difficulty concerns the birth of stars from an interstellar medium. Small dark globules and highly localized infrared sources have been observed which are presumably "protostars," stars (or at least stellar "placentas") in the very early stages of birth, fragmenting from a larger cloud of gas and dust and collapsing toward a stage where nuclear reactions will be initiated in their interiors. And certain stars are observed which, there is reason to believe, are only relatively recently born. However, how the protostars uncouple from their environment, and what happens between the observed putative protostar stage and the emergence of the fully-born star is unknown.

Though the critical factors can be listed, their precise contributions are uncertain and the problem is highly complex. Nevertheless, there is good reason to believe *that* the general "fragmentation-and-collapse" account of stellar birth is "on the right track," and that it is only the details of *how* this takes place that are not yet understood.

The situation is presumably the same with regard to the transition from laboratory synthesis of relatively small constituents of DNA (or some presumably primitive analogue) to the production of the giant self-replicating macromolecule itself: the gap is enormous and not yet understood. Here, however, there is a hitch that was not present in the case of the "gap" between protostars and stars: in the case of stars, observational evidence indicates that star formation is relatively widespread and therefore has at least a decent probability of occurrence given a proper interstellar medium and proper conditions which are themselves fairly common. But in the case of the evolution of life from inorganic molecules, we have only one planet to look at—one case in which such evolution has actually occurred. And in the absence of an understanding of a mechanism for evolution from relatively small macromolecules to huge self-replicating ones, we do not know what the probabilities of such an occurrence are. In spite of the overwhelming statistics that astronomers like to throw at us about the probable numbers of planets in the universe having conditions favorable to life (once produced), the probability of nature jumping the gap from chemistry to biology may yet be so low that earth's life may be unique or near-unique in the universe, vast as it is. The "filling in of details" in this case thus has ramifications beyond those of the star-formation gap: there, the probability of gap-jumping—the existence of a mechanism, however little understood, which is widespread in the universe—was not in question; here, it is.

Yet a third level of difficulty, this time by no means reducible to a matter of "filling in details," occurs in the case of the birth of galaxy clusters and galaxies. Luckily for theoreticians, observational evidence (such as the apparent black-body character of the 3°K microwave background radiation) indicates that the universe is, in the large, homogeneous and isotropic, and that such conditions held in the early universe, before galaxies and clusters of galaxies fragmented out. Where, then, did the *in*homogeneities postulated in our picture of the origin of galaxy clusters come from? (The theoretician no longer seems so fortunate.) In a universe in which quantum theory holds, fluctuations would arise spontaneously; these *might* be amplified in the relatively dense conditions of the early universe, and inhomogeneities of the proper size (galaxy-cluster-size and galaxy-size) *might* be selected out as ones which would collapse rather than be washed out. But thus far, no good theory of galaxy (or galaxy cluster) formation along these or any other lines has been forthcoming. However, the situation here is potentially more dangerous than mere lack of a good theory might suggest. Two independent considerations in particular make this so. First: evidence is now very strong that the nuclei of galaxies are regions from which enormous amounts of matter (and energy) are expelled, presumably periodically. Ambartsumian and others have suggested that the expelled matter is the source of new galaxies. The nuclei of the new galaxies might in turn carry the matter-producing capability. (There is even some hint of evidence that some small galaxies may have been ejected from larger ones.) On this view, the collapse theory of galaxy (or galaxy cluster) formation, already burdened with the problem of the origin of appropriately-sized inhomogeneities, would be rejected in favor of a "little-bang" theory of their origin. The second relevant consideration is this: why are clusters of galaxies still in existence if, as the cosmological inhomogeneity theory alleges, they were formed several billion years ago? If they are to have the gravitational stability required for such a long life, the clusters would have to have a certain mass to counterbalance their motions. The observed masses, however, fall short of the minimum required in every specific cluster for which mass-estimates are available, and by factors ranging from five to fifty or more. (Needless to say, estimating the total mass of a cluster of galaxies is a risky business; but could the observational uncertainties be of such a magnitude? and all in the same direction, of underestimation?) Is the missing mass present in "invisible" form—black holes, dead stars and galaxies, for example? There are difficulties with such proposals, and one must remember with soberness that once before when there was a "missing mass" problem (the advance of the perihelion of Mercury), the problem was solved not by finding the missing mass but by revolutionizing physics. Taken together with the problem of the origin of inhomogeneities, these two considerations cannot but lessen our confidence in the "collapse of cosmological inhomogeneities" part of our picture; they suggest that Ambartsumian's alternative cannot be dismissed out of hand. Indeed, were it not for the black-body character of the microwave background radiation, the observed helium abundance in the universe, and some other difficulties, one might be tempted to reconsider the Steady-State theory, with matter being continually (or at least sporadically) created in galactic nuclei rather than in intergalactic space. Nevertheless, although the history of science has often witnessed supposedly dead theories rise, phoenix-like, from their ashes, the objections against the Steady-State theory seem at present insurmountable; and Ambartsumian's suggestion is too undeveloped to be called a theory. (And how are we supposed to account for the huge amounts of matter and energy somehow produced out of the galactic nuclei? Would it be genuine creation *ex nihilo*? What would then become of the principle of conservation of energy? An alternative that has been suggested—not easy to swallow—is that matter comes through a "white hole" from another universe.) Hence, despite its difficulties, and in view of the slim observational evidence in its favor (distribution of globular clusters and of older stars), the cosmological theory of the origin of galaxies and galaxy clusters remains the best available. (However, that theory may be in for further trouble if small-scale anisotropies in the microwave background radiation continue not to be detected. Such anisotropies would be expected, on the cosmological theory of the origin of galaxy clusters, as relics of the original inhomogeneities.)

In this case, the nature of the difficulty may be summarized as follows: (1) contrary to the "lack of details" kind of problem, in this case the initial (or relatively early) conditions (the inhomogeneities or their relics) are not observed; (2) it is difficult, in the light of other considerations, to see how the appropriate conditions could be realized (how the appropriate inhomogeneities could arise); (3) there are other independent considerations which suggest an alternative possible explanation; (4) that alternative, however, is not as palatable as the theory available (in the present case because it has not been developed in detail, and because its development would seem to call for radical revision of other well-grounded ideas, or else for the introduction of radically new ideas for which there is little or no other warrant). This is one type of difficulty that might be classified under the heading of a *fundamental theoretical problem*—as opposed to the star-formation "filling in details" type of difficulty, which appears almost certain to be a *problem of theoretical incompleteness*.

These are far from the only known reasons for hesitancy about the picture I have drawn. One can of course expect a sensible caution about the cosmological portions of the picture, involving as they do such enormous extrapolations; and one can expect a tentative attitude toward the most recent attempts at synthesis. Heisenberg went to his grave opposing the quark theory, and we must remember that quarks are, after all, unobserved; and one would suppose that a scientific theory would have to be very good indeed if it is to maintain that the fundamental postulated entities are unobserv*able*. And as I remarked earlier, the symmetry between quarks and leptons is very incomplete. But there are also difficulties in parts of the picture that are of longer standing. The strong interaction borders on the intractable; the origin of planetary systems is still shrouded in obscurities; the origin of taxa higher than the species level remains something of an embarrassment for evolutionary theory; the problem of development in biology remains complex and very incompletely solved; the later stages of stellar evolution are still unclear, as indeed is the ultimate fate of the universe—whether it will go on expanding forever (as seems to be the slightly favored view at present) or will ultimately collapse again to a singularity, and, if so, whether it will "bounce" to produce yet another in a possibly continuous train of successive universes. These and their like can perhaps be said to be merely open questions, problems of theoretical incompleteness—matters of detail or extension of present knowledge and theory—rather than fundamental theoretical problems, overt threats to present theory (though some may have larger philosophical ramifications than others). But there are more serious difficulties too, which, though in some cases they differ significantly in general character from the problem of galaxy cluster formation, deserve to rank with the latter as fundamental theoretical problems, as dark clouds on the horizon of contemporary science. Whatever happens to our interpretation of quasars, something drastic may well happen to current physics. If quasars are "local"—relatively nearby—then we must account for their very large red shifts in some way other than as indications of great distance—possibly as effects of a large gravitational field (but then why are no other effects of the field observed?), or in terms of some entirely new law (but will that affect our interpretation of other red shifts as indicating an expansion of the universe?). If they are "cosmological"—very far away—we are faced with the problem of explaining energy production that appears to put even nuclear fusion to shame. In particular, if quasars turn out to be related to galactic nuclei (and there are strong indications that they are), then we again face the problem of the origin and nature of galaxies. There are difficulties with either alternative. In every problem in science there is the possibility of the unforeseeable; but in this case, unlike most others, we can as it were see that the unforeseeable is a *significant* possibility. Again, the failure to detect neutrinos from the sun has cast doubt on our theories of stellar evolution, and some have suggested that it indicates some shortcoming in our grasp of elementary particle interactions. Quantum theory and general relativity remain apart, and years of trying have not produced agreed-on progress toward their synthesis, or even general agreement as to how to try. For all its success in dealing with phenomena in a wide range of domains, the quantum theory has in the past decade or two been subjected to a revival of controversy as to its interpretation. Is a deterministic hidden-variable theory still feasible? Is precise determination of simultaneous position and momentum of a particle really impossible? One interpretation (Everett-Wheeler "relative state" interpretation) of quantum theory, which has the virtue of consistency if not of initial plausibility, has it that the universe splits into two independent universes on the occasion of every measurement, the two universes corresponding to the alternative possible outcomes of the measurement. With that, we have flirted, in this survey, with the possibility of parallel universes, successive universes, and now with possible universes being actualized. The past history of science has time and again produced new ideas that far outstripped prior imagination; speculation about other universes, though still on the borders, has entered the domain of science. We must be prepared for the possibility that there are indeed more things in heaven and earth than are dreamt of in our present picture of the universe. Even other universes.

But while we must keep an open mind about possible radical revisions of the picture of the universe and of life in it which I outlined earlier, we must also recognize that it is, I think it is fair to say, the picture within which the majority of physicists, chemists, and biologists today work. They work within it not in the sense that they accept it dogmatically, but in the sense that they believe it to be, overall, the picture best supported by present evidence, and that they believe the present task of science to be the development of the details of that picture, its further extension, and its testing and confrontation with alternative reasonable possibilities, especially in those parts of the picture which seem weakest. Much of the picture, indeed, seems unlikely to be rejected: what evidence could reasonably be expected that would lead us to deny, after all, that galaxies are stellar systems far beyond the Milky Way? that dinosaurs existed in the past? that DNA is at least implicated in heredity? And so on for a multitude of details which form the basis of much of our present picture. It is possible, of course—as philosophers since Descartes have been fond of reminding us—that these and indeed all facets of the picture may ultimately be rejected. But the logical possibility that we may be wrong, though it may be a reason for open-mindedness, is not itself a reason for skepticism or even for timid and indiscriminate caution. (This is the reason why I have emphasized that the difficulties I have been concerned with are *known* difficulties, which are, after all, the only ones we can hope to do something about.) And even in those regions of our picture for which we do have positive reasons for worry, we must not forget that, in the light of the available evidence, one picture may well be better than any of the available alternatives.

One final point in this all-too-sketchy survey of some aspects of the unity of the current scientific picture of the universe: there will be some who will claim that my account is "reductionistic," and who will claim that—for example—biology is *not* "reduced" to chemistry because all details of biological processes are not deducible from the chemistry of those processes. In this vein, since the details of the helium atom are not in general deducible from basic quantum-theoretical considerations as is the case with the hydrogen atom, one might as well say that physics has not been reduced to physics. If this is meant to imply that, to the extent that we cannot make such deductions, we do not have an *understanding* of the helium atom in terms of quantum theory (or of biology in terms of chemistry),then perhaps the fault lies in the deductive interpretation of understanding (or "explanation") which decrees that only deduction produces understanding. We do, after all, have such understanding despite the lack of precise deduction; and a more adequate account of "understanding" should allow for this. No doubt there is much in biology that is not understood in terms of chemistry; perhaps it can never be. But if this turns out to be the case, it will be because of specific aspects of the world, and not because of some prior philosophical strictures about what "understanding" (or "explaining") means.

## II

In the preceding discussion I argued that, as a result of the evolution of scientific thought, there has emerged a broad and coherent picture of the universe and of life in it, a view which, while incomplete and in some aspects open to serious question, is at present the best picture available. In the present section I will argue that this process of unification has not been restricted to the integration of beliefs about the world, but that there has also been a progressive tendency toward unification of those beliefs with the methods employed to attain well-grounded beliefs. That is, I will argue, the methods we consider appropriate for arriving at well-grounded beliefs about the world have come more and more to be shaped by those very beliefs, and have evolved with the evolution of knowledge.

Such a view of the intimate relation between knowledge and the methods of gaining knowledge flies in the face of the traditional sharp bifurcation of the two. For it is, and long has been, commonly assumed that there exists a unique method, the "scientific" or "empirical" or "experimental" method, allegedly discovered or at least first systematically applied in the seventeenth century, which can be formulated wholly independently of, and is wholly unaffected by, the knowledge which is arrived at by its means. It is as though scientific method is a set of abstract and immutable rules, like the rules of chess, independent of the strategies of the game but governing what strategies are possible.

Yet the most strenuous efforts of scientists and philosophers have failed to produce agreement as to precisely what that method is. Indeed, general philosophical theories about science according to which there is an eternal scientific method which, once discovered, needs only to be applied to generate knowledge, but which itself will not alter in the light of that knowledge, have proved to be either empty or false. Consider, for example, the view that science does not (or should not) admit concepts referring to what is "in principle" unobservable. The phrase "in principle" is a slippery one; but on any reasonable interpretation, what is "observable," even "in principle," changes with the development of new techniques, discoveries, and theories (think of the "direct observation" of the core of the sun by observing neutrinos). And on the other hand, perhaps the quark theory, if it is ultimately accepted, will have taught us not only something about nature, but also about what to do in explaining nature: about a role for the unobservable that was not allowed for by the straightjackets of philosophies of science that take observability to be something that is laid down forever. Similarly, what is "verifiable" and "falsifiable" can only be determined by the way things are, and our *beliefs about* what is verifiable and falsifiable can only be determined by our beliefs about the way things are. What for yesterday's science was considered unverifiable (hypotheses about the constitution of the stars, or about the origin of life) may today be a legitimate part of science; what some consider to be beyond the "line of demarcation" of the legitimately scientific (the unobservable, the unverifiable, the unfalsifiable) may at some stage, for good reasons, come to be a legitimate part of science (confined quarks, whose existence is unverifiable—unless, of course, we are willing to stretch the meaning of "verifiable" so that their existence is "verifiable" even though they are "unobservable"; but the philosophy of science has long been acquainted with the bankruptcy of such moves). Observability, verifiability, falsifiability, criteria for being a legitimate scientific problem (as opposed to a "pseudo-problem"), criteria for being a scientific possibility (as opposed to "metaphysics")—all these come, more and more in the development of science, to depend on the substantive content of accepted (well-grounded) scientific belief, and change with changes in that content. A sketch of some important developments in the history of science will indicate some ways in which this has come about.

In the seventeenth century, the boundaries between science, philosophy, theology, and mysticism were not drawn sharply. This fact must not be seen as evidence of some sort of intellectual schizophrenia on the part of the scientists concerned, or as an indication that they sometimes did science "badly." Consider Kepler: he was probably the first thinker to insist that every detail of our experience be accounted for precisely in terms of underlying mathematical laws. All experience was for Kepler interconnected by mathematical (for him, geometrical) relationships, and those interrelationships were clues to still deeper ones. In the light of this belief, Kepler felt compelled to ask not only such questions as, What is the precise relationship between the orbital speed of a planet and its distance from the sun? but also—*and in the same spirit*—What is the relation between the color of a planet and its distance? and, Given two planets forming a given angle at a person's birthplace at the time of his birth, what relationship does this fact have to his later life? and, What is the relationship between the "harmonies" in the motions of the planets and the harmonies in music (and in art and meteorology and. . .)? Angles formed by planets with locations on earth were as much "details of our experience" as, and had as much significance as, the angles swept out by a planetary radius vector in a given time. Within the general framework of his geometrical approach, Kepler knew no constraints on the kinds of questions to ask or the kinds of observable relationships it was appropriate to ask them about. Such constraints would later be introduced in the light of accumulating knowledge. For Newton, the planets were just further material bodies, obeying the law of inertia and exerting and responding to gravitational forces; their relative positions had nothing to do with either man or music. On the other hand, Newton still regarded theological considerations as relevant to his science: indeed, he saw his science as implying the necessity of periodic miraculous intervention by God in order to preserve the stability of the universe against the continual decrease in "quantity of motion"(momentum) due to collisions, against the disruption of the solar system through mutual planetary perturbations, and against either gravitational collapse in a universe containing a finite amount of matter, or a cancellation of all net gravitational forces in an infinite universe containing an infinite amount of matter (Bentley-Seeliger paradox). Laplace, having seen the resolution of problems about elastic and inelastic collisions and conservation principles that had plagued Newton's era, and himself claiming to have shown that planetary perturbations are self-correcting in the long run—but forgetting about the Bentley paradox—was able to inform Napoleon that science had no need of the hypothesis of God.

In such ways, as science progresses, constraints come to be imposed on the kinds of questions to be asked, and on the kinds of possibilities that can be legitimately envisaged, in science. But the development of such "rules of the game" is not all negative—not always a matter of cutting out questions and possibilities that had hitherto been considered legitimate. For the progress of science also, at various stages, through new discoveries, approaches, techniques, and theories, opens the door to new possibilities—new questions and new alternatives that were before ruled out as illegitimate or perhaps even self-contradictory or inconceivable. The work of Gauss and Riemann opened the way for thinking of a space with variable characteristics—a possibility which Newton, with good reasons at the time, rejected as self-contradictory in dismissing Descartes' similar suggestion. And no one needs to be reminded that quantum mechanics and relativity opened the floodgates for questions and theoretical concepts which would previously have been rejected out of hand. The idea that quantum theory contains its own theory of measurement, and even refashions the rules of logic in its own image, is completely consistent with the viewpoint I have suggested.

That viewpoint maintains that method not only determines the course of science, but is itself shaped by the knowledge attained in that enterprise. In many ways, scientific method is more like military strategy than it is like the rules of chess: the strategy shapes the course of the campaign, but is itself responsive to the lay of the land, and to the armaments that become available to it; and it adjusts to new situations and new devices. Science has not only tended to move toward unification of its substantive beliefs; it has also tended to move toward unification of belief and method. It has learned how to learn in the very process of learning.

CHAPTER 10

# WHAT CAN THE THEORY OF KNOWLEDGE LEARN FROM THE HISTORY OF KNOWLEDGE?

In recent years, philosophers of science have been increasingly concerned with questions about scientific change, and, in connection with those concerns, to rest their claims more and more on an examination of cases in the history of science. During the 1960s and early 1970s, those concerns tended to revolve around the question of *whether* scientific change, or at least major scientific change, is or is not "rational." It seems to me, as I shall argue in what follows, that that question is misguided in principle, at least as it is usually understood, and that it calls attention away from the most important and potentially most fruitful problems about the nature of scientific change. Furthermore, I believe, and will argue below, that the most fundamental reasons for investigating scientific change, and the sense in which and degree to which it is necessary to base such investigation on an examination of the history of science (as well as of contemporary science) have not been adequately grasped even by many who are sympathetic to the approach. Finally, I do not believe that the difficulties in the way of such an approach have been properly appreciated and taken account of, and in many cases have not been considered at all. In particular, I will examine here five basic objections or types of objections against the view that the philosopher of science, in attempting to understand the nature of science, must examine the rationale of scientific development and innovation, and must base that examination on a study of cases from the history of science (as well as from contemporary science).

My plan will be to begin by considering the first of these objections, as that discussion will place the remainder of the paper in a more general context of twentieth-century philosophy of science, and indeed of philosophy generally. This discussion will lead into a presentation of a view of the philosophy of science in terms of which the remaining four objections can be examined.

## I

The first objection is based on the distinction, made by logical empiricists among others, between "scientific terms"—terms occurring "within science," like 'force', 'mass', 'acceleration', 'catalyst', 'gene', 'superego'—and "metascientific terms"—terms like 'law', 'theory', 'hypothesis', 'explanation', 'confirmation', 'evidence', and 'observation', used in "talking about science," in characterizing scientific terms, statements, and activities. Thus " 'F = ma' is a scientific law" is a statement predicating the metascientific term '(scientific) law' of the scientific statement 'F = ma'. This distinction was frequently employed to define the central (though of course not the only) task of the philosophy of science: to understand science, in the sense in which the philosopher wishes to understand it, would be to understand what is involved in saying that a certain scientific proposition or argument is a law, theory, explanation, etc. More generally, philosophy of science was conceived as being concerned fundamentally with the "meanings" of metascientific terms or concepts.

For present purposes, the most important point about this view is the idea that, although the science of any given period should, insofar as it *is* science, exemplify (at least fairly closely)[1] the results of this metascientific inquiry, nevertheless *the analysis of the meanings of metascientific concepts does not depend on analysis of the content of science, in the sense that those meanings are not a function of that content, changing or evolving with the changes or evolution of the specific concepts, propositions, and arguments accepted or employed in science.* No matter how our explanatory theories about the world change, and no matter how well confirmed those theories might be, the meanings of such terms as 'explanation', 'theory', and 'confirmation', as well as the totality of other metascientific terms which together constitute the concept of 'science', will remain unaltered. What the philosopher of science is interested in are, for example, the characteristics that make *any* particular body of propositions, at *any* given period, a "(scientific) theory." Philosophy of science is thus concerned with the invariant characteristics of any possible theory, with the very "concept" of a theory (the very meaning of 'theory').

This view of the aim of the philosophy of science was, among logical empiricists, closely associated in practice with another view. Since philosophy of science was concerned with the concept of theory, with the characteristics of any possible theory, and since, according to the influential *Tractatus Logico-Philosophicus*, logic was the study of the total range of possibilities, it followed that the appropriate tool for the analysis of metascientific concepts is logic.[2] Again, the implication was, or was taken to be, that the "content" of science can be safely ignored, and only the "form"—the "logical form"—of scientific reasoning examined.

The logical empiricist conception of the central aim of the philosophy of science, reinforced by the tools of modern logic, thus rested on the assumption that the content of science in no way affects the meanings of metascientific terms; that is, it rested on the sharpness of the science-metascience distinction. And as a corollary of this assumption the historical changes which have taken place in the body of scientific concepts and beliefs were also irrelevant: the meanings which are the objects of the philosopher's scrutiny have an eternal ring about them, independent of the vicissitudes of ongoing science, the comings and goings of particular scientific ideas.

The explosive development of the history of science in the 1950s and 1960s raised serious doubts about this assumption and its corollary. In order to understand these doubts, let us suppose for the moment that the "meaning" of a metascientific term in some way determines the range of possible applications of that term; and let us express this by saying that there is some relation (perhaps of identity) between the meaning of the term and the criteria (not necessarily conscious or explicit) according to which it is applied to determine what can count as, for example, an explanation or a theory. The new discoveries in the history of science then appeared to many to indicate (a) that what counts as a legitimate scientific theory or explanation (etc.) at one stage of the development of science often differs, even radically, from what counts as such at another stage; (b) that the differing sets of criteria of application employed at the different stages could not—any of them, according to the more extreme views-be praised or condemned as more or less "rational" or "correct" than any of the others; and (c) that the criteria accepted at a given stage are intimately linked to the content of scientific belief at that stage. This last allegation seems to have been interpreted by these critics of logical empiricism as a claim that the criteria of what counts as (e.g.) an explanation at a given stage not only mark out the range of *possible* explanations, but even that they imply or strongly suggest the explanations that are *accepted* as correct at that stage.

The relativism consequent on this interpretation of (c), together with (a) and (b), is by now familiar. It was far too hasty and extreme a reaction to the logical empiricist view. For, even ignoring all the other objections that have been raised against the relativistic view,[3] the revelations of the history of science to which those critics of logical empiricism appealed could equally well have suggested a very different interpretation of (c), namely that *although the meaning of 'explanation' (or the criteria of what can count as an explanation) do, at any given stage, mark out a range of possible explanations, nevertheless the knowledge attained—the set of explanations which come to be accepted from among those possibilities—lead, at least under certain circumstances, to a change in the criteria themselves.* The "meanings" or "criteria of application" are not independent of scientific beliefs (as the logical empiricists held), nor do they imply them (as their relativistic critics maintained); the meanings or criteria of application of the metascientific terms are connected to the substantive scientific beliefs by what might, for the sake of brevity, be called a rational feedback mechanism (itself perhaps subject to reform in the light of new information which it itself helps reveal). "Meanings" or "criteria of application" determine ranges of possibilities; they themselves are subject to revision in the light of the possibilities which (on the basis of other criteria) come to be accepted as correct.[4]

This alternative view of (c) is compatible with acceptance of (b) and (c), if they are suitably interpreted. As to (a), what counts as a legitimate scientific theory or explanation at one stage could—accepting the alternative view of (c)—often differ, even radically, from what counts as such at another stage, though the notion of "radical difference" would have to be divorced from the infamous "incommensurability" doctrine of the 1960s: for there might now be a chain of reasoning connecting the two different sets of criteria, a chain through which a rational evolution could be traced between the two. And as to (b), we could recognize that, *given* the knowledge and criteria available at a particular time, certain beliefs about possibilities and truth *were* reasonable, even though alteration and improvement might be possible with the acceptance (even on the criteria current at that stage) of new beliefs.

Such a view would involve a denial of such absolute bifurcations as those between meaning and truth, analytic and empirical, science and metascience, method and its application: all would be subject to evolution along with the development of new knowledge. It would affirm that there are *no* concepts or beliefs that are immune to revision in the face of new knowledge: even levels of description as general (even "categorial") as those employing such concepts as "particle" and "entity" would be subject to revision or abandonment. Assumptions underlying much of the philosophical tradition stemming from Plato and Kant would thus be rejected. The view would imply that we learn *what* "knowledge" is *as* we attain knowledge, that we learn *how* to learn in the process of learning.

Such a view would itself, of course, have to contend with a number of objections. The most obvious, perhaps, is that it seems hard to see how criteria of rationality could be said to "evolve rationally" with the developing content of science unless there are higher-level, content-transcending criteria by which changes in criteria of rationality could themselves be said to be "rational"; thus we would be threatened again with collapse into either a "metascientific" or a "relativistic" approach. I do not think this objection is insurmountable; but the purposes of the present paper do not require that the view be defended, or its limitations shown, or its consequences elaborated. The important point, for present purposes, is that it constitutes a possibility which was not envisaged by either the logical empiricists or their critics (the former having also failed to allow for the possibility of the view presented by the latter). And for now the point of interest is the failure, in this regard, of logical empiricism: both the relativistic view and the alternative to it just outlined reject the logical empiricist conception of the aims of the philosophy of science as having to do with analysis of "metascientific" concepts; both imply that the alleged "eternal," content-transcending character of so-called metascientific concepts is a myth.

Much could have been said in defense of the "metascientific" approach: it could have been admitted, for example (but was not), that criteria of being an explanation (and so forth) vary from period to period; and it could have been agreed that these changes are dependent in some way on the changing content of science. But it could further have been argued that the philosopher of science is concerned with yet a *higher level* of criteria, namely *the criteria by which one identifies what are, in different particular traditions or at different stages of science, criteria for being an explanation (etc.).* These second-level criteria might now have been said to be invariant, and have been called "meanings," as opposed to the transient and evolving first-level "criteria of application." And so it could have been alleged that the critics had confused meaning and criteria of application, and it could have been reaffirmed that philosophy of science is concerned with meanings, in the sense now proposed, rather than with criteria of application. Indeed, the relativistic critics of logical empiricism might have been saddled with the charge that a claim like "The criteria for being an explanation at stage A are incommensurable with the criteria for being an explanation at stage B" presupposes what those critics seemed to be denying, namely, the existence of content-transcending criteria by which certain statements in the two traditions or stages could be identified by the historian as "criteria for being an explanation."

Numerous other such tacks might have been taken; but though they might have refined the logical empiricist conception of the aims of the philosophy of science, they would not have dealt with the fundamental issue. For example, if the specific defense of logical empiricism just suggested had been taken, then (1) we would still have to ask why philosophy of science should not be concerned with "criteria of application" as well as with "meanings." Indeed, it could be argued even more pointedly that, if we are to understand scientific reasoning, it is particularly the lower-level criteria—the criteria employed by scientists themselves in determining legitimate moves to make in raising and meeting their problems—that should be our central concern, and that attention to the "eternal" (logic, "meanings") was just what made philosophy of science in the logical empiricist vein seem so irrelevant to real science. But the issues are deeper still; for (2) the further question might have been raised as to whether the "second-level" criteria—the "meanings," on the suggested interpretation—themselves might not alter under certain circumstances with changes in the content of scientific belief and with changes in the ("first level") criteria of what can count as (e.g.) an explanation.[5]

The fundamental issue between the logical empiricist interpretation of the aims of the philosophy of science, on the one hand, and, on the other, the relativism of the 1960s and the alternative thereto which I sketched above, concerns the relations between the alleged "metascientific" concepts (or the criteria of their application) and the content of science. Is it really true that the conditions ("criteria") of being a scientific theory (for example) are independent of the substantive beliefs of science, and so do not change with alterations in those beliefs? Is there *any* level at which there is such independence of criteria from content? If so, what is the source of those criteria? (Traditional arguments for the existence of such a "level" have tended to claim that there *must* be one; but we have learned to be suspicious of such aprioristic arguments.) And for any "levels" which are not independent, but which alter with the evolution of scientific knowledge, what are the rules (if any) and circumstances according to which such changes of criteria occur? (And are those rules and circumstances themselves subject to change, and if so how?) *And the important point is that logical empiricism, simply assuming the sharp bifurcation between the "metascientific" and the "scientific," between the analysis of "meanings" (however interpreted) and the content of science, failed even to raise these questions*. They are questions that should have been raised from the start; but now they have become pressing. Our attention has been called to presumptive evidence that scientific change penetrates deep into the levels of scientific criteria; and a thorough reappraisal of the entire question of the relations between scientific beliefs and the allegedly defining concepts of science, and the ways in which those allegedly defining concepts operate and are operated on by the evolving content of scientific belief, needs to be undertaken. (It is for this reason that a full development and defense of the alternative sketched above is unnecessary here: it need only be recognized as an alternative, not only to logical empiricism but also to the relativism in which its critics were mired.)

On this argument, the necessity of examining scientific change arises from questions about the nature of the relationship between criteria employed in the acquisition of knowledge and the knowledge itself. These questions require an attention, first, to the content of science—both the possibilities which science is willing to contemplate, and the beliefs it is willing to adopt—and second, to the interaction between that content, as it evolves in the development of science, and the criteria, even at the highest levels, employed in arriving at that content. But there is another way of arriving at the same general approach, through reflection on peculiarities and problems about the character of contemporary science itself, rather than out of criticisms of philosophical views about science. I will now outline these peculiarities and problems and the approach they seem to call for.

## II

The picture of the universe offered by science today surpasses all imaginings of earlier thinkers. Space and time are held to be intimately linked, and the space-time is expanding, not in the sense that portions of matter are moving away from one another *in* space, but in the sense that the "curvature" of space-time itself is increasing. The universe is populated with objects, on the level of both the very large and the very small, which common sense and ordinary experience, or even the science of a few decades ago, could only describe as exotic, bizarre, or weird. The very distinction between matter and the space-time in which it is located is blurred and threatened even with obliteration; and at the level of the very small, the very notions of space, time, and matter, as traditionally conceived, may be inadequate. Elementary particles have properties so far from traditional and macroscopic expectations that they fully deserve to be called strange; even the terms 'elementary' and 'particle' seem not wholly appropriate at best, and the term 'interaction' covers cases where only one particle is involved. Distinctions like those between "particle" and "wave," "discrete" and "continuous," traditionally taken to be both exclusive and exhaustive of all possibilities, and perhaps even concepts as general as "entity" and "property", cannot do justice to the world as viewed by contemporary science.

The fact that the conclusions of science are, in many cases, open to specific objections, and that they are in any case subject to revision or abandonment in the light of further developments, does not lessen the problem; on the contrary, it only makes it worse. For many of those beliefs are so extraordinary that even their possibility, their existence as alternatives to be considered, would have been wholly inconceivable, or at best self-contradictory, to even the greatest or most imaginative of earlier thinkers. And at its frontiers—but by no means beyond the limits of legitimate scientific possibility—scientists trying to deal with the conclusions and problems of their subject are willing, and in some cases even compelled, to contemplate alternatives that are still more exotic and bizarre. No amount of reflection, however ingenious, on the range of logical alternatives could have been expected to anticipate the claims of relativistic cosmology or quantum field theory (to say nothing of the more speculative extensions and alternatives under active consideration) even as possibilities; no amount of examination, however painstaking, of our ordinary experience of the world could have been expected to lead to such conclusions or such possibilities.

Why, then, does science accept such conclusions, and why is it willing to consider alternatives so bizarre? One answer might be that the total body of evidence available makes reasonable a certain range of possibilities, and that in some cases the evidence for one alternative is strong enough to warrant its (tentative) acceptance. But when we examine the evidence put forth, we find that it is itself so unfamiliar, from the standpoint of common sense and ordinary experience, that we are only led to ask in turn why science accepts *that* sort of thing as evidence. Further patient, step-by-step explanation would then be required to delineate a series of successive departures, each made on the basis of putative reasons, from our expectations (whether founded in "common sense," ordinary experience, or whatever else), and finally eventuating in the necessity of accepting the evidence, alternatives, and beliefs of science. It would be as though we had a set of expectations or claims, with a delineation, for each expectation or claim, of a series of arguments (a "reasoning-series") leading at last to the assertions and alternatives of contemporary science. The philosopher of science would then be interested in the patterns of reasoning by which the successive departures from those expectations or claims proceed, with a view to generalizing and, if possible, systematizing those patterns.

Such reasoning-series would not necessarily make explicit reference to any views actually held in the history of science. In actuality, however, there would be considerable overlap; and where there is, the relevant historical cases almost invariably provide subtleties of reasoning that are unlikely to be found in the currently-provided reasoning-series; and those subtleties are of vast importance to the philosopher in his search for reasoning-patterns. Far from narrowing our insight by dealing only with "actual" rather than with "possible" cases, the history of science—as has been amply demonstrated by historians in the past several decades—exposes far more possibilities than we are able to generate otherwise.

Furthermore, the claims we might make in the name of "common sense" or "our ordinary experience" constitute only a small proportion of such claims that have been made. Departing from what we today might think of as common sense possibilities cannot begin to do justice to the bizarre character of contemporary science; for our "common sense" is undoubtedly much closer to those scientific conclusions and possibilities than anything earlier thinkers might have conceived. Yet the views of those earlier thinkers, at the time of their proposal, appeared plausible or convincing to their adherents, while the beliefs constituting our "common sense" claims often would not have. Arrival at *our* expectations (however based) from such beliefs is as much a part of the knowledge-attaining process as is the transition from our expectations to the claims of contemporary science. And the philosopher interested in developing a general theory of the knowledge-acquiring process must consider the reasoning-patterns involved in the former sorts of transitions also.

Finally, such consideration becomes absolutely essential when we ask, as we must, why what seemed so convincing to earlier thinkers differs so radically from what we take to be acceptable (on either a common sensical or scientific level) today. Have we simply learned more—while the rules of learning themselves, the reasoning-patterns involved in knowledge-acquisition, have remained constant? Or have we removed obstacles and misunderstandings in the way of applying rules of learning which, once grasped, will remain henceforth immutable? Or have the reasoning-patterns embodied in the acceptance or rejection of claims about the possible and the true evolved with the development of the structure of our knowledge-claims, so that we may expect similar evolution in the future? These questions—in particular the last—require us to attend to scientific change, as manifested in the history of science.

We have thus come full circle back to the point made in Section I above: by considering the peculiarities of contemporary science and the requisites for understanding them, we have arrived at the same conclusion as we found earlier by considering certain deficiencies in recent philosophy of science. The question of why science today believes the peculiar things it does about the universe, and why it is willing to consider the alternatives it does, requires attention to the question of how science has come to think in those ways.

## III

Against the background of this discussion of the methodology of the philosophy of science, we may now return to the consideration, promised earlier, of five objections against the idea that the philosopher of science, in trying to understand the nature of science, must examine the rationale of scientific development and innovation, and must base that examination on a study of cases from the history of science.

(1) The first of these five objections, that the philosopher of science, since he is concerned with the nature of science in general, cannot concern himself with the ephemeral content of science at all, has been discussed in Part I, above. However, we will encounter variations on the theme of that objection in the context of our examination of some of the objections to come.

(2) The second objection is a familiar one: that there is no such thing as a "logic of discovery"; there is only a "logic of justification." The processes by which ideas are arrived at in science are not "rational"; only when such ideas are already presented can canons of rational test be applied to them. The introduction of ideas in science is a matter of genius, of inspiration, of imagination, of creative originality, a matter for study by the psychologist or sociologist or historian, but not for the philosopher. Reasoning is to be equated with testing, not with creativity. A concern with the development of science can therefore be of no use to the philosopher of science.

In the past several years, a number of important distinctions have been made regarding the question of a "logic of discovery." The problem has been reformulated so as to recognize that the "reasons" involved in the introduction of new developments in science need not be conclusive ones, and that the "discoveries" involved need not be discoveries of *acceptable* ideas, but only of *plausible* ones. Given these reinterpretations, it is now clear that, while the question of why some specific person thought of an idea may not in general be a subject for philosophical investigation (though under some circumstances it might be), it is often possible to understand, against the background of the ideas and techniques available at a certain stage, why a certain idea was or would have been a reasonable one to suppose, and, further, that that idea was not only possible, but one worthy of serious consideration. That is, reasons are involved in the introduction (in the preceding sense) and serious consideration, as well as in the justification, of such new ideas in science. (Whether the reasons employed in all three of these contexts are, in all cases, of the same type, is a further question which need not be considered here.)

(3) A third major objection—one which has not been given the attention it requires—maintains that cases from past history of science, or general criteria extracted therefrom, cannot be used as supporting or counterevidence to theses about the nature of science in general, or about the nature of science as it exists today. The objection may be considered as consisting of two steps: (a) that even if consideration of cases from the history of science were

necessary for the understanding of science in general or as it exists today, appeal to such cases would not be sufficient to show their relevance to such understanding; and (b) that consideration of past cases is not necessary for such understanding.

(a) Scientific ideas in the past—so the first step in this objection goes—may well have been influenced by religious, economic, social, or psychological factors, or by presuppositions which, while we would recognize them as unambiguously "scientific," were nevertheless *employed* then in an "unscientific" way, dogmatically or even unconsciously. It may well be, for instance, that Galileo was not particularly interested in experiment, or that Kepler's "scientific" work was "inextricably intertwined" with elements that we today consider "mystical," or that Newton's interpretation of his "crucial experiment" concerning light and color presupposed, in an uncritical way that blinded him to alternatives, a particle interpretation of light. But even accepting these claims about historical cases, it does not follow that science does today, or always will, or must or should, proceed in similar ways. For it may be that we have *learned* the importance of experiment, and to distinguish "external" from genuinely "scientific" factors, and that we have become progressively more conscious and critical about fundamental assumptions. It is therefore necessary to establish not only that such claims about past science are correct, but also that they are relevant to the understanding of science in general, or that similar things hold about science today. That something was done by past science, or by past scient*ists*, is by itself insufficient to establish that science works that way in general or today.

(b) It is also unnecessary to examine such past cases. If we want to know whether present science makes presuppositions of the sorts mentioned above, the way to find out is to examine present science; and if we wish to know what "science in general" is like, we are much more apt to find out by examining up-to-date science than be looking at cases from earlier stages in which we ought to *expect* that irrelevant features would not yet have been sorted out.

We saw earlier (and will return to the point in objection (5), below) that there is a sense in which the science of any given period can be examined "in its own terms," independently of any appeal to the science of other periods. In particular, it is possible to examine current scientific theories and uncover the reasoning involved therein. But to focus only on contemporary science would be to ignore some of the most profound problems about the nature of human knowledge. Suppose it is true that we have "learned how to do science." What would have been involved in such learning? A progressive freeing of science from hindrances in the form of external influences and unexamined assumptions, gradually exposing a scientific method which, once understood, governs the further development of science without itself needing further elaboration? Or an evolution of method itself in conjunction with an evolution of the knowledge found by that method? And in either case, what is the character of the reasoning by which such changes are brought about? *An understanding of the criteria which are employed in contemporary science will not by itself reveal answers to these questions, any more than the consideration of the science of any other particular stage will necessarily provide conclusions applicable to science at other stages or in general.* To consider science at a particular stage, even the contemporary, is only to expose the criteria employed at that stage; but can it also give us an understanding of why we accept those criteria, and whether and how we can expect them to alter in the future? It certainly cannot if criteria evolve with the content of science—if we learn how to learn in the process of learning. And it is important to recognize that *the objection with which we are nowconcerned against the relevance of history of science relies on the very possibility that this may be true:* it rejects the relevance of past cases on the ground that what counts as scientific may have changed since the case in question. As I have said earlier, the present paper is not the place to defend a claim that this is in fact true, or to examine the extent to which it is true. For as long as it remains a possibility, our understanding of the knowledge-acquiring process is severely restricted by failure to investigate its possibility—that is, by limiting ourselves to criteria currently employed in that enterprise, and failing to examine the rationale by which those criteria may have evolved, and may be in the process of evolving.

In short, then, part (a) of this objection, to the effect that examination of any specific case in the history of science is an insufficient basis for more general conclusions about science, is valid. But on the one hand, the objection also holds, for the reasons given, against sole reliance on investigations of the reasoning involved in contemporary science. And on the other hand, the objection is misdirected. It is not our purpose to make generalizations about science on the basis of examination of single, isolated cases, or even, necessarily, on the basis of characteristics possessed in common by a number of cases. This is not merely because generalizations on the basis of single or few cases are always dangerous, but because the generalizations we seek are of a different sort. They have to do with the dynamics of rational change, which may make characteristics discernible at any one period of science altered or obsolete at a later stage, and therefore not suitable as bases of generalization. This is not to say that there cannot be or have not been any general characteristics of science, operating at all stages in the past; but whether there are any such, and if so the reasons why they have operated so far, and whether and why they can be expected to continue to operate, are questions that can be settled only by an investigation of science.[6]

Part (b) of the objection—that an investigation of the history of science is unnecessary—is valid only if we restrict ourselves to a very limited range of questions, the answers to which would leave us with a very limited understanding of the nature of science and of human knowledge. It is, as I have argued, possible to examine the science or specific areas of science of any particular period, to study its axiomatic foundations, the criteria employed in that area, at that period, in determining what are proper or fruitful areas for study, legitimate or important problems, promising lines of research, possible and acceptable answers to those problems, and so forth.[7] Indeed, the possibility of such examination is presupposed by any effort to study the dynamics of rational change in science. And there are many interesting and important problems, of philosophical concern, which can be illuminated by such investigations. But there are larger questions, having to do with the status of those criteria themselves—why they are adopted, and whether they are themselves subject to change, and if so why—which require attention if we are to understand the nature of human knowledge and its acquisition, questions which require an examination of scientific development.

(4) A fourth objection is the following: any attempt to understand science on the basis of actual cases (whether historical or contemporary) must be limited to description. But a description of what *has* happened in science cannot provide a basis for any claims about what might or will or ought to happen in a future science, nor about what could have or ought to have happened in the past. But an understanding of science is necessarily concerned also with such questions.

It is true that philosophy of science must be concerned not only with what has occurred or does occur in science, but also with what could happen, and even with what should happen, in science. If philosophy of science were limited to "mere description" (assuming for the sake of argument that a sense might be given to that notion which would be appropriate in the present context), it would be little different, except perhaps for a greater penchant for generalization, from the history of science. Nevertheless, in the senses of "could" and "should" which are relevant to the kind of investigation of scientific change outlined in this paper, it is not necessary to appeal to criteria of "possibility," or to "prescriptive principles," which have no grounding in actual facts. On the contrary, *there are senses of the expressions "E could have happened" (or "E could happen") and "E should (ought to) happen (have happened)" (a) which are the ones of crucial relevance for the understanding of the rationale of scientific change, and (b) the truth-conditions for which are to be found in what has in fact happened.* In other words, a description of what actually occurs or has occurred in science can serve as the basis for an account of epistemologically relevant alternatives and prescriptions. (Conversely, there are senses of these expressions which are correspondingly irrelevant to an understanding of the rationale of scientific development: contrast, for example, the sense in which study of recombinant DNA "can" and "ought" to be pursued because biology, as a scientific field, is, so to speak, "ready" to pursue its study and the problem is an important one for the further progress of the field, with the sense in which it "ought not" be pursued because of possible dangers to society.) In defending these claims, I will incidentally note that the historian of science, far from being concerned only with "mere description," is and cannot avoid being interested in questions about what "could" and "should" have happened in science, in the same senses as the philosopher of science. I will also make some observations (by no means exhaustive) about the differences between the history and the philosophy of science.

Let us first consider "E could have happened"; some of the most obvious interpretations of such statements are the following:

(i) In one sense, Thales "could have" devised the general theory of relativity, complete with all the mathematics and physics necessary for its formulation. This broad sense—call it "logical possibility," though that notion is not without its problems—has not proved very helpful in understanding the workings of science. For example, the fact that it sheds no light on the question of why general relativity was (and could have been) devised under the circumstances in which it was (and not, for example, by Thales) is a mark of its failure to come to grips with real scientific reasoning.

(ii) We also speak of what "could have happened" given the instrumentation, mathematical techniques, and physical ideas (whether accepted or not) available at the time in question. There are sometimes difficulties in applying this notion in specific cases: for example, it is sometimes not perfectly clear that, or to what extent, a certain idea was "available" at a certain time, or to a certain person or group at that time. On the other hand, there are clear cases in which such questions can be decided, or in which arguments can be given which are clearly relevant (whether decisive or not) to the question. And whatever the difficulties in formulating general criteria free of all loopholes (a fantasy in most cases), the existence of clear cases in which decisions can be made or relevant arguments presented without, or with only minimal, ambiguity, is sufficient to show that this concept has an application. And in those cases, the state of science at the time—the actual techniques and ideas available—determine a range of possibilities that "could have occurred." In this sense, we often speak of something that could have happened but did not, as having been the "natural" (expectable) thing to have occurred; at other times, we might speak of an event which, though it could have occurred, was not the reasonable or natural thing to have occurred. The historian of science is often interested in these sorts of cases (especially the former), and deals with them successfully and in an enlightening way. The philosopher of science, concerned with the reasoning which determines ranges of possible alternatives and selections from among those alternatives, is concerned even more crucially with such situations (especially the latter) than is the historian.

(iii) The preceding sense fades gradually into another, when we say that something "could have happened" at a given stage of science in the sense that the ideas and techniques concerned could have been developed with relative ease at that time. Again, the truth-conditions of such claims lie in the actual state of science at the time.

(iv) A further sense, important in the interpretation of science, is that in which certain ideas and techniques *were not* "available" at the time, and which "could" have been developed only later, and without which a certain idea could not be defended or even expressed clearly (or at all). Such is the sense in which the Cartesian "geometrical" program for science was not a "possible" one in the seventeenth century: it "could" not even have been stated clearly (and was not by Descartes) without the later developments of Gauss and Riemann. Such also is the sense in which Newtonian cosmology was shown to be "possible" only with the work of Milne and McCrea (1934). The historian of science, though he must often be aware of this sense, in general must not allow it to interfere with his judgment of what was "possible" (senses [ii] and [iii]) at the time with which he is concerned. For the philosopher of science, interested as he is in the development of criteria of possibility, this sense is of great importance: the philosopher of science must be concerned to understand the science of a particular period "in its own terms"; but he must also be sensitive to the limitations and errors in the approach of that period, limitations and errors which might not have been perceivable except in the light of later developments. As with senses (ii) and (iii), however, employment of this sense is firmly grounded in an understanding of the actual content of science at the stages concerned.

As to "what should have happened" or "what should happen," there is again a very broad sense, of no use whatever in understanding the nature and process of knowledge-acquisition in science (whatever *other* interests might lead to raising such questions), in which the concern is with certain moralistic presuppositions in terms of which judgment of an episode, or of science as a whole, is made.[8] There are, however, other senses that are useful to the historian and the philosopher, senses which are tied closely to the details of actual scientific procedure. For both the historian and the philosopher, to different degrees in different circumstances, and often for different purposes, are concerned with reporting and judging, for example, all of the following: weaknesses in the reasoning by which a particular scientist or group of scientists conceived their subject matter, raised their problems, argued in favor of certain lines of research as "promising," constructed alternative possible answers to their problems, chose acceptable solutions from among those alternatives, etc. (that is, we are interested in what they "could" have said or done—in any of senses (ii)–(iv) above—had they been more accurate or careful or aware, that is, if they had been doing their business—dealing with their own problems—well, as they, or in other cases as later scientists, conceived that business); vaguenesses or inaccuracies in their statements of premises or conclusions or arguments; presuppositions of their approaches, whether they were (or could have been) aware of them at the time; irrelevancies or other unnecessary ingredients in their arguments; consequences of their approaches which they may not have seen (and, if not, we are interested in why not). In all such judgments, formal rules of deductive or inductive logic are insufficient as criteria of what the person or group "should" have said or done; such rules must be coupled with a knowledge of the scientific situation at the time and what "could" have happened, especially in senses (ii) and (iii) above.

(5) The final objection that I will discuss here alleges that historical investigation always presupposes some point of view in the selection and interpretation of what happened in the past. Specifically, in the attempt to construct a philosophical interpretation of an episode in the history of science, the need to distinguish the "genuinely scientific" from the scientifically irrelevant in the episode requires that criteria of selection and interpretation be brought to bear. It should then occasion no surprise if the case, as thus reconstructed, appears to support philosophies of science in which those criteria are embedded. Other criteria of selection and interpretation would, however, produce a very different view of the case. Hence, because a philosophical interpretation of science must always be *presupposed*, it is impossible to *arrive at* such an interpretation through an investigation of historical cases.

Another version of this objection is that, in attempting to distinguish the "reasons" involved in an historical case from (for example) the psychological, sociological, economic, political, or religious factors involved, we cannot avoid bringing to bear on the case a conception of what is involved in something's being a reason, even though the very purpose of our investigation is purportedly to *arrive at* (but therefore definitely not to presuppose) such a conception.

This type of argument might be called a "philosopher's objection," designed to demonstrate the impossibility—the "logical" impossibility (based on the "very concept" of an investigation and the role of "criteria" therein)—of all inquiries of a certain sort. In the face of such objections, we might admit that philosophical lessons cannot be gleaned from a study of actual science, historical or otherwise (for such arguments can be raised against attempts to arrive at an understanding not only of historical, but also of contemporary, cases of scientific reasoning). And we might then feel compelled to return to a view that we must formulate our interpretation of scientific reasoning in advance of, and independently of, any investigation of actual cases of that reasoning. Such a view, a general form of what I have called the "metascientific" approach to the philosophy of science, can only eventuate in either relativism or a kind of Platonism. Such views, as I have already argued, not only ignore the strategies and tactics, the intricate maneuverings of attack and defense, that take place in the thick of the scientific battle; they also decree the impossibility that, with further victories and improvements in weaponry, not only the strategies and tactics may change, but also that corresponding changes might be wrought in what had previously been sanctified as part of the "very concept" of the scientific struggle, the very "meaning" of 'science'.

But the objection can be attacked more directly. It totally ignores the fact that inquiries into the history of science *do* take place, the results of which are clearly objective and successful, or at least more objective and successful than others. For although until fairly recently it *was* the case that most investigations of actual cases in science *did* distort those cases to fit a particular philosophical mold—a particular set of criteria of "demarcation" and interpretation of science—nevertheless in the past several decades the history of science has undergone a profound and far-reaching transformation. It is not merely that a vast quantity of new information about the history of science has been uncovered—information which has served to demonstrate the inadequacies of many earlier historical interpretations, as well as of the philosophical presuppositions or conclusions which were often associated with them. In addition, and perhaps even more importantly for present purposes, the very standards of investigation of the history of science have become far more sophisticated. Historians have become aware of the necessity of seeing scientific relevance not in terms of later science or of preconceived interpretations of science, but as it was seen by the thinker or group under investigation. And they have developed critical standards (usually, of course, implicit and uncodified) for deciding, independently of views of "what science is," whether that aim has been achieved. This is not to say that the aim *is* always achieved; but it does argue the falsity of any claim that there is no such thing as a "better" or a "worse" ("distorted") interpretation of a case in the history of science, and that there is such a thing as an interpretation based on evidence which is objective relative to philosophical theses about the nature of scientific reasoning. And a full systematic account of criteria of historical investigation which shows their objectivity (relative to the kinds of philosophical conclusions we are seeking) need not be provided for us to recognize *that* such investigation is possible.[9]

A number of other arguments could be advanced against the view of "meaning" and its relation to investigation which underlies the "philosopher's objections," but they may be left for another occasion. For present purposes it is enough to remind ourselves of the healthy skepticism that has arisen in the past quarter century regarding such *a priori* "proofs" of the impossibility of some undertaking, particularly in the face of the fact that we do seem to engage in it.[10] In such cases, it is the philosopher's objection that should be viewed with suspicion, not the enterprise.

We may, therefore, accept, at least tentatively, the possibility of the sort of investigation advocated in this paper: to examine cases in the development of science—including, of course, contemporary science, which, as I have argued, in many respects sets our problem—with a view to arriving, for each case, at a grasp of the scientific foundations of that case, including the interrelations between substantive ideas and criteria of what counts as an area of inquiry in science, what counts as an appropriate description of the items to be examined in that area, what counts as a legitimate and important problem about that area, what counts as a promising line of research, what counts as a possible and as a correct solution to the problems of the area, and related questions. (Even if there turns out to be more than one well-founded interpretation of the case, that in itself would be an important fact to take into account.) Our investigations must, however, take us beyond such analyses, to a comparative examination of the results thereof, with a view to asking what, if anything, there is in common between such cases, or whether (and to what extent) there has been an evolution in the criteria as well as in the substantive claims of science. We will want to know, of any features or criteria which are common to all science or which have been arrived at at some comparatively late stage of science, whether those features or criteria are necessary ingredients of the scientific enterprise (at least when that enterprise has come to be understood), and if so why they are necessary (they *can* be necessary without being in any usual sense *a priori*),[11] or whether they may possibly be altered or abandoned or replaced in the future, and if so then under what circumstances. And finally, we will want to see whether, on the basis of these investigations of scientific cases and scientific change, a more comprehensive and systematic account of the knowledge-acquiring (or at least the rationally justified knowledge-claiming) enterprise can be obtained. In all these investigations, our results will be tentative, and subject to alteration or rejection in the face of examination of further cases or more adequate examination of the same ones.

It is sobering to realize how little has been accomplished as yet toward a general treatment of these issues. The logical empiricist tradition, I have argued, tended to ignore them; their critics of the late 1950s and 1960s tended to react to logical empiricism in ways, described above, which led to relativism, i.e., to a denial that there is any overall rational development in science. Those relativistic views have, I believe, been effectively criticized; and, standing above all the criticisms, is the point that science is, after all, a paradigm case of the knowledge-acquiring process.[12] To deny that science and its development *can* be rational—a denial that seems to be the conclusion of the relativist position—fails to recognize that the terms "rational" and "knowledge" have a use. It is a condition of the adequacy of any philosophy of science that it show *how* rational change in science is possible, and a philosophy of science which, after asking *whether* scientific change can be rational, denies that it can be, must be rejected.

But in responding to the challenge of relativism, philosophers of science have, to a large extent, deviated from a concentrated attention on the above issues, focussing their attention on the question of whether science is "rational" at the expense of a conscientious examination of what has been considered rational, and of how and whether such considerations have evolved. And it is the latter sort of investigation that is of immediate importance and greatest potential fruitfulness in the attempt to understand human knowledge and human reason.

Yet once more a question arises, threatening us again with a variant of objection (4). For if our investigation is concerned with what have been *considered* to be "reasons" and "rational change," then—despite all our analysis of the notions of possibility and prescription given in response to objection (4)—will we not, in the end, have given a merely *de facto* account of science after all? Will we not have failed to account for the normative aspect of calling science as a whole a "rational" enterprise? Again the philosophical temptation, in the face of questions like these, has all too frequently been to argue that the prescriptive canons of rationality must be laid down (and themselves justified) in advance. Again I must disagree: not until we can see, in detail, the complex workings and interactions of the variety of criteria employed in science, and the ways in which those different sorts of criteria interact, in various types of circumstances, with one another and with the substantive knowledge-claims advanced in science, and how the scientific enterprise, even in the criteria it employs (and to which many philosophers have been too quick to accord the sacrosanct and inviolable status of "meanings"), might evolve—not until then will it be possible to see what is gained in the scientific enterprise, and why those gains are advantageous. An understanding of the honorific, as well as of the "descriptive," aspects of the concepts of rationality and knowledge should and must be the results and not the prerequisites of an investigation of the scientific enterprise. I have tried to argue also that they can be.

## NOTES

1. The logical empiricist view was that their results were "logical reconstructions" of actual science—logical models of science which did not have to correspond exactly to real science.

2. L. Wittgenstein, *Tractatus Logico-Philosophicus* (London: Routledge and Kegan Paul, 1961).

3. See, for example, I. Lakatos and A. Musgrave, eds., *Criticism and the Growth of Knowledge* (New York: Cambridge University Press, 1970); D. Shapere, "The Structure of Scientific Revolutions," *Philosophical Review* 73 (1964): 383–94; D. Shapere, "Meaning and Scientific Change," in *Mind and Cosmos*, ed. R. Colodny (Pittsburgh: University of Pittsburgh Press, 1966), pp. 41–85; and D. Shapere, "The Paradigm Concept," *Science* 172, 14 May 1971, pp. 706–9.

4. In speaking, in the remainder of this paper, of "criteria of application," I wish this concept to be interpreted broadly, so that any characteristics given in an alleged "definition" of a term (and to be attributed to members of its reference-class, i.e., to be possessed or exemplified by anything to which the term can be applied), as well as any characteristics assigned to all members of the reference-class but not as parts of the "meaning" or "definition" of the term, will be considered "criteria of application" of the term. (Even the requirement that the characteristics be possessed or exemplified by *all* members of the reference-class is too strong; however, I will ignore that point here.) I do not mean to imply that these characteristics are (always) appealed to in explicit and conscious tests of whether or not to apply the terms in question. These remarks may be extended to my use of the term 'criterion' in general. Part of the motivation for this broad sense of 'criterion' lies in well-known difficulties about the distinction between what is part of a "meaning" ("analytic")—and therefore, according to a long philosophical tradition, not subject to change on the basis of "factual" considerations—and what is not ("synthetic"). That distinction is not assumed in this paper, and this fact is reflected in my use of the term 'criterion' and its relation to the

(or any) concept of "meaning." And the point of not assuming such a distinction is, as will be seen in the sequel, to allow for the possibility that no part of a conceptual structure is intrinsically immune to revision on the basis of reasons.

5. My point is that, even if "meanings" were entirely divorced from any notion of "criteria of application," the question would still remain open as to whether they are immune to revision in the light of changes in accepted scientific belief.

6. Even if certain types of concepts or criteria were found to be common to all science up to the present time, we would still have to ask (a) whether the basis of acceptance of those concepts or criteria ensures that they may not change in the future, and (b) whether their acceptance thus far, and promise of continued acceptance in the future, is guaranteed by considerations of "logical necessity" or by the way we have found the world to be. We must, for example, allow for the possibility that, even if the same rules of logic have been employed in all science up to the present, that need not guarantee that those rules will remain unaltered in the future. Thus, although the issue of "quantum logic" still remains controversial, the possibilities it envisions need occasion no surprise in the light of the considerations advanced in this paper: we must leave open the possibility that a close analysis of some theory (perhaps quantum theory itself) might make reasonable certain changes even in the classical rules of logic in order to accommodate successfully the domain of that theory.

7. Logical empiricism has often been criticized for its reliance on logic and axiomatization, on the ground that those devices (a) require focussing only on a "completed" version of a science, and (b) "freeze" science into a "static" mold. While it is true that logical empiricism relied too exclusively on those devices, these two criticisms are incorrect. Logical treatment and axiomatization need not be restricted to a "completed" science, but can in principle be applied to any particular stage of development; and axiomatization of an area of science at a given stage does not imply that axiomatization of that same area at a later stage might not be very different. Indeed, such axiomatizations can be of help in revealing certain aspects of the cases in question and their evolution. They cannot, however, be relied on exclusively.

8. See the remark made above regarding recombinant DNA.

9. The words "objectivity (relative to the kind of philosophical conclusions we are seeking)" contain the germ of an account of how objectivity in historical inquiry is possible: criteria of selection and interpretation of historical data need not be "loaded" with a philosophical theory of science. This suggestion is an application of the more general analysis of the concept of objectivity in science developed in another paper, "The Concept of Observation in Science and Philosophy" (unpublished).

10. This attitude is due largely to the influence of L. Wittgenstein's *Philosophical Investigations*, 3d ed. (New York: Macmillan, 1968).

11. See note 6, above.

12. Whether science exemplifies *all* the methods of knowledge-acquisition employed in human activities is a question which cannot be dealt with until its knowledge-acquiring processes are fully understood. But the point is that any general epistemological conclusions about the nature of knowledge cannot ignore the kinds of investigations of scientific reasoning and knowledge-claims outlined in this paper. This is why the title of this paper is addressed to the theory of knowledge and not merely to the philosophy of science.

# PART III

# TOWARD A SYSTEMATIC PHILOSOPHY OF SCIENCE

CHAPTER 11

# THE CHARACTER OF SCIENTIFIC CHANGE*

## I

The prime intellectual achievement of modern science is a body of views of nature at once general in their conceptions and specific and precise in their explanations. Those views have, over the course of the history of science, become increasingly coherent, in the sense both of constituting a more and more unified perspective on a larger and larger body of detailed beliefs, and of providing an intelligible picture of the world we experience.[1] Although problems remain that can be expected to alter our present scientific picture, even in fundamental ways, some of its claims must qualify as knowledge and understanding of, or at least as well-grounded beliefs about, the way things are. They have been arrived at by an increasingly sophisticated and systematic process of investigating nature, a process roughly describable as being, or at least as having come to be, one of collecting evidence on the basis of observation and experiment, and of formulating hypotheses whose purpose is both to account for the observations and experimental results and to provide bases for further observation and experiment leading to new discoveries and broadened and deepened understanding. It is the responsibility of the philosophy of science to show, by an analysis which preserves the spirit of this achievement, how the achievement has been possible (allowing both for the possibility of knowledge at present and the possibility that current views might be wrong), and to interpret the processes by which that body of views has been arrived at.

The major traditional approaches to the investigation of the knowledge-seeking enterprise, as represented by such diverse thinkers as Plato, Kant, the early Wittgenstein, and the logical empiricists, have held one assumption in common: namely, that *there is something which is presupposed by the knowledge-acquiring enterprise, but which is itself immune to revision or rejection in the light of any new knowledge or beliefs acquired*. Many variations on this theme may be found; I will mention four. *First*: the view that that there are certain claims about the way the world is which must be accepted before any empirical inquiry is possible, or before further beliefs (well-grounded beliefs) can be acquired, which claims, being presuppositions of the knowledge-acquiring process, cannot be revised or rejected in the light of any results of that process. *Second*: the view that there is a method, 'the scientific method', by application of which knowledge or well-grounded belief about the world is obtained, but which, once discovered (by whatever means), is in principle not subject to alteration in the light of any beliefs arrived at by its means. *Third*: the view that there are rules of reasoning – rules, for example, of deductive or inductive logic – which are applied in scientific reasoning, but which can never be changed because of any scientific results. And *fourth*: the view that certain concepts are employed in or in talking about science which are not open to abandonment, modification, or replacement in the light of new knowledge or (well-grounded) belief. A variety of versions of this fourth view are important in current discussions. One such version maintains that the alleged immutable concepts are unavoidable ingredients in any fundamental scientific theory, as space and time were claimed to be, for different reasons, by Kant and Heisenberg. In other variations, the unchangeable concepts are held to be 'observational' ones upon which all scientific theorizing is supposed to rest. Still another variation maintains that certain 'metascientific' concepts (or 'terms'), like 'evidence', 'theory', 'explanation', which are used in talking about scientific concepts, claims, and arguments, must have meanings which are wholly independent of the specific content of ongoing science.

I will refer to all such views indiscriminately as 'presuppositionist' views of science, though the four sorts of views I have mentioned tend to hold further that the alleged presuppositions are *a priori* (or at least 'formal'), and that they constitute invariant characteristics of science, the essence of science, or what is means for an activity or claim to be scientific. There are presuppositionist views of science which are not apriorist or essentialist; when, rarely, I need to refer to them, I will distinguish their brand of presuppositionism by appropriate adjectives.

Whether in their propositional form (science presupposes certain *claims* about nature), either of their nonpropositional forms (what are allegedly presupposed are methodological strictures or 'purely formal' rules rather than claims), or their conceptual form (not claims, methods, or rules, but 'concepts' are presupposed), attempts to develop presuppositionist theories of science have faced appalling difficulties. Yet despite the avalanche of objections to particular formulations of these types of views, new forms of them have continually arisen from the ashes of previous incarnations. Reflecting on this remarkable persistence – or perversity – one cannot but ask for its reasons. Why have the repeated failures to generate a viable presuppositionist analysis of science not been taken more seriously as reasons for exploring nonpresuppositionist alternatives? Or, phrased more positively, why has it so often been felt, if only implicitly, that such an approach, in one form or another, *must* be taken, that there *is no* viable alternative? The answer to these questions, and also the weaknesses of the answers, can be brought out most clearly by examining some of the alternatives that have actually been proposed.

For periodically, opposing views have been offered, most recently in the years since the 1950's. On one front, many have claimed that the rules of deductive logic are in principle open to revision in the light of empirical knowledge,[2] and some, following earlier writers, have held that in fact quantum theory requires the abandonment of certain inference-rules of standard deductive logic.[3] Such forays, while immensely important and relevant to present issues, are relatively local when compared to the sweeping and highly influential attacks that have come from historians of science and historically-minded philosophers in recent years. Though drawing support from the difficulties of logical empiricism (especially the breakdown of the distinction between 'theoretical' and 'observational' components of scientific language), these writers have found their primary inspiration in the widening studies that seem to demonstrate that science does not, after all, have a single overarching methodology, nor even a unique method characterizing each area of science. On the contrary, according to their reading of the history of science, scientific change is not merely a successive alteration of substantive beliefs occasioned by new discoveries about the world; scientific change and innovation extend also to the methods, rules of reasoning, and concepts employed in and in talking about science. Even the criteria of what it is to be a scientific 'theory' or 'explanation' change; and thus the notion that 'theory' and 'explanation' are 'metascientific concepts', with meanings independent of scientific beliefs, is rejected.[4] The changes may be so radical in some cases as to alter the very conception of science, two scientific traditions being so different in some cases as to be wholly incomparable in any respect.

It is at this point that difficulties begin to appear that have led many to maintain that, if science is to be possible as an enterprise which progresses on the basis of reasoning about discoveries about the world, then there must be presuppositions of one sort or another of the kind discussed above. Let me state the case as baldly as possible in order to bring out the problem. Suppose we carry the historically-minded philosopher's insight to its fullest conclusion and maintain that there is absolutely nothing sacred and inviolable in science – that *everything* about it is in principle subject to alteration. (Notice, incidentally, that this doctrine is fully in accord with the spirit, if not the deed, of traditional empiricism – a point to which I will come back later.) Then included among the things that can change are standards or criteria of what it is to be a 'good reason' for change. But then how could criteria of rationality themselves be said to evolve rationally, unless there are higher-level standards or criteria of rationality, themselves immune to alteration, in terms of which changes of lower-level criteria of rationality could be judged to be rational? There thus seem to be only two alternatives: relativism, in which there is no real ground (*i.e.*, no ground other than the decree by fiat of a triumphant community) for saying that there is 'progress' in science, that one body of scientific beliefs is better than another; or else a presuppositionist theory according to which there is something, of the sort that can serve as a standard or set of standards or criteria for scientific rationality and progress, which is immune to the vicissitudes below, and which serves as the ultimate arbiter of those lower-level scientific disputes.

Many of the historically-oriented approaches of the 1960's compounded their difficulties by introducing further ideas tending to increase their commitment to relativism while obscuring the fundamental issue just described. For example, some of those views, taking a page from their presuppositionist competitors, went beyond the view that standards of rationality change piecemeal, holding that such standards (and presumably everything else in science) are determined by some fundamental viewpoint or 'paradigm'.[5] Much philosophical effort was devoted to arguing against these aspects of the antipositivist views; however, stripped bare of such overlay, the issue as I have described it above – the issue of how standards of rationality could be held to undergo rational change – is independent of these doctrines. It is also more fundamental, inasmuch as it is broader than a debate between two different presuppositionist views of science, one essentialist and nonrevisionist, the other antiessentialist and relativist.

Let us therefore return to that fundamental issue and its import for the interpretation of science. The problem, it seems to me, comes ultimately down to the following.

I. On the one hand, work in the history of science in recent decades strongly suggests that scientific change and innovation go deeper than mere discovery of new facts and simple successive alterations in the set of substantive scientific beliefs about the world, but extend also to the methodology, the distinction between a legitimate and an illegitimate scientific problem, what counts as a possible explanation, and a host of other aspects of the scientific enterprise that have often been said, in one way or another, to be independent of the processes of substantive scientific discovery and innovation.

For example, in generalizing the success of Copernican astronomy, Kepler maintained that all aspects of nature, down to the finest details, are interconnected by geometrical relationships which in turn are clues to still deeper relationships; that geometrical reasoning can lead to the discovery of truths that are hidden from mere sense experience[6]; and that all geometrical relationships are of equal relevance in the attempt to understand nature. Thus, in this latter connection, angles formed by planets with locations on earth were as much 'details of our experience' as, and had as much significance as, the angles swept out by a planetary radius vector in a given time. It made sense to him to ask not only, What is the relationship between the orbital period of a planet and its distance from the sun? but also, What is the relationship between the geometrical harmonies found in astronomy and those in music and anatomy and weather patterns, *etc*., and, Given an angle formed by two planets at a person's birthplace at the time of his birth, what relationships does that fact have to other facts about the person in his later life? Within the general framework of his mathematical approach to the explanation of experience, Kepler knew no constraints on the kinds of questions to ask or the kinds of observable relationships it was appropriate to ask them about. Incorporation of some aspects of Kepler's work into later views, and the success of those later views, would lead to the introduction of constraints rejecting certain Keplerian questions as scientifically illegitimate and irrelevant.

The preceding example illustrates the rejection of problems and corresponding possibilities as scientifically illegitimate; but there are also cases in which new discoveries, approaches, techniques, and theories open the door to new possibilities – new questions and alternatives that were before ruled out as illegitimate or perhaps even self-contradictory or inconceivable. Thus in the seventeenth century, Descartes proposed to account for the properties of matter in purely geometrical terms. To this program, Newton raised a fundamental objection: the parts of space are identical, so that their interchange, if it were possible, would bring about no change at all; space is, on the contrary, that with respect to which, in which, things move; and to speak of the points of space as themselves moving would be absurd (what could they move in?). Thus, to Newton, 'space' cannot have the property of movability that is characteristic of matter. Nor, indeed – though Newton of course did not say this explicitly – could it have inhomogeneities of any kind, since inhomogeneity is judged *with respect to* space. An assertion of space curvature, or of change of curvature, would not have been a *false* view; it was not even a possible view. It was absurd, inconceivable, self-contradictory. At the root of Newton's argument was the very intuitive idea that, if one is to speak of properties of an object such as curvature, one would have to do so in terms of some containing space. But since, to Newton, three-dimensional space was the ultimate reference-frame, not contained in any further space of greater dimensionality with respect to which inhomogeneities and changes of inhomogeneity could be referred, it itself could not intelligibly be said to have any sorts of inhomogeneities. All this changed with Gauss's 'Theorema egregium', in which he demonstrated that a two-dimensional surface has an intrinsic and invariant (with respect to choice of coordinate system) property of curvature which can be characterized and determined solely within the surface itself, without any consideration of any higher-dimensional space in which the surface is embedded. Furthermore, this property of curvature corresponded in all relevant ways to the property of curvature that had previously been characterizable only with respect to a containing space. Riemann extended Gauss's result to spaces of arbitrary dimensionality, so that it became possible to speak of the 'intrinsic characteristics' of a space of any dimensionality. It became possible, then, to do what was inconceivable and self-contradictory before: to consider physical space to have intrinsic characteristics without supposing it to be embedded in a higher-dimensional space; and further, to allow those characteristics to vary from point to point and from time to time. The way was paved for a revival of space theories of matter (neo-Cartesianism of Clifford, Eddington, Wheeler), but also, and ultimately more importantly, for the twentieth century space-time theories which attempted to relate the characteristics of matter and space-time.

A multitude of other examples could (and some will, below) be adduced to indicate that the boundaries between the scientific and the nonscientific – between, for example, the 'observable' and the 'unobservable', between the scientifically legitimate or possible and the scientifically illegitimate or impossible, between scientific problems and pseudo-problems, between scientific explanations and pseudo-explanations, between science and nonscience or nonsense – are not something given once and for all, but rather shift as our knowledge and understanding accumulate.

Indeed, the historians only reinforce what should have been evident in the first place: for no one can look closely at the contentions of modern science without being impressed by the profound depth of its departures from what anyone would have expected or even thought of. It is pehaps sufficient remainder of this fact merely to mention three further examples of such departures: the fall from grace of Euclidean geometry; the abandonment of the idea that there is a unique set of events in the universe simultaneous with a given local event; the abandonment of determinism as a *standard* for what a fundamental explanation should be.

II. Thus scientific change seems to be pervasive and deep, extending far beyond mere alteration of factual belief. Yet on the other hand, the philosophically-minded historian (or the historically-minded philosopher), in his rightful reaction against Whiggish history and positivistic philosophy, seems often to have gone too far in the other direction, forgetting that one also cannot look closely at modern science without being impressed by the fact that its claims *are* better than those of its predecessors.

Can justice be done to both these claims without returning to the 'absolutes' of the Platonic-Kantian-positivistic tradition or falling into the relativism of its latest critics? More specifically, rather than beginning with the idea that we must find the common essence of all science, can we begin at the other end, recognizing and examining the depth and extent of scientific change, trying to see whether such a perspective allows the possibility of scientific progress and knowledge, and only then, if and where necessary, admitting the existence of general features, requirements, or criteria? Naturally, in a brief exposition it is not possible to provide the extensive development and support which such a view requires, in the form of both case studies and associated philosophical analysis. Nevertheless, I will try to sketch a few of the directions which the full development of such a view might take.[7]

## II

The point of departure is the idea that all of the following, rather than being independent of the changes that take place in science, are intimately connected with the content of scientific belief: the determinants of what is to count as a body of information to be investigated and the way that information is described at any given stage; the various sorts of problems that arise concerning those domains; the methods by which the domain and problems regarding it are to be investigated; the range of possible answers to the problems; the criteria or standards for acceptance of some one answer as correct. Thus, for example, although the criteria of what can count as an explanation do, at any given stage, mark out a range of possible explanations, nevertheless the knowledge attained – the set of explanations which come to be accepted – lead, at least under certain circumstances, to a change in the criteria themselves. The 'criteria', if I may for the moment use that word uncritically (I will say more about it shortly), are not independent of scientific beliefs (as the logical empiricists held), nor do they imply them (as some relativists have maintained). The criteria of application of – for example – metascientific terms, just as of scientific terms, are connected to the substantive scientific beliefs by what might, for the sake of brevity, be called a rational feedback mechanism, itself perhaps subject to reform in the light of new information which those criteria themselves help reveal. 'Criteria of application' determine ranges of possibilities; they themselves are subject to revision in the light of possibilities which (sometimes on the basis of yet other criteria employed at that stage of science) come to be accepted as correct. That is to say, the 'rules' according to which we think about nature, and the vocabulary in terms of which we talk about it (even about what we 'observe') also develop with the evolution of the content of scientific knowledge and belief. Method, rules of reasoning, criteria (*e.g.*, of what can count as an explanation) go hand-in-hand with the beliefs arrived at by their employment, and are on occasion altered in the light of the knowledge or beliefs arrived at by their means. Constraints on scientific reasoning develop, being sometimes tightened and sometimes broadened, as science proceeds. And thus, although at one stage of science, what (for example) counts as a legitimate scientific theory or problem or explanation or consideration might differ, even radically, from what counts as such at another stage, there is often a chain of developments connecting the two different sets of criteria, a chain through which a 'rational evolution' can be traced between the two. We can then recognize that, given the knowledge and criteria available at a particular time, certain beliefs about possibilities and truth were reasonable, even though alteration and improvement were later possible, with the emergence of new knowledge and new criteria. (This is, of course, why appeals to historical cases cannot be taken uncritically as counterexamples to or support for theses about what scientific reasoning is.[8]) We still need to explain the sense in which such evolution can be said to be 'rational'; what I have argued thus far is that radical difference between scientific beliefs and criteria at two different epochs does not automatically preclude the possibility of connection, comparability, and progress.

Consider the example of description of the entities dealt with in the study of matter.[9] How does one deal with matter? Before the sixteenth or seventeenth century, the answer was debatable; nevertheless, a widespread view among working alchemists and many Aristotelian philosophers was that there is, fundamentally, only one kind of 'earth', existing in more or less perfect forms. Although the idea of separating an earthy substance from other earthy substances for the purpose of working on it alchemically was not inconsistent with this approach, nevertheless such 'removal of impurities' was at best only a preliminary and peripheral procedure. For even where removal of irrelevant material from earthy substances was an initial step in alchemical procedure, what remained was not what we would call a 'purified' substance, but rather an imperfect substance that still had to be brought, by various actions, to perfection. The later idea of chemical analysis as the segregation of substances and their breakdown into their constituents was, among these workers, far from central: a substance was imperfect of its kind not because it contained admixtures of material substances of other kinds, but rather because it was only a partial realization of a perfect or harmonious form of the one earthy substance. But increasing familiarity with a wide variety of forms of earthy substance, and increasing development of methods of dealing with those substances, led to an increasing attention, by the late sixteenth and seventeenth centuries, to the breakdown products of physical and chemical action, particularly heat. A two-century controversy arose as to whether the breakdown products were constituents of the original substance, or were merely created by the action of fire on it. By the middle of the eighteenth century enough had been achieved to make a strong case for the view that breakdown products were constituents, and for the view that understanding of material substances was obtainable through analysis of those products. Many ingredients had to go into the working out of such a view before its triumph in the work of Guyton de Morveau and Lavoisier[10]; but triumph it did, and it brought with it an alteration of the whole conception of the aims, problem-structure, and methods of the study of matter. The central problem in the attempt to deal with matter shifted from, How can earths be brought to perfection? to, What are the constituents of material substances? A reform of the nomenclature of material substances was inaugurated also: the language of chemistry was reformed so as to make the names of substances correspond to their composition and structure. (It is true that sometimes language 'pictures' objects and their relationships; but this picturing is not, as it was for the *Tractatus Logico-Philosophicus*, a necessary truth following from the very nature of communication; rather, it is something we make language do in the light of knowledge about the world, in order to help us gain further knowledge about it.) The idea of understanding matter through discovering its constituents and their structural and dynamic interrelations is not something that follows from, say, the very concept of 'understanding'; it had to be wrested from nature, tried and won by its successes. Nor has its career since the eighteenth century been without its vicissitudes and alterations. The notion of a 'constituent' has undergone significant modification, especially with quantum theory and quantum field-theoretic approaches and the wholesale importation of group theory into those subjects. The whole 'compositional' approach nearly died in the 1950's, when it appeared that a general quantum field theory was doomed because of the apparent nonrenormalizability of the type of theory that seemed appropriate for weak interactions, the seeming intractability of the strong interaction, and other considerations. Revival began in the mid-1960's, followed by spectacular vindication after 1974 with the victory of the charmed quark 'theory' and the rapid development of the color 'theory' of strong interactions, quantum chromodynamics. Even that success was not purchased without further alteration of compositionalist conceptions, however. The debates of the sixteenth, seventeenth, and eighteenth centuries had established that if, under appropriate conditions, certain products were separable from an entity, then those products were constituents of it. For various reasons, it seemed natural to convert this into an if-and-only-if proposition: if separable (in appropriately circumscribed conditions), then constituent; and if constituent, then separable. Already in the 1930's it had to be concluded that electrons emitted from a neutron or radioactive nucleus in beta decay could not have been constituents of the neutron or nucleus. Now, in the late 1970's, repeated failure to observe free quarks leads to the possibility that the second half of the if-and-only-if claim will have to be rejected (quark confinement). The relationships between constituency and separability have become vastly more complicated.[11]

## III

In such ways, shifts of aims, problems, methods, and vocabulary are linked to substantive beliefs about the world; aims, methods, and so forth in science are as much subject to discovery and evolution as are the facts and theories with which science deals. But there are further complexities of scientific change and innovation that can be brought out by looking at another sort of case.[12]

What is counted as 'evidential' and 'observational' in science is not something given once and for all, but evolves along with scientific knowledge. A striking and illuminating example is the claim, universally made by astrophysicists since the early 1960's, that it is now possible to make direct observations of the center of the sun. What is involved, of course, is a knowledge of the behavior of the neutrino – specifically (but only in part) of processes in which neutrinos are emitted, of the fact that neutrinos are weakly interacting and therefore can be expected to pass freely (without interruption or interference) from their origin through the body of the sun and interplanetary space, and of the kinds of receptors appropriate for intercepting such neutrinos, even if only rarely. For present purposes, what is important to note is the extent to which specification of what counts as observed or observable is a function of the current state of physical knowledge, and can change with changes in that knowledge. For (in many cases besides that of neutrino astrophysics) current physical knowledge specifies what counts as an 'appropriate receptor', the ways in which information of various types is transmitted and received, the character of interference and the circumstances and even the statistical frequency with which it occurs, and even the types of information there (fundamentally) are. The physics of the present epoch, for example, makes such specifications for a wide range of circumstances through the whole body of well-founded belief about elementary particles, their decays and interactions, and the conservation principles which govern such processes. In particular, for example, knowledge of the cross-sections for such particles, the probabilities of their interactions with other particles (or decays of individual particles) in given environments contributes to specifying the notion of 'interference' or 'interruption'. In particular problems, far more enters into specifying such notions as that of a 'receptor' than just what is contained in elementary particle theory; for example, in the solar neutrino case, knowledge of the environmental conditions (pressure, temperature, opacity) in the interior of the sun also plays a role, as does knowledge about various instruments.

But what, in such cases, is counted as 'knowledge' to build into the concept of observation? A vast background of information is available for this and similar purposes: it consists fundamentally of those beliefs which have been arrived at, concerning which there is no, or little, specific reason for doubt, or which have proved successful in various circumstances to which they have been applied. (Preferably, but not always, it is insisted that the beliefs that have been successful also not be subject to any, or serious, specific doubt.) It is such information that is brought to bear in determining what will count as observational; it also provides the bases for organizing information into areas for investigation, at the same time shaping the ways in which that information is described and the problems which make it an area to investigate ('domains').[13] It determines the kinds of problems that will be considered scientific, and of these, which are more and which less important. Arrived at by antecedent methods and having proved successful and free (or at least relatively free) from specific doubt, those beliefs sometimes lead to the rejection or modification of the methods which led to their discovery, and lead in turn to the adoption of new ones. They further delimit a range of possible solutions to problems, and produce standards and expectations regarding the characteristics which a solution to those problems must have. Expressed otherwise, the 'criteria' determining all these things are not something transcending science, but are an integral part of it; any strict separation of criteria from the subject matter to which they are applied must be rejected. And similarly, the unbridgeable bifurcation so often made in traditional philosophical discussions, between 'propositions' expressed in the indicative mood, and 'rules' or 'methodological strictures', to which the imperative mood is more appropriate, does gross injustice to scientific reasoning. The interplay between 'criteria', 'rules', 'methodological strictures', and the 'cognitive content' of science is constant and tight. Those ideas certainly cannot constitute a final and rigid categorization in terms of which science should allegedly be construed.

While there are certainly considerations in science which play the role of 'criteria' in one way or another, that term can also be misleading in other ways than those suggested above. For a second-line defense of a science-metascience distinction might run as follows. It might very well be (so the objection to the present account would go) that what counts as 'observable' might depend on and vary with the content of accepted scientific belief; nevertheless, at any given stage science insists that whatever it is at that stage that is not counted as 'observable' be excluded from the realm of the scientific. And thus an insistence on 'observability' in this more abstract sense is what is preserved throughout the alterations of scientific belief, as an essential characteristic of the scientific enterprise. The really 'metascientific' principle involved is something like this: "Whatever is observable at stage *X* of the development of science is scientific; whatever is unobservable at that stage is nonscientific." (Arguments similar to this one can be constructed for such traditional candidates for distinguishing characteristics of science as 'verifiability', 'confirmability', 'falsifiability', *etc.*). But such a line of defense will not hold: there are many cases in which observability (in the sense of 'what counts as observable' at a given stage of science) has been rejected as a governing rule of what counts as 'scientific'. A contemporary case is one of the most impressive: rather than sacrifice the highly successful quark theory in the face of failure to observe free quarks, theorists have appealed to a property (quantum number) of 'color' which explains why quarks can combine in the first place.[14] 'Color', introduced originally to keep the quark theory consistent with the Exclusion Principle, is postulated to be unobservable: in the mathematical theory of the color force, that force will be smallest at small distances and larger at greater distances. An isolated quark would be colored, and since nature conspires to hide her true colors, therefore free quarks are unobservable. To try to pull a quark away from its fellow quarks in a hadron, we would have to introduce enough energy to produce a new quark-antiquark pair, which would combine with the previously-existing quarks to produce new hadrons, thus effectively preventing the appearance of a free quark. Color has become the central concept of the most promising theory or proto-theory of strong interactions available today; indeed, quantum chromodynamics may ultimately provide the basis for a truly unified quantum field theory. (It is not yet clear, however, whether *QCD* really implies full quark confinement.[15]) For present purposes, however, the moral of the story is that science determines not only *what counts* as observable (or verifiable or falsifiable or whatever), but also *whether* observability (*etc.*) is to play the role of a criterion of scientific admissibility. In the present case, the unobservability of free quarks and color is not sufficient ground for rejecting them as 'unscientific'; on the contrary, there is good reason for retaining them, in the following respects: (1) their success (and the success of the quark theory generally) in accounting for a vast range of hadronic phenomena; (2) the fact that their unobservability is a natural outgrowth of (or at least is compatible with) a theory or type of theory which has been introduced and proven successful for reasons independent of the particular hypothesis in question; and (3) the promise of that theory or type of theory for further successes in the future (particularly the promise of unification with the theory of electromagnetic and weak interactions). It is true that science has learned the importance of observations; but it has also learned that there may be circumstances in which insistence on observability may be an injudicious dogma. It has been so in the past.[16]

However, questions still remain with regard to the view I have presented. At least, it might be asked, wouldn't we in general *prefer* a theory according to which the basic entities and properties discussed in the theory are observable to one in which this is not the case? If someone were to observe free quarks or separate 'colors', or were to construct a theory which was in other respects equal in explanatory power to the present *QCD*, but in which free quarks and separate colors were observable, would that not be a source of relief or triumph? Theories according to which separate colors would be observationally distinguishable have been proposed (Han-Nambu), but have not been embraced, automatically or otherwise, with the enthusiasm that has been accorded the now-standard theory. And at the present stage of investigation, with quantum chromodynamics having developed so successfully in its present form, observation of free quarks (or separate colors) would be more of a concern than a triumph, or – what is more likely – would be accomodated as a matter of course in a *QCD* in which quark confinement can be violated, being rare but not prohibited.

Still, one might feel, in cases where a theory admits unobservable entities or properties, the preferability or admissibility of such a theory must be justified by extenuating reasons, such as (1)–(3) above – whereas (we might feel) such justification is not required in the case of theories whose basic entities and properties are observable. I have argued elsewhere that claims of 'observability' in science are far from immune to the need for justification: circumstances can and do occur in which those claims are dubitable where claims of other sorts are not. But apart from that, there is an important truth behind the felt objection that observation is somehow privileged.[17] It is of course true that, at a given stage of scientific investigation, certain general sorts of considerations become characteristic of scientific reasoning. We have learned, for example, that 'observation' plays a central role in science; and, too, we have learned that considerations of the sorts exemplified by (1)–(3) above also play a role, and in particular – among other things – exercise a controlling effect on observational considerations. We have also learned in a general way when and when not to let such considerations enter in, what weight to give them in particular circumstances. And we have learned that the observability requirement must be taken very, very seriously: that a theory must be very good indeed if it is to violate that requirement.

There are a great variety of such general types of considerations: the knowledge-seeking enterprise is not, nor has it ever been, governed by some *single* sort of consideration like verifiability, falsifiability, or observability. It is, as was suggested at the beginning of this essay, the business of the philosophy of science to delineate this complex network of general types of considerations and their interplay. However, there are two points about this network of 'rules' that make the primacy of observability not the necessary truth it might seem at first thought.

The first point is that the philosopher of science must also recognize the changing character of this network over time. Primacy of observability – what primacy there is – was learned through the successes of emphasis on that requirement and the failure of its rationalistic, intuitionistic, revelationist, mystic, and other rivals. And the tempering of that requirement by other considerations was also learned through the failures of an exclusive insistence on observability and the successes of broader theorizing. It seems reasonable to expect that there will be further alterations in the character of this general network in the future. On the other hand, there may well be features of it that will not be altered. It may well be, for example, that observability – if I may for the moment speak as if that were a precise requirement – will always remain a primary requirement in science. We have no reason at present to question its status, subject to the tempering effect of other considerations. After all, for something to be revisable does not imply that it will ever be revised; we may simply have learned how to learn about the world. On the other hand, it is not therefore a necessary requirement, any more than it is an *a priori* one.

The second point is that this network of 'general sorts of considerations' does not consist – as my discussion so far might misleadingly suggest – of a set of clearly- and unambiguously-formulable ideas or even clearly- and unambiguously-applied methods. The process of inquiry consists partly of employing concepts that are unclear, ambiguous, and sometimes even inconsistent; this is as true of 'metascientific' concepts like 'observation', 'evidence', 'theory', 'confirmation', and 'explanation' as it is of concepts like 'mind', 'matter', 'element', 'particle', 'infinitesimal', 'force', 'energy', 'temperature', 'delta function', and 'singularity' at various stages of scientific history. And beyond this is the openness and alterability of concepts however clear at any given stage ('life', 'electron', 'neutrino', 'island', 'ocean', 'schizophrenia'). Greater clarification and closure are often achieved with further investigation (and we also learn in what clarification consists); but that clarification depends not only on 'conceptual analysis', but also, and even primarily, on beliefs about the way the world is. For this reason, there is always an openness to concepts in science, an openness to alteration in the light of new beliefs about nature. While it is the responsibility of philosophers of science to elucidate the general characteristics of scientific reasoning, we cannot expect some 'absolute precision' (mythical in any case) in our analyses. What clarity we can achieve we must of course seek (and it is sometimes possible to be relatively clear about the respects in which clarity cannot be achieved); but we must be prepared to acknowledge irreducible vaguenesses, ambiguities, alternative interpetations, open possibilities of a variety of sorts – and this not because of our failure to discover the true principles of scientific reasoning, or even the ones which are applied at a given epoch of scientific history. Rather, there is no such truth, or at least there is none yet. At any given stage (at least so far and in the foreseeable future) there are a great variety of types of considerations. There may be disagreement as to what they are, how they are to be interpreted, what they apply to, what weight they are to be given when they are applied, and a host of other questions. And the considerations may not even be formulated. Such lack of agreement on 'methodological' matters is an integral part of the scientific enterprise: arriving at reasoned clarifications and eliminations of alternatives in respect to them is as much a characteristic of the enterprise as is debate about alternative substantive beliefs and elimination of some of these in favor of others. Indeed, the two aspects of the enterprise are inseparable.

Thus, even in extracting the requirement of observability and its primacy from the dynamic process of science, the philosopher of science has not located a precise rule, unambiguously (much less slavishly) followed in science. Although in some cases of sophisticated science what is considered observational is extraordinarily precise, there may be only 'family resemblances' between what is considered observable in one area and what is considered so in another. Not all such usages are equally clear, and some degree of openness is in any case generally apparent or implicit. There is no single, unambiguous, perfectly clear requirement to be found, even though advances in clarity may well be forthcoming in the future. A stage may even be arrived at beyond which no further ambiguity or unclarity reveals itself, no reason to doubt that the concepts at which we will then have arrived are perfectly adequate for dealing with the world. But neither the necessity nor the impossibility of arriving at such a stage can be established in advance.

Philosophy, and philosophy of science in particular, has long suffered from lack of agreement about the issues with which it deals and even as to whether the various issues and alternatives are formulated in a clear and precise way. Has the time not finally arrived when, instead of viewing such disagreement as a sign of failure and need for further work, we should recognize it as a datum and seek its explanation? The above discussion proposes such an explanation: the single truth, and the precision, simply are not there; whether we ever arrive at such a stage, we have not done so yet. From this perspective the fundamental fault of traditional philosophy lies with its concept of 'meaning', and that in a dual sense: its doctrine that meanings are, or should be, perfectly precise (*e.g.*, be formulable as a set of necessary and/or sufficient conditions) before they can be of value in a cognitive enterprise; and secondly, its doctrine that considerations about meanings are wholly independent of considerations having to do with substantive beliefs about nature. If we are to have a concept of meaning *at all* which serves to illuminate aspects of the knowledge-seeking enterprise, it is imperative that such a concept reject these two doctrines and incorporate the points made above.

This discussion makes it possible to carry the analysis of the role of 'criteria' in science one step further. Earlier in this section I discussed, as a possible objection to the present view, a distinction between a 'principle of observability' and particular 'criteria' of 'what counts as observable'. The objection claimed that there is a 'general metascientific principle' of observability, "Whatever concepts deal with what is observaole at stage $X$ of the development of science are scientifically admissible; whatever do not are non-scientific." I have argued that this principle is not universally adhered to. Now, the objector might hold that it is *what counts* as observable (the reference of 'observable', or, perhaps more relevantly, the criteria which determine what counts or should count as part of the reference) that is subject to change. The term 'observable' in the statement, however (the objection continues), has a meaning which is wholly independent of what is counted as observable or the (changing) criteria of what is to count as such. (The 'meaning' on this view would, of course, not determine the reference.) The objector would thus purchase, as the domain with which he is concerned as a philosopher, a guarantee that general features of science are invariant. But he pays a heavy price: for, even apart from the question of whether such a view of 'meaning' is defensible, if philosophers are to be concerned with meanings in that sense, they divorce themselves from concern with the actual reasoning-processes that take place in science, by which scientists actually *decide* what is observable in any given case. A philosophy of science along such lines condemns itself to being irrelevant to actual science. An alternative that is perhaps more appealing is to view 'meanings' as just another, 'higher' level of 'criteria' – criteria for identifying what count as criteria of observability (or, in more standard terminology, a concept of meaning according to which meanings determine reference.) Only then why should *those* higher-level criteria be immune from the effects of scientific change? [18]

The view being developed here has no need of 'meanings' which are independent of 'criteria'. For it takes seriously the suggestion that the 'meaning' of a term – whether a term like 'electron' or one like 'explanation' or 'observation' – resides in the way it is actually used by scientists, in the considerations they employ in deciding whether to apply or withhold the term. That use (or, expressed in a potentially misleading way, that criterion or set of criteria) can alter with new developments or differ somewhat in different circumstances; and no use at any particular stage or in any particular context need be 'perfectly' precise, unambiguous, and closed in order to be functional. (It may, of course, be remarkably so, as in the case of 'observation' in neutrino astrophysics.) But there will in general be resemblances, of ancestry and descent, between uses in successive stages, and of cousinship between contemporaneous uses in different contexts. In the clearest cases, those relationships are determined by the reasons underlying the alteration. (Rejection of certain characteristics of electrons and introduction of new ones; new types of explanation, as illustrated by the Gauss-Riemann case; the history of the concept of observation leading up to the view of the 'observability of the center of the sun'; new discoveries forcing hitherto unrecognized distinctions – distinctions between types of neutrinos, islands, or schizophrenia – or relationships – as determined by Faraday between galvanic and current 'electricities'.) There may be common features over an extended period of scientific development, but none has the privileged status of a 'meaning' of the term which *must* remain throughout. On the other hand, if one wishes to speak of the 'meaning' of a term over an extended period in which there are alterations, one need speak only of such resemblances: there need be no common core of meaning, as long as there is a traceable relationship of changes introduced for reasons. The view I am suggesting thus takes seriously the Wittgensteinian proposal that "the meaning is the use": the meaning of 'observable' is given in terms of what is counted as observable at a given stage (or the criteria in terms of which that is decided, to the extent that those are clear), and – here going beyond Wittgenstein – in terms of the reasons for considering those things to be observable. It also makes literal, and gives an historical twist to, the Wittgensteinian idea of 'family resemblances' as a more realistic view than that which insists on a common, essential core which is the 'meaning'.

I do not wish to deny that in many cases in science, contemporary as well as historical, there have been separations of types of considerations. One finds that at one time or another, methodological considerations, or 'meanings' or 'criteria', or ideas of what an explanation is or should be, *have* been held to be separate from substantive beliefs. And indeed, at the time they *were* separate, in that the body of accepted scientific belief, and the way those nonsubstantive considerations were formulated at the time, were not sufficiently developed to indicate any clear way of connecting them. But over its development, science has tended to find ways of removing such segregations, of breaking down bifurcations between methods or rules or meanings and the body of substantive scientific beliefs. The advantages of such unification – the reasons why scientists have come to seek ways of achieving it – and the means by which they sometimes do achieve it, remain to be investigated.

## IV

To return now to the scientist's use of background knowledge as the base on which to advance further, to new discoveries and new understanding: as I remarked above, science builds on what it has found it can trust, what it has least specific reason to doubt and what it has found most broadly applicable. Such acquired trustworthy belief can be applied to the creation of new types of observations which can be used in the testing of hypotheses about which we require more trust. This is not to say that what we trust and build on is sacrosanct. If the predictions of our hypotheses disagree with our observations (as they have in the case of the solar neutrinos), it is not necessarily the hypothesis under examination (in that case, hypotheses about the processes of stellar energy production) that is at fault; it may be any portion of the theoretical and instrumental ingredients which form the background of the experiment. As a purely logical point, this assertion is not new, of course. But perhaps because the logical point was the one thought to be most important philosophically, the crucially relevant aspects of scientific reasoning in this connection have been largely ignored or at best described on a metaphorical level. Rather than emphasizing the *logical* viability of alternative ways of resolving problems, what we need to study closely is the rough rationale that often exists in science for the circumstances in which and the order in which we subject the ingredient accepted ideas to suspicion and, correlatively, the order in which we seriously consider new alternatives. (The fact that the order is rough does not preclude the possibility of saying some relatively precise things about it.) Such orderings have been learned in the course of the development of science. They are determined by the background of well-founded beliefs relevant to the situation in question, most importantly by what we have learned to take as well-founded reasons for doubt or suspicion. We begin by suspecting those ingredients which are, in the light of our previous well-founded beliefs, most likely to be at fault, and least costly to give up. If the difficulty persists, the threat penetrates deeper and deeper into the structure of accepted beliefs involved in the purported observation and its background. Thus, in the case of the solar neutrino experiment, the disagreement between observation and theoretical prediction has persisted after re-examination of the solar models employed (and which we had specific reasons for thinking might be at fault), the chemistry of argon, and a host of other parts of the background of the experiment, until even the completeness and accuracy of our understanding of elementary particles has been brought into question. Such ordering is, of course, rough, partly because it is often not completely clear whether and to what extent some ingredient of the problem situation is or is not open to doubt, and partly because the assumptions involved in the hypothesis and the experiment testing it may not be explicit, and may in fact never have been formulated. Progress in science consists partly in sharpening reasons for doubt and partly in making deeper assumptions explicit and therefore realistically open to examination and use; and thus the ordering of priorities as to what alternatives to investigate seriously sometimes becomes less rough.

The making explicit of hitherto-unexpressed assumptions takes place in many ways, and performs numerous functions in the scientific enterprise. Thus the introduction of group-theoretical ideas into physics made possible the precise formulation of ideas (symmetry principles) which were in many cases not previously even mentioned, presumably because they would have seemed too obvious; it also made possible the discovery of deeper relationships than had previously been suspected (relation of symmetry principles and conservation laws); and finally, it made it possible to reject certain of those ideas (*e.g.*, failure of certain symmetries to hold for weak interactions) and to formulate previously unforeseeable and unformulable ideas of great explanatory power (broken symmetry). Nothing in science is intrinsically sacrosanct, whether substantive beliefs or methodological rules; but part of the process of making ideas potentially subject to doubt and modification lies in formulating them explicitly and in fruitful ways. Often this 'fruitfulness' arises out of a formulation which builds knowledge into language. (This is one of the 'cognitive' functions of reforms of scientific language like that in chemistry in the eighteenth century.)

In connection with the points I have been making, there is a close analogy between science and mathematics, and in particular with investigations of the foundations of mathematics and – ironically – with the relationship between mathematics and metamathematics. The analogy with metamathematics upon which logical empiricism based its conception of the philosophy of science was indeed unfortunate. Once, one could hope with Hilbert that there would be something final and unalterable about metamathematics: notions of 'system' and 'proof' would be made explicit and precise; and metamathematics would stand over mathematics proper like a dependable guardian, providing its secure (because primitive) foundations, establishing the consistency of particular mathematical systems and legitimizing its use of infinitistic methods by its own uncompromising reliance on trustworthy finitistic ones. That is no longer the case, and not only because of Gödel. Metamathematical notions are as subject to revision, rejection, extension, and generalization as are mathematical ones.

Throughout the history of mathematics, there have been attempts to constrain working mathematics by the imposition of strictures laying down what was to count as 'legitimate' mathematics (or, in another vein, as mathematics which could legitimately be applied in science). The story is a long one, from Greek worries about irrational numbers to insistence on limitations to equations of no higher degree than third, to demands that mathematics be restricted to the 'constructive' or (at least for metamathematics) to 'finitistic' methods. Each and all of these attempts to impose restraints required the rejection of substantial and important portions of mathematics; and in the end, it was not the mathematics that was rejected, but the restraints. In their attempt to extend their concepts and understanding of mathematical structures, working mathematicians have in the end been willing to introduce whatever methods might be required to achieve those ends (meanwhile appreciating, of course, that those methods might have their own problems and limitations as well as advantages). Thus despite Hilbert, and despite what one might suppose from Gödel's results regarding the impossibility of an absolute consistency proof for number theory, Gentzen could produce such a proof by going beyond the conditions of Gödel's proof (though without contradicting its basic result) by introducing a rule of transfinite induction, a rule which is at least difficult to interpret as 'finitistic', and in any case goes well beyond prior conceptions of what was to count as such. Such methods are widespread today. Though there are differences between mathematics and physics (perhaps not as great as they have been purported to be), the parallels with physics in these respects are striking. If obervationalist strictures dictated the rejection of the color concept, so much the worse for observationalist strictures. And a close examination of the arguments of those who, following Mach and Duhem, opposed the most important new ideas in physics in the first three decades of the present century, illustrates the same point: it was the philosophical strictures, not the physics, that were abandoned (or, more exactly in this case, reformulated in the 1920's so as to appear not only as supportive of, but even as the epistemological bases of, those new physical ideas).

In their impact on philosophy, Russell's mathematical logic and Hilbert's program for mathematics and metamathematics proved to be enormously pernicious red herrings. Philosophers were led to believe that they had, in logic and 'metascience', the keys to formulating what is essential and unalterable in human knowledge and the knowledge-acquiring enterprise. Focussed thus on the presumptively unchanging, their attention was diverted from the way science was altering the most profound aspects of our belief-structure. It is time that the latter achievement be placed at the center of philosophical scrutiny. We mut concentrate our efforts on understanding the depth and character of scientific change; and the burden of proof, rather than the privilege of assumption, must be placed on those who would maintain that, despite the direction of science in the past century, there is something which it necessarily and irrevocably assumes.

## V

In discussing the use of scientific beliefs as bases for further engagement in the knowledge-seeking process, and in talking about the questioning of those beliefs, I have consistently spoken of 'specific doubts'. The kinds of doubts relevant in scientific reasoning are specific ones. (This, too, is something that had to be learned.) Any claim about the world, however trustworthy it may have proved, *might* turn out to be wrong; specific doubts about it *might* arise. But contrary to a tradition in the theory of knowledge dating back to Descartes, *the possibility of doubt is not a reason for doubt*; the possibility that we may turn out to be wrong even in those beliefs which we at present have least (or even no) specific reason to doubt is not itself a reason for ceasing to trust those beliefs, for ceasing to use them as the bases for further extension of the knowledge-acquiring enterprise. A Cartesian or philosophical doubt is one which applies indiscriminately to any belief whatever; such doubts play no role in the scientific enterprise, and are at best only misleading reminders that specific doubts, that is, doubts that are specifically relevant to some particular knowledge-claim, might always arise.[19]

This point provides the key to an answer to what I earlier called the basic question with regard to a view of science like the one I have proposed – the question of how, if all standards of reasoning or criteria of rationality are subject to change, scientific change can be said to be rational. Here we see clearly the obscurity produced by words like 'standard' and 'criterion': they conjure up visions of something standing above and independent, and make us think that unless there is something of that sort, we can have no 'reasons' at all, since we have denied the very existence of standards or criteria. But I have argued that it is the content of scientific belief that shapes our 'standards' or 'criteria' of what count as reasons or as reasonable in science, and that such 'standards', far from standing above and independent of science, are as much a part of the activity of science, of the processes of discovering and coming to understand, as are the substantive beliefs themselves; indeed, in many ways they are indistinguishable from the latter. The activity of science is not well described in terms of two distinct *kinds* of statement, substantive and criterial; what occurs is better conceived in terms of statements playing different roles in different circumstances. And if 'criteria' are so inextricably involved in the activity of science, then what better 'standards' or 'criteria' could we employ – at least where we are able – than those beliefs, methods, and so forth that have proved successful and have not been faced with specific doubt? In the attempt to find some basis for considering certain things to be observable, or for distinguishing between those hypotheses to consider and those not to consider, and so forth, what else should one expect to use and to build on, wherever possible, if not such beliefs?[20] That our reasons, or the beliefs and methods which constitute their bases, may be wrong, and are sometimes shown to be, does not invalidate the practice of using the best information (and *hence* the best reasons, the best conceptions of legitimate problems, the best methods, *etc.*) we have available – 'best' in the sense I have mentioned, of having shown themselves successful, and not having become subject to specific doubt. (Of course we may sometimes have to fall back on ideas with lesser credentials; or we may find it convenient to do so when a problem can be handled more easily using ideas or methods that we have good reason to believe are inadequate. And we must remember that it is often debatable whether we actually have good reasons to doubt, how serious the doubt really is, and similar matters. A fuller presentation of the present view would take explicit account of these points.)

As I remarked at the beginning of this essay, an adequate philosophy of science must show how it is possible that we *might* know, without guaranteeing that we *must* know. The view I have been proposing does justice to this condition. For it makes *possible* – but not *inevitable* – a scenario like the following: that with regard to some domain, we might someday arrive at a theory concerning which, try as we might, even for hundreds of years, we can generate no specific reasons for doubt. Further, the theory might be easily applied (for example, if it is a mathematical theory, its equations are solvable so that its application to particular problems is straightforward); it might be highly successful; it might be of very broad scope; and we might be unable to find any alternative theory that does as good a job. What else would count as discovery of the way things are?[21]

Yet the preceding question is not a completely rhetorical one, because there is a traditional objection to views of truth similar to the one I am advocating. The objection is to doctrines that claim to "define" 'truth' in terms of reasons for belief. It is alleged that the reasons justifying our belief that a claim is true only provide *evidence* for the truth of that claim, which might be false no matter how overwhelming the evidence: it does not follow from the existence of overwhelming evidence that a claim is true. Hence, according to the objection, the *meaning* of "*X* is true" is something different from the evidence for believing that *X* is true. Therefore, while a 'pragmatic conception of truth' can express a *criterion* for accepting a claim as true, the *definition* or *meaning* of 'true' must be sought elsewhere. As usual, the definition sought is to be divorced wholly from any considerations having to do with how we come to accept or reject beliefs as true; it has to do only with "what it would be for a statement to be true," independent of our evidence for that statement. And so, again, the problem of 'meaning' is set up so as to be wholly irrelevant to the decision-making practice in which 'criteria' are involved.

But what is the *point* in saying that the truth may be other than what we believe, no matter how compelling the evidence? Does that point in any way require us to suppose that there is a 'meaning' of 'true' in the esoteric and irrelevant sense alleged? Reconsider the scenario sketched above, in which we have a very highly successful claim (or set of claims) about the world, and have no (specific) reason to doubt that claim. To say that the truth may nevertheless lie elsewhere than in that claim is *simply* to say that doubt may arise even about that highly justified claim. Recognition of that possibility exhausts the content of the idea that 'truth' may be something other than what is justifiably believed. And there is no need to convert that recognized possibility that we may be wrong into an analysis of what it would be for our claim to be *really true*, an analysis totally irrelevant to the context of coming to accept a claim as true, irrelevant to the way the concept of truth plays a role in the knowledge-acquiring process. And hence my original question, What else would count as discovery of the way things are (than arriving at a highly successful claim regarding which we can find no specific reason for doubt) is rhetorical after all, once the philosophical distortion is removed.[22]

The point I have just been making conspires with two others to indicate that, and precisely how, the present view provides a 'realistic' interpretation of science, an interpretation that justifies the use of such words as 'discovery' and 'knowledge' in characterizing the aims and at least some of the achievements or potential achievements of the scientific enterprise. The first of these further points (the second in all) is this: that the distinction between 'realistic' and 'idealized' treatments of problems, domains, concepts, and theories is not a philosophical overlay on science, an inference or interpretation made about science which is fundamentally irrelevant to the scientist's actual thought and work. On the contrary, that distinction is deeply embedded in ongoing science: unless the scientist is 'talking philosophy', he gives specific reasons for considering a treatment to be an 'idealization' or 'abstraction' or 'simplification' or 'model', specific reasons for saying that things really could not be the way he is considering them to be, or that things are or even must be a certain way.[23] And again: where no such reasons exist (and where no equally good alternative account is available), the mere possibility of doubt regarding the 'realistic' account is not itself a reason for doubting that that account is realistic, *i.e.*, true.

The third point supporting a realistic interpretation has to do with the very roots of the attempt to find out about the world. Most contemporary theories of learning agree that learning proceeds from interaction with an environment.[24] Indeed, the basic human condition out of which the processes of learning and, eventually, inquiry arise is one of differentiation of the self from a world of externally existing objects. What does this mean? It means a differentiation of the self from things that may have properties which do not always manifest themselves, and which may not yet have been discovered, and which therefore may have unexpected effects on us. It means, further, that those things can interact with other things as well as with us. Inquiry, the attempt somehow to deal with that world, to come to terms with it in some way, is intimately associated with what Piaget calls 'adaptation', with its complementary subaspects of 'assimilation' and 'accomodation'.[25] Inquiry begins with the distinction of ourselves from a real, independently-existing world and the need to deal with it in some way.

Vague though the question inevitably is, one cannot help but ask where our ideas for dealing with the world 'originally' come from. 'Originally' is a long way off; let us for the present consider only the sorts of theoretical ideas first introduced by the Greeks. It seems that experience is capable of suggesting an immense variety of possible generalizations. We find that we cannot make things move (free fall and buoyancy apart) except by pushing them; and we generalize this into a law of nature (Aristotle). Things might conceivably have gone otherwise, with generalization from the experience of things continuing to move for a while even after they have been pushed, and of their continuing to move longer and further, other things being equal, the smoother the surface they slide on (inertia). Again, we might build a view of nature on the observation that the grains in a pile of sand are indistinguishable except on close inspection (atomism), or we might elevate into a cosmological principle the observation that a colored liquid diffuses so uniformly through a cup of water that every drop, no matter how small, seems to contain the same proportions of the two liquids (Stoics). Such views, suggested by a limited range of experiences, must then face the task of further development and application, in the course of which they may be confronted with the kinds of doubts that can lead to modification or rejection. Would two different starting points necessarily converge to the same view under the stresses of doubts and alterations? (How would the sciences of two different planetary civilizations compare?) The view I am proposing offers no guarantee that they will converge, any more than there can be a guarantee that we will ever arrive at a theory regarding which no further doubts will arise. The present view is a realistic interpretation of science only in that it shows how it is *possible* that we can make discoveries and existence-claims and arrive at knowledge; it does not – and it should not – provide any guarantee that our knowledge-claims are correct, or that we (and all other intelligent beings) are bound to arrive at a single view. Thus the present view differs from traditional ones which maintain that, in order to defend scientific realism, we must provide such guarantees.

In a very important respect, the analysis of knowledge and the knowledge-seeking enterprise is distorted by views of science as continually rejecting the false beliefs of the past. Although there are certain cases in which beliefs are rejected as utterly without foundation, as wholly useless for prediction or for otherwise dealing with the world, such cases are more the exception than the rule, and other types of cases should not be uncritically assimilated to them. Another common type of case, far more important in the attempt to understand the nature of the scientific enterprise, are ones in which a new, more comprehensive theory shows past theories to hold only under limited circumstances, and (in some cases at least) in which the new theory also shows why those past theories were successful in that limited domain. It has often been noted that special relativity showed, not that Newtonian mechanics was straightforwardly false, but rather that it held only approximately in certain limited conditions. Although there are some important differences between the ways the two pairs of cases are related, the relation between inertial and Aristotelian views of motion can be fruitfully viewed as similar in important respects to that between relativistic and Newtonian mechanics. From this perspective, we need not suppose Aristotle's view of motion to have been simply 'refuted' and replaced by the 'correct' inertial view; what the latter did was to show that Aristotle's view holds only within a limited range of experience, and (together with the Newtonian law of gravitation) why it holds in that domain.[26] While we must not go to the opposite extreme of ignoring the role of doubt (and thus of truth and falsity) in the rejection of old theories, a perspective which emphasizes the role of greater comprehensiveness in scientific change calls attention to the fact that, for all their limitations on a broader scale, older theories often *are* successful to at least some extent. Such limited success is, after all, only to be expected, because so many of those theories were suggested by experience. (Indeed, how it would be possible for a hypothesis or theory to be *utterly* false constitutes a difficult but largely ignored problem for empiricist philosophies. On the other hand, the difficulty of the problem should not lead us to dismiss entirely the view that some ideas can reasonably be said to be 'utterly' without foundation.)

## VI

In the course of this essay, I have spoken often in terms of certain general features of science and its ancestors in the knowledge-seeking enterprise. Such general characterizations might suggest that the account I have presented still retains after all, in those respects, *a priori* and unrevisable principles, rules, or methods. It is now necessary to remove any such suggestion. I shall consider each of the following in turn:

(1) the assumption of an independently-existing world of objects;
(2) the goals of inquiry, and of science in particular;
(3) the principle that the views we construct must not be inconsistent, and, more generally, the rules of deductive logic; and finally,
(4) the principle of revisability itself, and correlatively, the notions of doubt and necessity.

These are all large and difficult topics, and require far more treatment than can be given here. In particular, specific arguments that specific claims (or methods, *etc.*) are necessary must be met by specific rebuttals; and that cannot be done in the present context. Nevertheless, I will advance some general considerations arguing that in none of the respects (1)–(4) is there any implication of in-principle unrevisability about science. More specifically, I will argue that in none of these respects is either of the following the case:

(a) that we have, in that respect, something which is presupposed by science, either in the sense of being a precondition of inquiry (or of science in particular) or in the sense of being an essential defining characteristic of the knowledge-seeking enterprise (or of science in particular);

(b) that we have, in that respect, something which, while learned in the process of inquiry, can, once discovered, be seen – in advance of any further inquiry – to be unrejectable thenceforward, to be a necessary truth (or a method or rule of reasoning that must always be respected).

I now proceed to discuss these issues with regard to the above-mentioned four respects.

(1) In connection with what I referred to earlier as the basic human condition lying at the root of the knowledge-acquiring enterprise, I spoke of

the differentiation of the self from a world of independently-existing objects. Does the knowledge-seeking enterprise, and its modern version, science, therefore require the assumption of an independently-existing world of objects, an assumption which can never be surrendered in the face of any results of inquiry?

There is certainly good reason to believe that the distinction between self and independently-existing world is learned. Developmental psychologists are in general agreement that the distinction is not present in infancy. And since the history of thought about the world is the history of individuals, we may presume that the idea of an independently-existing world was learned in past history also.

But perhaps it might be held that, once the belief in an independently-existing world is arrived at – or at any rate once we become engaged in inquiry – we must maintain such a belief. But what is it that, in so doing, we are supposed to be maintaining, how does it function in the process of inquiry, and in what ways is it unrevisable? Surely it is not the claim that 'objects' exist that is supposed to be unrevisable: for even apart from the multitude of earlier alterations in that concept, the very applicability of such a term in post-quantum physics is very questionable, except perhaps in an extremely attenuated, near-useless sense. Whatever the general character of what we believe to exist at a given epoch in the history of inquiry, that character may later have been so radically revised that its relationship to what was believed earlier is unrecognizable until we examine the intervening evolution of thought which led to the later beliefs.

But do we at least presuppose that *something* exists independently of ourselves, even though we are unable to specify the character of that something in an unalterable way? But even if there were such a belief, it would in no way guide us as to which specific beliefs to hold or to reject – it would play no role whatever in the scientific enterprise, and thus the sense in which it would be a 'presupposition' of that enterprise becomes lost.

At this point we cannot but conclude that our belief in the existence of an independently-existing world is neither more nor less than the sum of our specific substantive beliefs, together with our recognition that doubt may always arise with regard to any of those beliefs. It is not as though we held *both* a body of individual beliefs *and* a further belief that there is a world; rather, the sum of our individual beliefs, plus the recognition that they may have to be abandoned in the face of specific reasons for doubt, is what *constitutes* our belief in an independently-existing world. The sum of our specific beliefs makes up our belief in a 'world'; and our recognition of the possibility of doubt lies, as we saw earlier in our discussion of 'realism' (Section V), at the heart of our belief in the 'independent existence' of that world. The possibility of doubt, arising initially from our finding that things may have properties as yet undiscovered, or be attributed properties which they do not have, initiates inquiry; and the failure of specific doubt to arise in connection with beliefs which allow us to 'deal successfully' with that world is the basis for supposing those beliefs to be true of that world.

(2) I now turn to a group of question regarding the goals of science. I have spoken of the inquiry which arises from the distinction between self and world as an attempt to develop some way of 'dealing with' that world, of 'coming to terms' with it somehow. Does that attempt constitute the establishment of a goal defining and delimiting the scientific enterprise, a goal which is unrevisable as long as we continue to engage in science? The very vagueness with which that 'goal' has had to be formulated should put us on our guard. For at best, in the beginnings of the inquiry-and-learning enterprise (whether individual or historical), goals, and the means of achieving them, and what constitutes success, are poorly defined. Being vague, they can be described only in vague terms; what clarity they have has been achieved only gradually. Man has learned what goals to seek in inquiry, and what constitutes success in that enterprise, in the very process of learning about nature. He has learned what sorts of considerations to take seriously, which sorts of doubts to take into account as guides in scientific reasoning, and even that 'success' and 'doubt' are considerations at all, in the very important sense that those ideas are given content only through inquiry.[27] He has learned not only about nature, but also how to learn and how to think about nature. Can the aims of the primitive animist, in his construction of myths about the world and his place in it, be unambiguously described as 'understanding' or even as 'control' or 'adjustment'? In the same sense as we understand those concepts? Can he even be said, in all cases, to have *had* aims or goals, except in a very loose sense? Or consider again the contrast between 'perfectionist' alchemical approaches and 'compositional' ones, where we have seen alteration in both the aims of science and what counts as a successful view. We had to learn, through the process of debate and experiment that culminated with Guyton de Morveau and Lavoisier, a different and more powerful way of 'dealing with' nature. We learned what to aim for in dealing with matter in the same activity in which we established that things have constituents and in which we learned how those constituents are to be studied.

Despite their vagueness, expressions like 'dealing with' or 'coming to terms with' or 'adjusting to' the world do have a use, as do a multitude of other expressions by which we manage to talk about long stretches of the knowledge-seeking or scientific enterprise. For such expressions call attention to analogies between the activities which we characterize as scientific and certain activities of an earlier age. Those analogies are, in some cases at least, the products of the historical process by which those modern activities have descended from the earlier ones. But we must not conclude that, because some such terms *can* be applied over long historical stretches, we have therefore found some common core of all scientific inquiry. Still less must we suppose that, because (we think) we have found such a common core, there must be a precise analysis of its constituent concepts which is the very 'meaning' of 'inquiry' or of 'science'. Even insofar as the possibility of general characterization of inquiry or science exists at all through the use of such general concepts, that possibility does not imply the existence of an 'essence' of inquiry or science, either in the sense of precise characteristics which any activity must have in order to count as 'inquiry' or 'science', or in the sense of precise characteristics which are unalterable. At best it calls attention to a relationship of ancestry-and-descent. And the important task is to try to recognize that what is being discussed through such terms is subject to development, to radical alteration, some aspects being rejected and new ones added, and to try to understand the mechanisms by which such changes take place.

Nor do we find unalterable requirements of the scientific enterprise when we turn to more specific aims. For example, we find ourselves living in a world in which, on occasion at least, similar circumstances occur, with similar consequences; things and events fall into types, characterized by similar properties and behavior. Generalizations prove possible, from ones on the level of simple categorizations of objects to broad-scale generalizations like the Aristotelian and inertial conceptions of motion, or like atomistic and continuum theories of matter; from ones that we find to hold without exception, to ones that hold only under limited circumstances, to ones that we find useless and reject. We find, further, that many of our generalizations can be systematized – can, for example, be seen to be special cases of or otherwise related to others, or that (as we have found) the recurrent properties and behavior of types of objects can be seen as effects of deeper properties of constituent parts – that a great deal of nature can be understood in terms of 'elementary constituents' and the 'forces' between them. (And we have learned to consider *that* to be a form of 'understanding'; and, as was suggested earlier, we are still learning in what respects 'compositional' approaches fail as well as succeed in providing understanding.) Further, we find that our ability to 'deal with' the world is enormously facilitated by generalization and systematization of our well-grounded beliefs; for example, we are enabled to apply our prior knowledge to new situations, both to direct us in our decisions and actions and to guide us in the search for further knowledge. (And once more, we learn to call such organization of our beliefs 'understanding'.) Indeed, the processes of generalization and systematization have worked so well for us that we seek to extend them further. (It is thus that normative rules arise in science; but they, too, are hypotheses based on their success.) It was, for example, something about the nature of systematization (its achievability through mathematics) and its value (particularly the possibility achieved through that geometric system of going beyond sense experience and inferring the three-dimensional structure of the universe) that constituted the major lesson learned from the Copernican as opposed to the Ptolemaic system –as Copernicus, Kepler, and Galileo all realized.

Nevertheless, although generalization and systematization are, or rather have become, aims guiding us in our inquiries, their application is hypothesized on the basis of our past successes in making such generalizations and systematizations, and on the understanding and practical utility they bring. Any particular generalization or systematization, and therefore the whole aim of generalization and systematization, is in principle open to frustration. For example, we live all the time in science with domains that have not been related to one another. Though we continue to try to relate them, there can be no guarantee that we will succeed; and we may ultimately come to admit the existence of distinct and irreducible domains. There is no question here of a requirement that science unify or be considered to have failed.

(3) Very similar remarks apply to the principle of consistency. The benefits conferred by consistency are manifest, and are connected with the advantages of systematization. Inconsistency would prevent our being able to make decisions as to action or belief in particular circumstances, while consistency contributes to the greater systematization of our beliefs. Yet despite the perennial reproaches of Berkeleyan philosophers, techniques and ideas have been employed with great success in the history of science despite the fact that those ideas were inconsistent as presented (infinitesimal calculus until Cauchy and Weierstrass, or, in another respect, until Robinson; Dirac delta function before Schwartz; certain idealizational techniques[28]). Although we have learned to prefer consistency to inconsistency because of its achievability and success, and so continue to seek it, there can be no guarantee that we must always find a consistent reinterpretation of our inconsistent but workable techniques and ideas. We may have to live with them, learning in the process that, whatever its advantages, consistency cannot be a requirement

which we impose on our ideas and techniques on pain of their rejection if they fail to satisfy it. Perhaps if we were to encounter a great many cases in which we found inconsistent ideas useful, and were unable either to reinterpret them in a consistent way or to find alternative techniques that were equally good, our general preference for consistency – our 'adherence to a principle of consistency' – would be considerably lessened. It is because of our success in clarifying and making consistent such concepts as 'infinitesimal' and 'delta function', and because consistency contributes to our ability to deal with the world in a theoretically and practically satisfactory manner, that we continue to employ that principle wherever possible.

The principle of consistency is thus applicable in scientific thought, but does not serve as a requirement of scientific acceptability. Other logical principles (*e.g., modus ponens*) play the role of helping us make explicit what is implicit in our beliefs. But the question remains of whether rules of logic are alterable, in the sense that they can be replaced by alternative rules. For the purposes of the present paper, the question is whether they are alterable in the light of substantive scientific beliefs, as we have seen methods, domains, problem-fields, ranges of possibilities, and so forth to be.

A large variety of interpretations of quantum mechanics attempt to make sense of that theory by considering it to require some modification of logic (most often the distributive laws). Opponents of such interpretations sometimes object that what is modified is the algebraic structure of physical theory, which is not what is meant by 'logic'. In the light of our earlier discussion, however, such an objection cannot be decisive. For even if logic has hitherto had certain characteristics, including being separate from substantive scientific belief in the sense of being immune to revision in the light of such belief, it might well be altered in order to achieve greater integration with other areas of thought, especially if such alteration and integration introduces greater clarity into those other areas. We might then speak of an improvement in our understanding of what logic is as well as of an improvement in our understanding of quantum mechanics.

The subject of the revisability of logic in the light of scientific beliefs is too vast and complex to pursue in this essay. However, one very important point about the relation of that topic to the views presented in this paper should be noted. If logic is revisable, as an integral constituent of our body of substantive beliefs about the world (or if it is in the process of being made a more integral part, as quantum logicians suggest), then the view being advanced in this essay allows the following possibility. We might arrive at a situation in which, according to the logic at which we have arrived at that stage, there *could* be no alternative to the set of substantive beliefs at which we have also arrived in conjunction with arriving at that logic. In that case we would have arrived at claims that, we would claim further, are 'necessarily true'. (Our present philosophical distinction between 'physical (or theoretical)' and 'logical' necessity would no longer have application.) And further, we might be unable to find any reasons for doubting the truth of our beliefs, and therefore for doubting the truth of our claim of necessity. It would still remain the case that reasons for doubt, and for modifying the interlocked system of logical and substantive beliefs, might arise. But, as has been repeatedly observed throughout this paper, that possibility is itself no reason for doubt. In an important sense, indeed the only sense that can reasonably be assigned, we would then have arrived at a body of beliefs which we would have every reason to believe are true, and even necessarily so.[29]

(4) There is, of course, no guarantee either, on the view I have been presenting, that a state such as that just described will ever be arrived at. However, the very possibility (given, of course, that logical and substantive beliefs can be integrated in the appropriate way[30]) brings us to the final point of this paper: the status of the 'principle of revisability' itself, the idea that, with regard to any of our beliefs, rules, methods, *etc.*, specific doubt can always arise, and that we must be prepared for such contingencies. Is this principle itself revisable, or is it an essential – perhaps even *the* essential – characteristic of science? In order to deal with this question, we must examine more closely the role of this 'principle'.

Like so much else that we have considered, the role of doubt in inquiry, too, is learned. We have learned that, in the knowledge-seeking enterprise, universal doubt, doubt that applies indiscriminately to any belief whatever, is irrelevant; only doubts specific to a particular belief constitute reasons for doubting that belief. But we have also learned – more in the past century and a half than in previous times – that, however firm the credentials of any particular belief might be, it may not be true after all: doubt may yet arise which will force its revision or rejection. Feelings of certainty, impressions of self-evidence, arguments claiming to demonstrate necessity of a particular belief or the impossibility of alternatives, claims of inconceivability of alternatives – all these and similar grounds for claims of necessity have failed in a wide variety of cases. Views that, had they been thinkable at all beforehand, would have been dismissed as impossible, inconceivable, or even self-contradictory, have been incorporated into our most profound theories of nature or at least admitted as scientific possibilities (Gauss, Riemann, and Clifford on Newton's rejection of the Cartesian program). And yet the sobering effect of those lessons does not constitute a proof that doubt can be expected in the case of all our beliefs. Indeed, it does not constitute a specific reason for doubting any particular belief. It is wholly unreasonable to doubt beliefs that have proved enormously successful and concerning which no specific reason for doubt has arisen; and where such specific doubts do arise, we should apportion our doubt in strengths commensurate with the seriousness of the reasons for doubt. That the sun will rise tomorrow, that my car is in the parking lot where I left it a moment ago, that dinosaurs existed in the past, that galaxies are distant systems of stars, that Newtonian mechanics, classical electromagnetic theory, or quantum electrodynamics capture a great deal that we can trust (even though they might hold only in certain limited domains) – such beliefs and many more, even though they do not carry their immunity from any possible revision imprinted upon them, it is unreasonable to doubt, precisely because of their enormous success and the nonexistence of any (very serious) specific reasons for doubt. That is the sort of thing we have *learned* to consider 'reasonable' – all of us, that is, except some overly fastidious philosophers in their studies, anachronistically insisting that there must be such imprinted guarantees.

But what this amounts to saying is that the 'principle' of revisability is not a principle that provides specific guidance in any of our inquiries. That principle is only an attitude fostered by past failures, a self-admonition to expect that specific reasons for doubt might arise. Yet it is unreasonable to insist on constantly so admonishing ourselves with regard to beliefs about which doubt rarely or never actually arises; and the admonition might reasonably diminish or even disappear in a far more general way were the circumstances envisioned above ever to be realized. For if we were to arrive at a stage at which what we believe we also have reason to believe is necessary, and if, after thousands of years of effort we were unable to generate specific reasons for doubt with regard to those beliefs, then the *reasonableness* of taking any further tentative attitude toward them would diminish asymptotically with time. (This conclusion holds even if we do not have reason to believe that the beliefs at which we have arrived are necessary.)

In this essay I have attempted to indicate some central features of a view of science, and of the knowledge-seeking enterprise generally, according to which that enterprise involves no unalterable assumptions whatever, whether in the form of substantive beliefs, methods, rules, or concepts. I have also tried to state as explicitly as possible some major problems which such a view must confront, and have tried to show how they may be dealt with. A great many aspects of the view and its problems require further elaboration and argument; some topics, like the question of the revisability of logic, have only been touched on. There is much more to be said, too, about the details of the way or ways in which substantive beliefs force or otherwise affect the choice or rejection of methods, rules, and so forth, and conversely, the ways in which methods, rules, and so forth affect the formulation and choice or rejection of substantive beliefs. For there need not be, and surely is not, only one form of such relation. Further, I have said nothing about the question of alternative formulations of scientific ideas and methods. Are there – indeed, must there always be – alternative and equally satisfactory ways of formulating domain descriptions and explanatory theories? Or is there a tendency, prompted by reasons within the scientific enterprise, to try to tighten our views in such a way that not only competing alternatives are eliminated, but also alternative formulations which appear, at some stage at least, to be 'equivalent'? Finally, despite the attempts made at various points, there is a need for fuller discussion of the relevance of the views of this paper to traditional and contemporary philosophical theories and problems.

I have discussed some of these topics, to some extent at least, in other papers; others must be dealt with on future occasions. However, I hope that what I have said here will convey some idea of a view of science which, while attempting to avoid the absolutes of Platonic-Kantian-positivist ways of thinking, also endeavors to escape a relativism in which there is no such thing as knowledge and progress. It attempts, that is, to do justice to the variety of patterns of thought and method in the history of science while at the same time allowing for the possibility, and occasional actuality, of true knowledge-acquisition and progress in science. Through examination of those patterns of reasoning, it concludes that the processes of scientific discovery are concerned not merely with factual claims, but also with the introduction, alteration, and rejection of problems, possibilities, methods, rules, standards, and even the most general 'metascientific' concepts in terms of which we think of science. In these respects, it is perhaps the first truly uncompromising empiricist philosophy ever proposed. It makes it possible to see science as a creative enterprise aimed at discovery and understanding – but an enterprise in which creativity takes the form not only of acquisition of further well-founded beliefs, but also, sometimes, of alteration of our concepts of what 'discovery' and 'understanding' themselves consist in. It makes it possible to understand how we can learn and even come to know, while at the same time we can learn how to learn and how to come to know.

## NOTES AND REFERENCES

* This paper was written during a sabbatical year (1978–79) in the incomparable environment of the Institute for Advanced Study, Princeton, New Jersey. I am indebted to a great many people for discussion and comment on the ideas contained herein. Among those whose criticisms and suggestions contributed directly and valuably to the thoughts expressed are: Peter Achinstein, Stephen Adler, John Bahcall, Freeman Dyson, Michael Gardner, Ivor Grattan-Guinness, O. W. Greenberg, Hannah Hardgrave, Nancy Maull, Yuval Ne'eman, Richard Rorty and the members of his seminar on 'Relativism' at Princeton University in the Fall of 1978, Neal Snyderman, John Stachel, and Frederick Suppe. I am particularly indebted to Morton White for numerous valuable discussions of these and related topics during the year.

I also thank the National Science Foundation for support (under Grant SOC76–19496) of earlier phases of the research leading to this paper.

[1] This is true despite the superficial appearance of fragmentation in modern science and of tight coherence in earlier views. I have surveyed some cases of fragmentation and unification in earlier science in 'Scientific Theories and Their Domains', in F. Suppe (ed.), *The Structure of Scientific Theories*, University of Illinois Press, Urbana, 1974, 518–565, and in today's science in 'Unity and Method in Contemporary Science', a revised and expanded version of a paper published under the title 'Unification and Fractionation in Science: Significance and Prospects', in *The Search for Absolute Values: Harmony Among the Sciences*, International Cultural Foundation, New York, 1977, pp. 867–880.

[2] W. V. Quine, 'Two Dogmas of Empiricism', in *From a Logical Point of View*, Harvard University Press, Cambridge, 1953, pp. 20–46.

[3] H. Putnam, 'Is Logic Empirical?', in R. S. Cohen and M. W. Wartofsky (eds.), *Boston Studies in the Philosophy of Science*, Vol. V, Reidel, Boston, 1969, pp. 216–241. A representative sample of discussions of the topic of 'quantum logic' is found in C. Hooker (ed.), *The Logico-Algebraic Approach to Quantum Mechanics*, Vol. I, *Historical Evolution*, Reidel, Boston, 1975; see also B. van Fraassen, 'The Labyrinth of Quantum Logics', *Boston Studies in the Philosophy of Science*, Vol. XIII, Reidel, Boston, 1974, pp. 224–254.

[4] But see my 'What Can the Theory of Knowledge Learn from the History of Knowledge?' *The Monist* **60** (1977), p. 492, for a difficulty in this view.

[5] T. S. Kuhn, *The Structure of Scientific Revolutions*, University of Chicago Press, Chicago, 1962. For criticisms of Kuhn's views, see I. Lakatos and A. Musgrave (eds.), *Criticism and the Growth of Knowledge*, Cambridge University Press, Cambridge, 1970; D. Shapere, 'The Structure of Scientific Revolutions', *Philosophical Review* **73** (1964) 383–394; Shapere, 'Meaning and Scientific Change', in R. Colodny (ed.), *Mind and Cosmos*, University of Pittsburgh Press, Pittsburgh, 1966, pp. 41–85; Shapere, 'The Paradigm Concept', *Science* **172** (1971), 706–709.

[6] Copernicus's system, while far more unified than the piecemeal, makeshift Ptolemaic one, also implied, in virtue of its geometrical structure alone (as the Ptolemaic system did not), the three-dimensional arrangement of the solar system – the order and relative distances of the planets. It was the unity of the Copernican system, and still more its implication of knowledge beyond what is revealed by the senses (*i.e.*, by what we see as the projection of the planetary images on the celestial sphere) that most impressed Galileo and Kepler, and, indeed, Copernicus himself. Together with developments in

mathematics, this achievement led to the view that mathematics is the key to a realistic understanding of the physical universe, and probably had much to do with the rise of rationalistic epistemologies. *Cf.*, D. Shapere, 'Copernicanism as a Scientific Revolution', in A. Beer and K. Strand (eds.), *Copernicanism Yesterday and Today*, Pergamon, New York, 1975, pp. 97–104.

[7] More detailed discussion of some of the issues raised in this and the following section are contained in the article cited in Note 4.

[8] 'What Can the Theory of Knowledge Learn from the History of Knowledge?', *loc. cit.*, 488–508; *cf.* esp. pp. 497–503.

[9] The treatment of this case as given here differs in certain respects from that in my earlier paper, 'The Influence of Knowledge on the Description of Facts', in F. Suppe and P. Asquith (eds.), *PSA 1976*, Philosophy of Science Association, East Lansing, Michigan, 1977, pp. 281–298.

[10] The chief factors involved were: (a) the possibility of compositional analysis, and the idea that knowledge of composition gives understanding of matter; (b) the concept of an element as a breakdown product, and the existence of elements; (c) the concept of weight as being centrally relevant chemically; (d) Lavoisier's oxygen theory of acids, metals, and calxes; and (e) the reform of chemical nomenclature to reflect composition (as conceived in Lavoisier's oxygen theory).

[11] Indeed, continued failure to observe free quarks may eventuate in the abandonment of compositionalist approaches in fundamental theory: the concept of a particle may have to be viewed as appropriate only in an 'asymptotically free' limit, so that *fundamental* theories will no longer be characterized in terms of their particle content.

[12] The discussion of observation and the case of neutrino astrophysics given in this section is elaborated more extensively in a forthcoming paper, 'The Concept of Observation in Science and Philosophy'.

[13] 'Scientific Theories and Their Domains', *loc. cit.*

[14] The concept which has come to be known as 'color' was introduced in the form of 'parastatistics' by O. W. Greenberg ('Spin and Unitary Spin-Independence in a Paraquark Model of Baryons and Mesons', *Physical Review Letters* **13** (1964), 598–602; reprinted in J. J. J. Kokkedee, *The Quark Model*, Benjamin, New York, 1969, pp. 122–126). The parastatistics and color interpretations are quantum-mechanically equivalent.

[15] Claims that free quarks have been observed are made periodically, most recently (as of this writing) – and to be taken very seriously because of the experimental competence of the investigators – by G. S. La Rue, W. M. Fairbank, and J. D. Phillips, *Physical Review Letters* **42** (1979), 142–145.

[16] Examples are the antagonism to the development of statistical mechanics by Ostwald and his fellow 'energeticists', and the opposition by Mach and Duhem to much of the new physics that was being developed in the early part of this century.

The arguments given in the preceding discussion show that observability is not a necessary condition of scientific admissibility; the question of whether, and under what circumstances, it is a sufficient condition is examined in the paper referred to in Note 12.

[17] Difficulties about the concept of observation in science, and in particular about the 'theory-ladenness' of observation-claims, have led many philosophers to shy away from the concept, so that the kinds of reasons they adduce for accepting or rejecting a scientific theory sometimes appear to have to do with everything else *except* observation

and experiment. But this is a mistake: the central role of observation and experiment in the objective testing of theories cannot be denied. The problem is not *whether* observation plays such a role, but rather *how* science manages to maintain 'objectivity' despite the 'theory-ladenness' of observation. *Cf.* 'The Concept of Observation in Science and Philosophy'.

[18] Instead of speaking, as I have, of 'meanings' and 'criteria' and their interrelations, philosophers often speak of 'meanings' as being concerned with the *essential* as opposed to the nonessential criteria for applying a term. That way of putting the distinction, however, assumes that certain criteria *are* essential, and that is the very point at issue. It is gratuitous to assume, in the case of 'metascientific' terms, that because scientists of different periods differ as to the criteria for being an explanation, therefore there *must* be some meaning (essential criteria) common to all those uses but distinct from the 'nonessential' ones about which they disagree. And the further assumption, necessary to that view of meanings, that those essential meanings must be independent of any factual considerations, is likewise gratuitous. Although science does aim at finding out what things are fundamentally like (their 'essences', if one cares to use that term), and reshapes its uses of terms to accord with what they are like, those 'essences' are arrived at by investigation and are subject to alteration in the light of further investigation.

Critics of 'criterial' approaches to the analysis of meanings agree with my view that all descriptive criteria have to do with 'extralinguistic matters of fact', *i.e.*, that those criteria are subject to debate and alteration. However, those critics agree with adherents of the 'criterial' approaches, as I do not, that there must be something common throughout such disagreements and changes. Their solution is to say that that common feature is achieved not through a core of essential criteria, but through sameness of *reference*: despite differences as to what we say about the thing, we are still talking about the 'same thing'; that we are is guaranteed in terms of causal relations to some original 'baptismal' occasion, all later ascriptions of properties being to that referent. (This view is presented at length in S. Kripke, 'Naming and Necessity', in D. Davidson and G. Harman (eds.), *Semantics of Natural Language*, Reidel, Boston, 1972, pp. 253–355. Important contributions have also been made by Donnellan and others; see S. P. Schwartz (ed.), *Naming, Necessity, and Natural Kinds*, Cornell University Press, Ithaca, 1977. As with Kripke, Putnam applies the view not only to proper names, but also to natural kind terms (*e.g.*, 'Meaning and Reference', *Journal of Philosophy* 70 (1973), 699–711, reprinted in Schwartz, pp. 119–132.) As applied to the analysis of science and scientific change by Putnam, this view seems motivated by the desire to defend realism and to avoid Kuhn's conclusion that successive 'paradigms' are 'incommensurable' despite their use of the same terms. But for that purpose, the causal theory of reference is both ineffective and unnecessary: comparability of theories cannot, as Putnam wishes, be guaranteed by the dictates of a philosophical theory of reference. In the case of terms like 'electron', we have no way of telling whether several theories (say, those of Lorentz and Dirac and contemporary quantum electrodynamics) are 'talking about the same thing' except in terms of the chain of reasoning connecting their respective views of electrons: one traces the connections between earlier and later uses in terms of the reasons why certain properties ascribed by earlier workers were abandoned or modified and others added. To say, beyond this, that they were *also* referring to the same thing, is to add nothing. Thus Putnam's version of the 'causal theory of reference' amounts to nothing more than a denial of Kuhnian incommensurability, not an argument against it. One might, of course, argue that terms like 'electron' are not "natural kind terms", but are rather "theoretical terms", so that the causal theory of reference is not meant to apply to them. But it is debatable, to say the least, whether such a distinction can or even should be upheld. Certainly it should be foreign to a realist interpretation of science. (However, Putnam now seems to have abandoned realism; see his *Meaning and the Moral Sciences*, Routledge and Kegan Paul, Boston, 1978, Part Four.)

On the other hand, the view advanced in this paper focusses on such chain-of-reasoning connections, the 'criteria' and changes of criteria involved in talking of electrons – or of gold or lemons or alcohol, for that matter. In this respect, it is closer to the 'criterial' views. However, it agrees with the critics of those views in holding those criteria to be matters of 'empirical fact'. Most fundamentally, my view disagrees with the assumption behind both of those views: that in order for there to be continuity between two discussions (say, between two stages of scientific development), there must be something in common, whether 'essential criteria' or 'rigid reference'. I take these emphases of my view to be in accord with views that hold that 'meanings' are to be understood in terms of criteria of assertability rather than of necessary and/or sufficient conditions of application –as long as criteria of assertability themselves are not taken to be 'essential'.

The causal theory of reference, in the form presented by Putnam, at least, suffers from one further defect. The original ('baptismal') use of the term 'electron' (Stoney, 1894) 'referred' not to a particle at all, but to a unit of charge. When that unit of charge was found to have mass, the term came to be applied to the particle, which was then said to have ('carry') the charge. Since 'charge' and '(massy) particle' are also presumably descriptive terms, I find it difficult to understand what it is supposed to have been (something which was neither charge nor massy particle) to which both Stoney and Thomson were 'referring.' (To interpret this case as a mere 'rebaptism' would ignore the real –scientifically important – relations between Stoney's use and the use in which 'electron' was applied to a particle.) The term 'alcohol', and a great many others in science, have the same kind of history and raise the same kind of difficulty. Putnam's 'referent' thus assumes the character of a Lockean 'something-I-know-not-what', and serves as little function.

[19] With regard to specific doubt, we must distinguish between (a) the fact that a belief *is* doubted (whether or not it is correct to doubt it, *i.e.*, whether or not reasons for doubting it actually exist in the body of well-founded scientific beliefs of the time) and (b) the fact that such reasons for doubt exist (whether or not they have been noticed). It is with the latter that the notion of 'specific doubt' is concerned; however, in the actual process of scientific reasoning, it is of course the former kinds of doubts that serve as *hypotheses* that (and to what extent) a certain belief is objectively doubtful.

As to philosophical doubts, it might be that some such doubts apply not universally, but only to large classes of beliefs. In such cases, philosophical doubts could be characterized as doubts which apply equally to a proposition and its negation. Since in the scientific enterprise we are still in the process of learning what sorts of doubts to consider scientifically relevant, it is to be expected that new classes of doubts will come to be considered irrelevant.

I speak constantly of 'scientific beliefs'; but how does one tell what the beliefs (and in particular the well-founded ones) of a particular scientist, group, or period are? Although there are difficulties here, what counts as a scientific belief, and as a well-grounded one, becomes clearer with the progress of science. Deciding what were the beliefs held by seventeenth century thinkers, and whether and to what extent those beliefs were well-grounded, is far more difficult than it is for sophisticated areas of science in the twentieth century. In the (best) latter cases, like the case of the solar neutrino problem, what is accepted is quite explicit, and the relative extent to which different beliefs are accepted (and hence the rough ordering of consideration when difficulties arise) is also relatively clear. There may, of course, be unformulated, 'implicit' beliefs at any time; those, however, as my way of referring to them suggests, play an explicit role in scientific reasoning only in hindsight, after they have been formulated.

In any case, the view I am presenting does not look on beliefs as 'habits' of behavior, or 'customs', as in Peirce and Dewey, but as objective claims (including claims regarding methods, rules, standards, *etc*.).

[20] Traditional sense-datum theories held that the data on which we build our beliefs are given in experience and are not open to doubt, even in principle. On the view presented here, we accept, as that on which we build further knowledge-claims, that which we have found no (specific) reasons to doubt. There is no experience, and no linguistic description of experience, that is intrinsically immune from doubt, *i.e.*, concerning which it can be known in advance that no circumstances of doubt could arise. A number of trenchant criticisms are advanced by Williams against the traditional 'foundationalist' epistemology, according to which knowledge can be attained only if it builds on intrinsically certain 'givens' (M. Williams, *Groundless Belief*, Yale University Press, 1977). Further details of the difference between the view of the present paper and foundationalist epistemologies (and sense-datum theories in particular) are found in 'The Concept of Observation in Science and Philosophy'.

[21] In its insistence on the revisability of beliefs, and on the role of specific as opposed to universal doubt, the present view is reminiscent of that of Peirce. However, see Note 19. Also, I place more emphasis on purely theoretical aspects of the knowledge-seeking enterprise than do most pragmatists: even though I will argue shortly that the roots of that enterprise lie in the attempt to 'deal with' the world, my view allows theoretical understanding to become a goal in itself as science develops. Further, it should be clear that I reject Peirce's view that the method of science *must* converge on truth: "The opinion which is fated to be ultimately agreed to by all who investigate is what we mean by the truth" ('How to Make our Ideas Clear', in P. Wiener (ed.), *Values in a Universe of Chance*: *Selected Writings of Charles S. Peirce*, Doubleday, New York, 1958, p. 133). This view (and the associated definition of truth) is of course inconsistent with Peirce's 'fallibilism'. The view I have presented avoids such difficulties while at the same time preserving a notion of truth, and the *possibility* of our arriving at it, in terms of success and the absence of specific doubt.

[22] It is thus that problems like the Gettier 'paradoxes' arise only from the framework of concepts and approaches created by philosophers; and it is thus that they have no further relevance. For the Gettier problems, see E. Gettier, 'Is Justified True Belief Knowledge?', *Analysis* **XXIII** (1963), 121–123, and G. Pappas and M. Swain (eds.), *Essays on Knowledge and Justification*, Cornell University Press, Ithaca, 1978.

[23] D. Shapere, 'Notes Toward a Post-Positivistic Interpretation of Science', in P. Achinstein and S. Barker (eds.), *The Legacy of Logical Positivism*, Johns Hopkins University Press, Baltimore, 1969, pp. 115–160. The relevant discussion is in Part II.

[24] Chomsky's theory of language acquisition is an exception. However, I cannot see that

Chomsky has established either that there are innate ideas or that his theory requires that there be such.

[25] Piaget's ideas of adaptation, assimilation, and accomodation are scattered throughout his works; a concise and clear exposition of them is found in J. Flavell, *The Developmental Psychology of Jean Piaget*, Van Nostrand, New York, 1963, Chapter 2.

[26] The difference between the relation of special relativity to Newtonian mechanics on the one hand, and the relation of the principle of inertia to the Aristotelian conception of motion on the other, must not be underestimated. Perhaps the most important difference arises from the fact that Aristotle's view of motion was quite ambiguous, being open to a number of different interpretations as to the relations between force, resistance, and speed. The principle of inertia did not simply select one of those interpretations and generalize it; it rejected the conception of the problem as being that of the relations between those three quantities. *Cf.*, D. Shapere, *Galileo: A Philosophical Study*, University of Chicago Press, Chicago, 1974, esp. pp. 39–45, 57, and 73.

[27] It should not be supposed that, in saying that we have learned that 'doubt' is a consideration in scientific reasoning, I have contradicted my earlier statement that the recognition of the possibility of doubt initiates inquiry. In the case of 'the roots of inquiry', the 'doubt' consists only of the vaguest awareness that objects may behave in unexpected ways. It is quite another matter to learn that, in trying to learn about the world, we must take into account specific doubts, based on our prior well-grounded belief.

[28] Such inconsistent idealizational techniques are discussed in my 'Notes Toward a Post-Positivistic Interpretation of Science', *loc. cit.*, Part II. It is shown there how such techniques can function effectively, and that, in particular, their inconsistency does not allow us to infer any proposition and its denial.

[29] If our interlocked system of logic and substantive belief is necessary, how could we conceive the possibility that we might yet be wrong, that specific doubts might yet arise? In order to make sense of my scenario, must we not assume the existence of a transcendent logic in terms of which we can see our system of logic-and-substantive beliefs as only one of many possibilities? No: the universal (philosophical) possibility of doubt is not an assertion of an alternative. The view that 'possibility' must be understood in terms of a superlogic which defines a 'range of possibilities' (state-descriptions, possible worlds) is wholly superfluous here. The only 'alternative possibilities' one needs in this scenario are the previous views, rejected now as impossible in the light of the view at which we have arrived. The remaining possibility that we might yet be wrong – that specific doubt may yet arise to prove our system not necessary (or even true) is only an attitude fostered by previous error.

[30] It should be noted, however, that even within our present logic, a scenario could occur which is quite analogous to the one just discussed. For we might arrive at a stage at which our understanding of nature was so deep that we could lay out all possibilities (all state-descriptions, to use a Carnapian metaphor) explicitly, and show all to be false but one. In this case, however, we would be left only with 'physical necessity', rather than 'logical necessity'. (In the scenario previously discussed, it will be recalled, that distinction was lost.) The discussion of (4) holds under either scenario, and thus is really independent of the assumption that logic can be integrated with our substantive beliefs.

# DISCUSSION ON THIS CHAPTER[1]

GARY GUTTING: In order to help me understand what you are saying, I would like to focus on one of your strongest claims, where you say that nothing in science is intrinsically sacrosanct, whether substantive beliefs or methodological rules. When you put it that way, someone might think that you mean that in five hundred years science might have changed so much that it becomes basically what we now call playing football. But that is not so on your account because in the changes of methodological rules, you have a notion of a legitimate successor. If you put in a new methodological rule, it has to be a legitimate successor to the ones that went before. If that is so, though, I wonder whether you do not have to admit that there are *kinds* of beliefs or *kinds* of rules that are, in fact, essential to science – maybe not any specific rules but certain types – so that you would at least have a continuity of the enterprise.

SHAPERE: What counts as a legitimate successor, though, is determined by the content of science at a given time, by its rules, methods, substantive beliefs, and their interplay – so that, with the further evolution of science, what counts as a 'legitimate successor' may be different from what counted as such earlier. I give no general, unchanging 'notion' of legitimate succession, nor do I need to do so to assure continuity of the enterprise – even in the form of limits on *kinds* of beliefs or rules essential to science, as you put it. The continuity of science can be assured without assuming common essential features as long as there are chain-of-reasoning connections; and even what counts as a 'reason' might change, so that the character of the chain could too.

As to whether science could become football, I think your worry is motivated by a feeling of certainty that science just *couldn't* become football, and that we know this in advance, so that science must have an essence that differs from that of football. But the cash value of that feeling cannot amount to more than the following: given everything we believe about science and football today ('our meanings' of 'science' and 'football'?), we have no reason to think that science will ever become football. What other reasons could we hope to give? Yet we also know that science, or, more generally, thought about the world, can change, even drastically. We can predict some possibilities and dismiss others, but we can't be sure, any more than Democritus could have imagined that atomic theory would take its present form and be studied in the way it is now and for such different purposes from those he had in mind. It's foolhardy at best to convert the reasons we *have*, the characterization of science as it is now, into an essence. At worst, it distorts the nature of the scientific enterprise. But on the other hand, there is no way to really worry about possibilities so dim that we have no reason to think they will occur, and every reason to think they won't. Such possibilities are *mere* possibilities – only philosophers' considerations that are really no considerations at all.

LARRY LAUDAN: Dudley, I completely share your view that there is an important and largely ignored symbiotic relationship between methodological standards that it is appropriate to bring to bear on science and the development and evolution of science itself. It's just a fact that appropriate methodological standards change through time. What I am troubled about is the belief that we can retain a notion of the rationality of science by dropping even at the metalevel any assumption that there are persistent metacriteria for the choice of criteria.

SHAPERE: Yes, as I said, that is the central problem.

LAUDAN: It seems to me that you smuggle such criteria in, because in all of the examples you give, your argument is basically this (please correct me if I'm wrong): we encounter a case in which there is a conflict between a successful scientific development on the one hand and a set of accepted methodological standards on the other; and you argue that we take the successful science to some extent as veridical and use it as an instrument for criticizing the methodological criteria.

SHAPERE: It can happen that way sometimes. Of course, the opposite can happen too: the methodological standards might have been more successful, less open to doubt, than the particular substantive beliefs under consideration. There are, in fact, many ways in which the interaction can go, and if I tended in the paper to emphasize cases in which it's the methodological rules that are dropped or modified in the face of substantive considerations, that's because that's the direction in which the 'symbiotic' connection has most often been denied.

LAUDAN: Now to set it up the way you have, however, is to *assume* that we have, independently of our standards, a conception of what *successful* science is. And it seems to me that what you are doing is smuggling in permanent, transcendent criteria of scientific adequacy by talking about successful science. If you are not doing that, how would you resolve a case in which there is a conflict between a scientific theory and a set of methodological procedures when we are *not* prepared to abandon the methodology and pay lip service to the science?

SHAPERE: Notice that I didn't say we pay 'lip service' to the science. But no, Larry, I haven't been smuggling. Thinking about the history of science, you can see very quickly, I think, that we have had to *learn* what constitutes success in science, just as we've had to learn everything else. We arrive at a body of beliefs and standards, and among the latter are standards of success; there may be many of those, too. Nor need we suppose that there's some one unchanging super-standard of success in terms of which we decide which standards of success are successful. For example, at some stage one standard of success may be dominant, but we find we can't fulfill it, while another 'lower' standard gets satisfied quite frequently and well, and people begin paying more attention to it. That's just one way in which standards can change without assuming a transcendent, unchanging criteria of success. There are others; the case of alchemy and chemistry, which is in the paper but which I couldn't read today because of time, is a good one to think about in this connection.

As to your final question, just as there are cases where methodology bows to substantive considerations, so there are cases where substantive considerations bow to methodological ones. That happens any time evidence is rejected (and reduced to the status of 'purported' or 'alleged' evidence) on the ground of being unsound in the light of well-developed, highly successful methods (consider the rejection of certain statistical results).

ERNAN McMULLIN: I am sympathetic to the attack on presuppositionist sorts of philosophy of science, but I am a little troubled by the effort here to get away from presuppositions entirely. I shall put it a little differently than Gary and Larry put it. It is a simple Popperian sort of point. If you take your starting-point, which is to look in a more or less descriptive way at what counts as scientific rationality in an epoch, the first question you have to answer is how to make your criteria of demarcation stick. How will you separate off nonscientific rationalities of that epoch? How will you disqualify some and accept others? If there is not some sort of 'ballpark' intentionality of science, which can be used in some broad sense at least as a screen, then it would seem to follow either that you are begging the question in that epoch *or* that you leave yourself open to the sort of point that Gary made, which is that you are relying entirely on a kind of organic continuity, from one generation to the next. This would leave open the possibility that the scientific rationality of the future might look *entirely* different from the scientific rationality of today.

SHAPERE: I *do* leave that open; but of course that possibility isn't a positive (specific) reason for worry that it *will* look 'entirely different' (whatever that means) from the specific rationality of today.

I want to make two points in connection with what you've said. First, as to the need for 'ballpark criteria', note that we *do* have these for at least sophisticated areas of contemporary science. These criteria pick out ('Whiggishly') aspects of past history as scientific and exclude others. But then we begin to study the people and activities we've picked out, and we find that they considered other things to be relevant to their 'scientific' thinking (*i.e.*, to what we Whiggishly pick out as their scientific thinking). And after a while we may begin to see what they had in mind. So we begin to reappraise our idea of what they thought to be 'scientific'. After quite a bit of this, we begin to think also that some people (like Aristotle and the alchemists), whom we hadn't considered to be 'scientists' or scientifically relevant from our Whiggish perspective, can't be separated off any more, because of their connections with what other people whom we have been considering scientists were doing. And finally, perhaps we start getting some understanding about how and why things changed to get us to our present view of science, to our present 'demarcation criteria', such as they are. Now why can't things happen that way, without assuming inflexible criteria and getting into all the problems of question-begging they would get you into? And although the story I've told is oversimple in some respects, doesn't it describe fairly well how the subject of history of science has actually developed? Historians of science certainly think they can avoid such question-begging, at least in an asymptotic sort of way, and that they have avoided the question-begging of earlier 'Whiggish' historians like Mach. I think they've taught us quite a bit about avoiding the difficulties you've brought up. In fact, the interpretation of scientific change that I've given is, I believe, fully in conformity with that approach to finding out what 'past science' is without assuming any kind of permanent 'intentionality' of science, while avoiding certain conceptual confusions into which some historians have fallen (*e.g.*, implicit relativism). Indeed, the view was designed with that in mind, among other things.

My second point with regard to your question, Ernan, is this. Your view, and that of the two preceding questioners, still is that there *must* be something transcendent – some superstandards or supermethods or superwhatevers that stand above scientific change and are immune to revision, that ensure continuity within science and demarcation of science from other sorts of human activity. It's significant, of course, that you talk, not of *specific* beliefs, rules, standards, *etc.*, being transcendent in this way, but rather of *ballpark* ones, or (as Gary Gutting said in a similar vein) "*kinds* of beliefs or *kinds* of rules ... essential to science ... maybe not any specific rules but certain types – so that you would at least have a continuity of the enterprise." That's symptomatic both of an assumption and of a problem: people have had their fingers burned in the attempt to find *specific* things defining and delimiting science (that's the problem), so they say that, since there *must* be *something* common and essential (that's the assumption), it must be vague or general. (And of course it must be made *very* vague and general if it is to do the work required of it, of applying to all 'science'; but then, once made so vague and general, it cannot do its assigned job after all.) But why not question the assumption – the imposed requirement that there *must* be something in common? I claim that we can get the same work done (making it possible to speak of science as *a* subject) without that assumption. The whole attitude is an example of what I call the 'Platonic Fallacy': the idea that if things are called by the same name, or otherwise classified together, there *must* be something in common between them. In this case we have a great many activities classed together as 'scientific', so there must be something common to them defining and delimiting science.

Suppose you try to propose such superwhatevers. Don't they themselves have to have some kind of justification? And since by fiat you have made our best-grounded beliefs about the way things are irrelevant to justifying them, what are you going to appeal to? If they are supposed to be descriptive of what science is, or prescriptive of what science may become, you will have well-known problems either way. If they are supposed to be methodological, the only justification you could hope for is that they could be shown to be necessary in order that science be able to proceed. And we know how arguments of that sort have fared in the past, too.

EUGENE LASHCHYK: I have a problem distinguishing your present position from the one which you think you have rejected, and that's Kuhn's position.

You argue that Kuhn is wrong because he claims that meanings, criteria, observations are paradigm-dependent. Now you suggest that you follow Wittgenstein, that actually what we have to do is analyze meaning, *etc*., in terms of use; but what is usage except a kind of context? We have to study it in its context. Your present position looks to me like the position you want to reject, which you take Kuhn to hold.

SHAPERE: The differences were suggested in passing in the paragraph in Section I of my paper beginning "Many of the historically-oriented approaches of the 1960's . . . ," but I'll explain more fully. The fundamental difference is that, for Kuhn – and of course he's notoriously hard to pin down – there is one overarching presupposition or set of presuppositions ('paradigm') that somehow governs everything else throughout a whole scientific field, community, or tradition. It determines rules, methods, standards, problems, substantive beliefs on specific questions, *etc*. (Note that Kuhn has here committed the Platonic Fallacy: in order for a subject to be *a* subject, there must be a common essence – even if it cannot be formulated.[2] Only for Kuhn there is a common essence only within a paradigm, not for what we call science as a whole.) On my view, there is no such thing, nor need there be. Different sorts of considerations may be relevant in different situations, and when things change in science, they generally change in a more piecemeal way. The changeover in chemistry that is usually said to revolve around Lavoisier and to have constituted a 'revolution' was a far more gradual process, involving numerous factors[3] which, when they were all drawn together, had a very great cumulative effect. The same is true of *most* other so-called 'revolutions'. (I say 'most' because there is nothing in my view that precludes the possibility of introducing very radically new ideas without the way having been paved. If one means by 'continuity' in science the view that the past always paves the way for new ideas, then I not only reject the idea that there must always be a common *essence* if a subject is to be *a* subject; I also reject the idea that, for a subject to be *a* subject there must at least be continuity within it – in that sense of 'continuity'.)

Of course I recognize that, historically, there often have been unquestioned assumptions guiding science; the idea that all explanation must be 'mechanical' is an example. But I would argue that science has learned to try to get away from such dogmatism. That doesn't mean trying to get away from all assumptions whatever, but to try to make assumptions explicit and to subject them to examination wherever possible. (Recall the case of symmetry principles as discussed toward the beginning of Section IV of the paper.) Thus, even to the extent that science ever *was* done in a Kuhnian way (and that extent is exaggerated greatly by him), we have learned that that was not a good way. (Here's an example of why a study of historical cases alone cannot tell us everything about what science is now.) Certainly elementary particle theory today is 'dominated' by nonabelian gauge theory approaches to the study of fundamental interactions and the attempt to unify them. But reasons – specific scientific reasons – exist for taking such approaches, and for suspecting (or in some cases believing quite strongly) that there may be limitations to them, and alternative approaches are not drummed out of the profession as long as their promise can be shown. Or at least, if they are, there is a widespread feeling that we ought by now to have learned better.

As a first approximation, the difference between Kuhn and myself on this fundamental point can be put by saying that whereas Kuhn's view is a *global presuppositionism* (the same overarching presupposition or paradigm is brought to bear in every activity or idea in the paradigm tradition, and no contrary presuppositions are allowed as 'scientific'), mine is a *local presuppositionism* (different presuppositions may be brought to bear in different situations, and where there is a more general and widely-applied approach, as in current elementary particle theory, it too is subject to being questioned). But that is only a first approximation, since when one is careful about the word 'presupposition' (and related words like 'interpretation'), one finds that the point of such statements as "All thinking (and/or describing, claiming, arguing, *etc.*) requires presuppositions" or "All thinking (or whatever) involves interpretation" is much more innocuous than is often supposed.[4]

This fundamental difference between Kuhn's view and mine is the source of many others. For example, Kuhn's 'global presuppositionism', together with his view that there are many distinct 'paradigms', lies at the core of the relativism which his view implies – the implication that there is no such thing as progress in science, and that no one fundamental theory (paradigm, set of fundamental presuppositions) is better than another. Besides differences stemming from this basic one, there are of course many others. My view is uncompromisingly nonessentialist; Kuhn's, while it is nonessentialist with regard to all science, is essentialist with regard to each paradigm tradition. Kuhn does not acknowledge reasons for adopting a new paradigm, only reasons ('anomalies') for becoming dissatisfied with an old one. (However, when Kuhn sheds his philosophy and gets down to cases, as in his fine discussion of the background of quantum theory in *Criticism and the Growth of Knowledge*,[5] his good sense gets the best of him.) On my view – and in this I oppose Popper as well as Kuhn – there can be reasons leading to the adoption of new theories, however fundamental. Again, the case of chemistry illustrates this very well; but so do many others.

MICHAEL GARDNER: You present a dilemma that either there are unchanging standards, historical standards, or else that we can only evaluate the work of a given period by its own standards. And it seems as if you then try to escape this dilemma by saying that what *we* should do is evaluate the standards and work of past eras and evaluate the rationality of historical changes in these standards by reference to *our* standards. Is that what you mean when you say that certain Cartesian doubts do not invalidate the practice of using the best information, thus reasons, thus conceptions of legitimate problems we have available? Do you mean that we should use our standards to evaluate past standards?

SHAPERE: There are several ways in which the ideas of a certain period can be evaluated. One is in terms of the criteria and well-founded beliefs available at *their* time; another is in terms of later ideas which showed those to need modification or rejection. There is of course no inconsistency in using both of these ways; each serves a purpose. It seems to me that old 'Whiggish' histories of science made the mistake of using *only* the second type of criterion. On the other hand, there is a grave danger – and a tendency – on the part of many historians to use only the first, which leads them to a form of relativism.

GARDNER: That brings me to my objection. When you are doing the second thing – using the rational standards of the present day to evaluate what has been going on in the past, you are taking a Quinean line, I suppose. The philosopher should accept the scientific principles and the conceptions of the methods that are now accepted by the scientific community. One of the things that scientists now believe, I guess, is that the methods they use really are the best and that they always have been the rational way of doing science. It's just that Aristotle didn't know that. So when you propose accepting the standards of today, since those standards implicitly assert that they have always been valid, why don't you end up with *ahistorical* standards – ahistorically valid, not ahistorically known?

SHAPERE: Of course we judge past beliefs, methods, standards, and so forth in terms of the beliefs, *etc.*, which we have found most successful and free from specific doubt, just as we apply those beliefs in the further investigation of nature. But to say that therefore those present beliefs and so forth are 'ahistorically valid', that "they have always been valid," is potentially misleading. After all, even though our present beliefs, *etc., may* never be revised, may never become subject to doubt, such doubts may arise, and revision of our present beliefs, *etc*., might carry with it a revision of our assessment of the past.

It's also misleading, because ambiguous, to say that the beliefs and methods, *etc*., that we use today "always have been the rational way of doing science." Partly that's misleading for the reason I've just given: we may be wrong too. But besides that, this way of treating rationality, in terms of our present-day perspective, ignores the other important sense of 'rationality,' according to which people in the past could be justified in believing what they did even though later they turned out to be wrong, when new information became available that they didn't have. (Actually there is only one sense of 'rationality' involved, according to which, very roughly, one is rational if one bases one's judgments on the best-founded beliefs available. But it's still useful to distinguish between judgments of rationality from our perspective and judgments of rationality from that of a past thinker or group.)

THOMAS NICKLES: Dudley, the general thrust of you position is reminiscent of Quine, but you seem critical of a Quinean sort of view at a couple of points. It would help me to see just where you differ from Quine, and why.

SHAPERE: There are a number of points of agreement between Quine's views and mine: that nothing in our system of belief is immune to revision (I would include methods, *etc*., in addition to beliefs here, and perhaps Quine would agree); that our system changes piecemeal rather than as a whole; and that science attempts to work "a manageable structure into the flux of experience"[6] (that way of putting the point seems more practice-oriented than I think appropriate for more sophisticated science, and I prefer an even more noncommittal locution like 'deal with'). My agreements are primarily with theses advanced in 'Two Dogmas of Empiricism' and elaborations of those theses in later works; but there is much in his later work that I do not think is compatible with those theses and with which I am not in sympathy. However, even in the early paper there are a great many differences between my view and his. They stem, I think, from the difference in our approaches to the analysis of the knowledge-seeking enterprise. Quine's conclusions are not drawn primarily from a close examination of science and scientific change, but rather from the viewpoint of a logician. Consider his Duhemian view that no statement is ever tested in isolation, so that, in the face of a recalcitrant experience, we have a wide latitude of options as to how to adjust our system of relevant ideas. (Quine claims that the *whole* of science is tested in any such experiential situation, but I can see no basis for that view.) What do we do in such a situation? Quine speaks of a network of beliefs in which some are centrally and some peripherally located. "The edge of the system must be kept squared with experience; the rest, with all its elaborate myths and fictions, has as its objective the simplicity of laws." (*Ibid*., p. 45) We "disturb the system as little as possible" (p. 44), even if this means rejecting the recalcitrant experience. But all this, and Quine's glosses on it in other writings, says precious little at the very point where I think the real problem for the analysis of the knowledge-seeking enterprise begins. For that problem is, *Given* that, in any experiential, observational, or experimental situation, we bring to bear a great many background beliefs, what is the character of the interplay between those beliefs and experience (or observation or experiment)? How, if at all, for example, is the objectivity of observational or experimental testing preserved in the face of such application of background beliefs? With regard to the Duhemian situation in particular, how do we decide what adjustments to make in the face of 'recalcitrant experience'? By what reasoning do we decide to make the adjustment here rather than there, and in this way rather than that? Quine has essentially stopped at the logical point of underdetermination of belief by experience: that is the real focus of his interest in the Duhemian situation, as is shown by his later writings. But from the perspective of the philosophy of science, he has taken only a small step, failing to go beyond the underdetermination thesis to give the crucially-needed careful analysis of how science reasons to arrive at a readjustment of its belief-structure.[7]

It is not even clear to me that Quine thinks there is really much in the way of reasoning, beyond the underdetermination thesis, to analyze. Experience, it seems, not only underdetermines the structure of our belief-system; it is actually impotent in helping us decide how to adjust that system in the face of recalcitrant experience. Doing so is a matter of minimal disturbance (that being 'our natural tendency' (*ibid*., p. 44) or simplicity; but very often he suggests that adjustments are made in terms of our 'likings', 'preferences', 'interests', or 'purposes', without any indication that there may be reasons for holding those preferences or having those purposes, *etc*. Because of all this, I cannot see his view as a truly empiricist one. It is more a conventionalist view, bordering on a relativism that resembles Kuhn's views more than mine. Quine's later emphasis on the 'indeterminacy of translation' in addition to the 'underdetermination of experience' seems to me to exacerbate this tendency.

Quine rejects a realist view of science: "I continue to think of the conceptual scheme of science as a tool, ultimately, for predicting future experience in the light of past experience. Physical objects are conceptually imported into the situation as convenient intermediaries – not by definition in terms of experience, but simply as irreducible posits comparable, epistemologically, to the gods of Homer." (*Ibid.*, p. 44) I disagree with that view on the ground that it does not do justice to what I think can be shown about what science does. Science tells us what things exist (or, more precisely, what it claims exist), for example; and its commitments are decided on the basis of scientific considerations. To suppose that we must throw its statements into a 'canonical' logical form in order to make its existence-claims explicit by seeing what it quantifies over or needs to quantify over is not only superfluous, but bizarre.[8]

EDWARD MacKINNON: My difficulties are of a slightly different sort than the others in the sense that with a few minor qualifications, particularly in the interpretation of Kant, the view you are presenting is what I would hold. The problem that bothers me is, what do you do with it? The difficulty I find with this is that it is still very much on the level of general strategy arguments as opposed to actually providing the detailed analyses of historical examples and showing how they work. If one wants to uncover a continuity in science, you have different types of things that might be done. But if one does not want *a priori* principles or immutable presuppositions, then what is needed is a much more detailed structural account of the nature of science and its development in terms of things like the distinction between content, mathematical form, depth level, and phenomenological levels, and laws of science and laws about the laws of science (invariance principles and conservation laws) – enough of a detailed structural account so that one could go through certain developments and say that there was *this* sort of continuity here and that sort of continuity there and, in the overall picture, some depth continuity perhaps. I just wonder whether you think this has to be done and whether you are trying to do something like this?

SHAPERE: Yes, I am very concerned with the detailed examination of cases, and the view I have presented is based partly on such examination. The discussions of cases in this paper are encapsulations of long and detailed studies of the cases in question, and I have dealt with a number of other cases in other published and unpublished papers and in some of my courses. (Some aspects of the philosophical motivations behind my examination of cases, and the approach I take to such examination in courses as well as in research, are described in some detail in my paper, 'Scientific Thought', forthcoming in *Teaching Philosophy*.) But while clearly necessary for gaining an understanding of the scientific enterprise, examination of cases isn't sufficient. There are philosophical issues that have to be dealt with too, of the sorts I've discussed in this paper. Those issues are not things one has to clear up *first*, before one can get on with the real job of looking at the cases – as though they constituted the advance plan of our investigation. (Is that perhaps what you had in mind in referring to 'strategy arguments' in my paper?) On the contrary, their solution is an integral part of the *goal* of the investigation: they are not the mere plan of the war; they are what the war is all about. Neither the philosophical issues nor the case studies can be successfully attacked independently of the other. Among other things, one must take a philosophically critical attitude toward every way in which one is tempted to think about the cases, trying thereby to guard against philosophical categorizations that might be misleading, unclear, ambiguous, inappropriate, question begging, or otherwise faulty, while at the same time trying to develop an account that will be more adequate, both philosophically and historically. The distinctions and categorizations you present in your question, for example, might very well be quite appropriate sometimes. But I think we would have to be suspicious if they were put forward uncritically as the appropriate conceptual tools with which to approach the analysis of any case in the history of science up to the present day and beyond. I trust you don't intend them to be taken in that way.

NICKLES: Several people, partly inspired by Larry Laudan's book, *Progress and Its Problems*, are coming to view scientific problems and problem solving as a, or perhaps *the*, central focus for general methodology or philosophy of science. I am not very clear how important problems are in your account.

SHAPERE: The analysis of scientific problems is certainly very important. Problems are, after all, a central aspect of the scientific enterprise. But problems don't just exist 'out there'; nor do we just make them up. There are reasons why they arise. There are reasons why certain problems are considered to be legitimate scientific problems and others are not. There are reasons why certain problems are considered to be important and other problems, even though legitimate, are considered unimportant or as problems which, while important, can't be dealt with at the present time. There are reasons why a certain range of possibilities, under certain circumstances anyway, is considered to be a range of possible answers to a certain problem. There are reasons behind expectations about what will constitute an answer to a given problem. And so forth. Well, if there are reasons which form the bases on which problems arise, *etc*., certainly we mustn't forget that those reasons are there. One might even insist that since the problems have their bases in those reasons, the latter are more important than the former – more fundamental, promising of deeper insight. I think that there is danger in focussing on problems as the central aspect, since it might lead one to concentrate on classifying problems while neglecting a close analysis of the reasons behind them. I wouldn't want to ignore either one, or, for that matter, any of the other important aspects of science.

LYNN LINDHOLM: I would like to introduce what I think is another problem, the difficulty in handling pluralism in describing what is rational about scientific reasoning. As many people here have pointed out, there are different forms of discovery, different forms of rationality. Some people do their research by attending to metaphysics; other people look at the facts. In many of the conference sessions, speakers have been pushing us to appreciate the various forms of discovery and various forms of doing scientific research. What hasn't been appreciated is that it is difficult to understand scientific progress on such an account. One can't just allow a thousand flowers to bloom, as Feyerabend advocates, because if I describe William Harvey as the inductivist researcher, Galileo as an aprioristic researcher, and someone else as a person who attended to simplicity and thereby made his discoveries, I can't then give an overall idea of why the tradition is progressing. We want to do that as well as to answer the questions about individuals. This particular problem bothers me a lot, and prevents me from appreciating the push toward pluralism as much as I would like to. Can we have pluralism without anarchy?

SHAPERE: I agree with your assessment of the situation: the problem you point to is a real one, and it indicates that the push toward pluralism is being overdone. At bottom, that push is the result of not being critical enough of generalizations based on what has happened in science in the past. Certainly, a great variety of methods *have* been used in the past, usually confusedly. But (even ignoring the confusion) the fact that they've been used, even by great creative scientists, doesn't make them sacred. A method may have been appealed to in the past precisely because scientists of that period had nothing particularly better to rely on; and it may have proved useful and successful because there was *something*, though not everything, to it. What happens is not that we preserve the many methods, but that we find what there is of value in each of them and what is not, when each is applicable or most useful and when not, and so forth, rejecting some methods and sometimes introducing others. We learn, for example, when we are entitled to generalize and about what (inductivism), under what circumstances to rely on general theories (apriorism (?)), what to count as simple, and other like things. We learn how to combine these methods: for example, how to use general theories to tell us what generalizations we are entitled to make and what we are to count as simple, or how to use what we have learned to count as simple to help us decide what general theories to accept. Thus we begin to systematize our 'methods' with our beliefs about the world. Our best beliefs are put to work in rejecting, clarifying, modifying, and systematizing our methods, and those methods are in turn applied in increasingly sophisticated ways that lead to new beliefs in terms of which we may again refine our methods, and so on.

But these are precisely the points I have made in the paper: method, like substantive belief, is learned. We learn what methods will help us learn, and how and when. With increase of knowledge (*i.e.*, of well-founded, doubt-free beliefs), our methods grow more sophisticated, and we are able to learn more, and more effectively: the rich get richer. And we try to systematize our methods as we systematize our beliefs: systematization, too, we have found, is a highly successful 'method'. Thus, for example, we try to become clearer about when the principle of observability may be overruled by other methodological considerations and when not (recall the discussion of quarks and color in Section III). Though the past benefits of systematization, and our past success in achieving it, lead us to seek it further, we may or may not succeed in our further search for it. But even if we were to fail – and, after all, we are not yet *perfectly* clear about *exactly* when observability may be overruled, for instance – even then, there would be no basis, historical or otherwise, for the claim that pluralism of method is or should be a fundamental and inviolable principle of science. (We would at most have learned that such pluralism is a fact to be lived with.) On the contrary, insistence on that principle flies in the face not only of the learned affirmation by science of the value of systematizing methods and beliefs, but also of its learned denial that there are any such inviolable principles.

## NOTES

[1] *Editor's note*: This discussion includes material from a separate, panel discussion, which could not be published.

[2] See p. 44 of T. S. Kuhn, *The Structure of Scientific Revolutions*, Univ. of Chicago Press, Chicago, 1962.

[3] See Note 10 to my paper.

[4] Details will be given in my paper, 'The Concept of Observation in Science and Philosophy', *Philosophy of Science*, December 1982.

[5] See pp. 256ff of Kuhn's 'Reflections on My Critics', in I. Lakatos and A. Musgrave (eds.), *Criticism and the Growth of Knowledge*, Cambridge Univ. Press, Cambridge, 1970.

[6] W. V. Quine, 'Two Dogmas of Empiricism', reprinted in *From a Logical Point of View*, Harvard Univ. Press, Cambridge, 1958. See p. 44.

[7] Again, see my 'The Concept of Observation in Science and Philosophy', *Philosophy of Science*, December 1982.

[8] Beyond what I have stated in Section V of the present paper, my arguments on this point are given, though in a somewhat rudimentary way, in Part II of 'Notes Toward A Post-Positivistic Interpretation of Science' and Part IV of 'Scientific Theories and Their Domains'. (See Notes 23 and 1 above for complete references.) They will be considerably elaborated and extended in a forthcoming paper, 'On the Role of Conceptual Devices in Science'.

CHAPTER 12

# THE SCOPE AND LIMITS OF SCIENTIFIC CHANGE*

Most philosophies of science have supposed that there is something about the scientific enterprise that is either presupposed by that enterprise, and is thus immune to revision or rejection, or else is discovered in the course of that enterprise, and is seen, from that point on, to be immune to revision or rejection. Let us call such beliefs – ones which are allegedly immune to revision or rejection – *necessary claims* about science. They need not be "propositional," in the sense of being overt claims about the way things are. Indeed, few philosophers any longer maintain that there are any such necessary propositional claims, the nearest survivor being the view that there are "observational" statements which, insofar as they are truly observational, cannot be modified or rejected. Rather, the focus of modern philosophies of science has been on such allegedly necessary claims as the following: that there is a method, "the scientific method," by application of which knowledge about the would is obtained, but which, once discovered (by whatever means), is in principle not subject to alteration in the light of any beliefs arrived at by its means; that there are rules of reasoning – rules, for example, of deductive or inductive logic – which are *applied* in scientific reasoning, but which can never be changed because of any scientific results; or that there are "metascientific" concepts, like 'evidence,' 'theory,' 'explanation,' which are used in talking about scientific concepts, claims, and arguments, which have meanings which are wholly independent of the specific content of ongoing science, and which, collectively, define what science is and always will be.

Views of this sort, maintaining that there are necessary components or presuppositions of the scientific enterprise, have faced enormous difficulties in the past. Apart from specific objections to specific variations on the theme, the general view that there are necessary claims in or about science seems to go against the trend of science in the past century or so, where we have witnessed the successive overthrow of one allegedly "necessary" claim after another. More recently, historians of science and historically-minded philosophers have argued that deep change is characteristic of the knowledge-seeking enterprise throughout its development. They argue, in effect, that such change extends not merely to profound alteration of our substantive beliefs about the world, but also to conceptions of the goals of science, the demarcation between science and non-science, the distinction between legitimate and illegitimate scientific problems, the methods of science, the standards of adequacy and acceptability of scientific solutions, and, in general, to everything that is a constituent of the scientific reasoning-process. Indeed, one might summarize the view proposed (however implicitly) by these historians and philosophers as asserting that there is nothing in scientific reasoning, not even conceptions of "reasoning" itself, that is in principle immune to revision.

There is much that is appealing in this view: besides being more in accord with the history of science, particularly its more recent history, it promises to liberate us from the last shackles of apriorism and essentialism. It promises to provide us with a view of science that is uncompromisingly empiricist, in that it suggests that not only do we learn about the world through experience, but we also learn how to learn and think about it in the same way.

And yet despite these appealing aspects, the view that there are no necessary claims in science faces severe difficulties. Among the many such difficulties is the following. Suppose we try to maintain that there is absolutely nothing sacred and inviolable in science – that *everything* about it is in principle subject to alteration. Then included among the things that can change are standards or criteria of what it is to be a "good reason" for change. But then how could criteria of rationality themselves be said to evolve rationally, unless there are higher-level standards or criteria of rationality, themselves immune to revision, in terms of which changes of lower-level criteria of rationality could be judged to be rational? There thus seem to be only two alternatives: relativism, in which there is no real ground (no ground, that is, other than the decree by fiat of a triumphant community) for saying that there is "progress" in science, that one body of scientific beliefs is better than another; or else a presuppositionist theory according to which there is something, of the sort that can serve as a standard or set of standards or criteria for scientific rationality and progress, which is immune to the vicissitudes below, and which serves as the ultimate arbiter of those lower-level scientific disputes.

It is readily apparent that this difficulty is a fundamental one, and perhaps accounts in part for the reluctance of so many philosophers in the past to try to develop a theory according to which there are no necessary scientific claims. In this paper, however, I will argue that the difficulty can be overcome through a deeper understanding of what is involved in something's being a "reason" in science. It will, of course, not be possible to do full justice to this important question in the brief time available today. Nevertheless, I hope the general lines of my argument will be clear.

My point of departure is the intuition that, in any argument concerning a subject-matter, those considerations will be relevant as reasons that have to do with that specific subject-matter. The question, "What does that have to do with the subject we're discussing?" is a challenge that the consideration adduced by our opponent is irrelevant, that it does not constitute a reason either way in our dispute. If this intuition sounds somewhat tautologous, let me at present merely caution that it will provide the key for bringing out some of the central aspects of what happens in scientific reasoning, and, indeed, of what scientific reasoning is. I will in any case return to discussion of its basis later. For the present, as steps toward understanding the implications of that intuition, let me turn successively to two points: first, what is involved in being a "scientific subject-matter," and, following that, what is involved in "having to do with" a subject-matter in science.

The science of a particular epoch can be seen as the investigation of various areas or domains. A "domain" can be defined roughly, for present purposes, as a body of information which is problematic in certain respects, and the items of which we have reason to believe are related in the sense that a unified account of them (with respect to their problematic aspects) can be expected. Domains, in this sense, can be as broad as the subject-matters of fields like electromagnetism, genetics, or organic chemistry, or as narrow as the specialized interests of individual research workers. I have discussed the mechanisms of domain formation and functioning in some detail in previous papers.[1] For present purposes, what is important about them is that the division of science into domains or subject-matters is not something that is given by immediate experience in any final way, but is something acquired through painstaking investigation, and subject to further alteration. The relationships involved are discoveries, the fruits of accumulated knowledge. That there is a subject of electricity to study had to be found out; that it constitutes a unified subject-matter and not several distinct ones (static electricity, current electricity, animal electricity, *etc.*) had to be established, by careful experiment and argument, by Michael Faraday. That that domain could be unified with that of magnetism was learned through a long and tortured process of investigation. Domains can be split as well as unified. But overall, science attempts to make clear and precise the various interrelationships between the items it studies, and it is indeed a mark of highly sophisticated science that those interrelationships are highly articulated.

Nor is the association of "items" into domains merely a matter of grouping together independently-describable information. On the contrary, reformulation of the language in which we talk about items often accompanies their grouping or regrouping into subject-matters or domains. Chemistry in its modern sense became a subject when older vocabulary for talking about matter – vocabulary based primarily on the sensory appearances of substances – was replaced by a new descriptive language in which substances were named in accordance with the elemental substances of which they were compounded. The possibility of that reform of nomenclature underscores the role of prior knowledge (or belief) in the formation of domains and their description. In the case of the eighteenth-century chemical reform, a number of factors entered in to make the reform feasible: the possibility of analysis into constituents (rather than those alleged constituents being created by the process of analysis itself); the idea that knowledge of composition gives understanding of matter; the concept of an element as a breakdown product, and the discovery that there are a relatively small number of breakdown products common to a wide range of chemical substances; the concept of weight as being centrally relevant chemically; and Lavoisier's oxygen theory of acids, metals, calxes, and combustion. With the further development of an area of science, the formulation of domains tends to become more and more built on previous knowledge or well-founded belief.

The relations between the descriptive language of domain items and the "observation-language" in terms of which hypotheses about the domain items are tested is not straightforward and simple. Hence it is necessary here to emphasize that that observation-language, too, tends, in sophisticated science, to rest on a vast store of prior well-founded beliefs. As an example, consider the claim, universally made by astrophysicists since the mid-1960's, that it is now possible to make direct observations of the center of the sun. What is involved, of course, is a knowledge of the behavior of the neutrino – specifically, but only in part, of processes in which neutrinos are emitted, of the fact that neutrinos are weakly-interacting and therefore can be expected to pass freely to us, without interruption or interference, from their origin through the body of the sun and interplanetary space, and of the kinds of receptors appropriate for intercepting such neutrinos. And it is prior knowledge that specifies what counts as an "appropriate receptor," the ways in which information of various types is transmitted and received, the character of interference and the circumstances and even the statistical frequency with which it occurs, and even the types of information there (fundamentally) are. The physics of the present epoch, for example, makes such specifications for a wide range of circumstances through the whole body of well-founded belief about elementary particles, their decays and interactions, and the conservation principles which govern such processes. In particular, for example, knowledge of the cross-sections for such particles, the probabilities of their interactions with other particles (or decays of individual particles) in given environments contributes to specifying the notion of "interference" or "interruption." In particular problems far more enters into specifying such notions as that of a "receptor" than just what is contained in elementary particle theory ("observation" is not laden merely with "theory"); for example, in the solar neutrino case, knowledge of the environmental conditions (pressure, temperature, opacity) in the interior of the sun also plays a role, as does knowledge about various instruments.

So the formation of domains or subject-matters, and their description, and the language in which we express observational tests of hypotheses about domains, are highly dependent on "background information," on accumulated knowledge or presumed knowledge which is brought to bear on that formation and description. Analogous considerations and cases could be given to show how other aspects of the development of science are also shaped by background belief: the problem-structure of the science of a given epoch, for example, or the range of possibilities envisioned at that epoch. But what about that "background information" itself? What constraints govern its conception and employment? What justifies the claim that certain beliefs "have to do with" a subject-matter and can enter into considerations about it?

In relatively early and unsophisticated stages of science, the line between "scientific" and "unscientific," between the scientifically relevant and irrelevant, is not clear, or is even nonexistent: there is, at that time, no clear basis for exclusion of certain sorts of considerations as scientifically irrelevant. (There may, indeed, be grounds for supposing them to be relevant, even though later science will exclude them.) In the seventeenth and early eighteenth centuries, there was no clear line between what we today would distinguish as "scientific" and "theological" considerations. Newton's arguments in favor of absolute space and time were heavily tinged with theological considerations. Furthermore, he believed that there were at least three ways in which the laws of physics implied the necessity of God's interference in the world: first, in order to keep the motions of bodies from dying down in the face of momentum loss through impacts; second, in order to preserve the order of the solar system against the disruptive effects of gravitational perturbations; and third, in order to resolve the inconsistency of a universe in which, if matter were distributed finitely through space, all bodies would fall to their mutual center of gravity, but in which, if matter were distributed infinitely through infinite space, all gravitational interactions would be cancelled. Yet by the middle of the eighteenth century, confusions about *vis viva* and momentum had been resolved, and Newton's first reason had dissipated; and Laplace showed, or at least claimed to show, that gravitational forces between planets are self-correcting in the long run. Apocryphal or not, the story about Laplace's reply to Napoleon's query about the place of God in Laplacean science ("We have no need of that hypothesis") captures the fact that, to Laplace and the increasing majority of his physicist successors, theology had been excluded as irrelevant to science. (The third Newtonian argument, the Bentley-Seeliger paradox, was neglected until the twentieth century.) In a similar vein, Keplerian questions about astrological influences, or about why there are exactly six planets – questions Kepler himself admitted because they were equally as "geometrical" (in his interpretation of that concept) as questions about the shapes of planetary orbits and about the relations between orbital speed and distance of a planet from the sun – came to be excluded by the success of the view that matter exerts causal influence only by impact and (possibly) attraction and repulsion at a distance. These examples show how certain types of considerations come to be excluded as scientifically irrelevant; but there are also, sometimes, cases involving the introduction of radically new types of considerations. Thus the work of Gauss and Riemann made it possible to treat physical space as having intrinsic characteristics without supposing it to be embedded in a higher-dimensional space, and further, to allow those characteristics to vary from point to point and from time to time. Newton and Leibniz had rejected the Cartesian idea that matter can be understood in purely geometrical terms precisely on the ground (among other reasons) that it was inconceivable for space to have such characteristics.

In general, then, as science develops, successful beliefs produce constraints on what can count as scientifically relevant, sometimes broadening, sometimes narrowing the range of the scientifically relevant. The line between the scientific and the non-scientific – between what can count as a scientific consideration and what cannot – is thus an acquired characteristic of scientific inquiry; it is not innate, essential to and definitory of the scientific enterprise itself. In its earlier phases, science relies on considerations that may later be excluded, or fails to rely on considerations that will later be alleged, on the basis of what we have learned, to be relevant. (Many of the latter cannot even be formulated until a certain stage in the sophistication of scientific language is reached.) What earlier were – at best – vague and loose relevance-relations are rejected, tightened, or altered in more profound ways involving revisions in our ways of thinking and talking about a certain subject-matter;

and indeed, we learn even *that* we should seek such tight relations of relevance. Thus, metaphysical, theological, political considerations have all played roles in the development of scientific ideas, as have otherwise ungrounded analogies or symmetry considerations. Such considerations have later been found to be irrelevant, in the light of what we have learned about the world; or, as in the case of analogies, they have been transmuted into or foresworn in favor of relations which are supported by a background of successful beliefs. Thus stellar classification was originally (last quarter of the nineteenth century) based purely on the colors of stars, and theories of stellar evolution on the analogy of stellar colors with the progression of colors in heated objects. The success of such an approach (in a highly modified form) has led to the conviction that there is *evidence*, not mere analogy, in favor of the view that astronomical objects of all sorts are thermodynamic entities. And symmetry considerations have passed from being purely aesthetic demands to being very precise, integral, and testable constituents of elementary particle theory.

This is not to say that science, or any particular area thereof, has wholly dispensed with the need for now-recognizably non-scientific or questionably justified considerations such as analogies. Even in so paradigmatically advanced an area as particle theory, Yukawa relied on analogy with the extant quantum theory of electromagnetic interactions in constructing his exchange-particle theory of strong interactions. And later physicists appealed to non-Abelian gauge theories of the Yang–Mills type, successful already in quantum electrodynamics, as a model for construction of theories of the weak and later of the strong interaction. (It is true that, in these cases, certain difficulties had to be overcome before the analogy could prove applicable: appeal to "analogy," in sophisticated science, is increasingly rarely a matter of the mere noting of resemblance.) And many physicists continue to hope for a similar theory of gravitation.[2] Yet the clear hope of science, based on its past successes, is to exclude the need for such considerations: to develop its body of well-founded beliefs to such an extent that it will include within itself the grounds for all such reasoning. Science attempts, that is, to *internalize* the considerations that are to count in its further deliberations: to establish clear and tight relations of relevance, to draw the boundaries between the scientifically relevant and the scientifically irrelevant as sharply as possible, and to make the relevant considerations so comprehensive and precise that they are sufficient for raising and resolving all scientific problems. That is a procedure that has proved successful in the past, so successful that it has become a normative principle governing what is to count as a "reason" in science. (That pattern of development is indeed characteristic of the way normative principles are evolved in science; they, too, are introduced in the light of what we learn.)

But this brings us back to our starting-point. For what we have found is that science attempts, as far as possible, to develop in such a way as to exemplify the intuition that what is relevant as a reason in science must have to do with the subject-matter at hand. The explicit development of domains is a direct instance of this intuition: the attempt to formulate, as clearly and unambiguously as possible, what the subject-matter of an area of investigation is. Similarly, the evolution of the distinction between scientifically relevant and irrelevant considerations, a distinction made in the light of our successful beliefs, works in the direction of making explicit and unambiguous which beliefs may be brought to bear in the further building of our scientific conceptions. Among those successful beliefs, some will be established, to some degree at least, as relevant to specific domains, or to specific instrumentation which may in turn provide the bases for tests of specific hypotheses about specific domains. Thus, what I have spoken of as the "internalizing of considerations" in science is simply a specification of what is to count as a "reason." (Or, to put the matter in another way, what counts as a "reason" is something that itself evolves in the light of what we learn.) And again, it is fully in accord with the intuition about reasons with which I began – namely, the intuition that those considerations are to be considered relevant as reasons that have to do with the subject-matter at hand.

What is the status of that intuition itself? It, too, has been acquired in the course of inquiry; it is not the product of a purely intellectual analysis of the essence of "reasoning." The recognition of the need to delineate our subject-matter clearly and sharply, and to specify as exactly as possible what is and is not relevant to that subject-matter, has not been present in the knowledge-seeking enterprise from its inception. Science has gradually become aware of it, and seen its success as an approach. And that success has elevated it to the status of a guiding principle of scientific inquiry.[3]

In short, then, science builds on the basis of its successful beliefs. But "success," while in general necessary, is not (in general) sufficient to qualify an idea for being "built into" further scientific development – into, for example, the formulation of new domains, or the new formulation of old domains, or the introduction of new possibilities. For a belief can be successful while there still exist specific reasons for supposing that it cannot be true, that things could not really be that way. Such beliefs we call "idealizations," and I have examined their role in the scientific enterprise elsewhere.[4] In general, idealizations do not serve as "background information" for the construction of new beliefs (or methods, *etc.*). For that purpose (in general) there must – ideally – be no specific reason for doubting the belief. (I qualify these remarks by saying "in general" because, in certain circumstances – such as the lack of anything better to go on, or the questionability of its idealizational status – an idealization *can* be so used. I should also note that, as in my other writings, I contrast "specific reasons for doubt" with "universal or philosophical 'reasons' for doubt," the latter – like "A demon may be deceiving me," or "I may be dreaming" – applying indiscriminately to any claim whatever. Such philosophical doubts, as science has learned, play no role in the scientific or knowledge-seeking enterprise.)

Thus a "reason" in science consists of a belief (a) which has proved successful, (b) concerning which there is no specific reason[5] for doubt, and (c) which has been shown to be relevant to the specific domain in which it is being applied as a "reason."[6] These characteristics hold as ideals, of course: in practice we must usually rely on beliefs that have not proved unambiguously successful or unambiguously free from doubt, and whose relevance may sometimes be rather questionable.

As an example, consider again the investigation of stellar energy production and stellar evolution *via* solar neutrinos. Since the primary theory of energy production for stars in the sun's mass range is the proton-proton chain, the highly successful apparatus of elementary particle physics is brought to bear. The applicability of particle physics, together with more specific knowledge, leads to the conclusion that cleaning fluid (perchloroethylene, $C_2Cl_4$) is an appropriate medium for the reception of the particular solar neutrinos being sought. By inverse beta-decay, the isotope chlorine-37 is converted to argon-37, and thus, in turn, facets of the chemistry of argon become relevant in deliberations about the solar neutrino problem.

I have thus far left the notion of "success" unexamined, and in spite of its centrality, will be able to discuss it only briefly here. I have argued extensively elsewhere that we learn what success is in science. Within the very general context of "dealing with experience," many different conceptions of the goals of science, and of what constitutes "success" in "dealing with experience," are found in the history of science. The chemical revolution of the eighteenth century, for example, carried with it a change in conception of the goal of matter-study, from (as one view among many) the idea of bringing matter to perfection to the idea of understanding matter in terms of its constituents. That change of goal brought with it changes in conceptions of what it is for a view of matter to be "successful." Standards of success are among our beliefs, and there are a variety of ways in which they can change without the assumption of a transcendent, unchanging criterion of success. For example, at some stage one standard of success may be dominant, but we find we cannot fulfill it, while another "lower" standard gets satisfied frequently and well, and people begin paying more attention to it. This was, essentially, the case in the chemical revolution. (In such cases, it might seem more appropriate to speak of having "learned what failure is," rather than of having learned what "success" is; but the two kinds of learning are, after all, not separable.)

In the light of the account I have presented, we can see why science need not appeal to a transcendent and irrevocable principle of rationality in order to account for the occurrence of rationality and progress within scientific change. For what better standards or criteria could we employ – at least when we are able – than those beliefs (and methods and so forth) that have proved successful and have not been confronted with specific doubt, or at least specific doubt which has either not been removed, or else which has been shown to be not compelling enough to worry about? In the attempt to find some basis for considering certain things to be observable, or for distinguishing between those hypotheses to consider and those not to consider, and so forth, what else should one expect to use and build on, wherever possible, if not such beliefs? No further sorts of reasons are available to us, and none further are required, in order to account for the rationality and progress of the scientific enterprise. That our reasons, or the beliefs and methods which constitute their bases, may be wrong, and are sometimes shown to be, does not invalidate the practice of using the best information (and *hence* the best reasons) we have available – "best" in the sense of having shown themselves successful, and not having become subject to specific and compelling doubt.

And thus the difficulty mentioned earlier regarding the view that there are no necessary claims in or about science, all being in principle revisable in the face of experience is, if I am right, dissipated. That objection, recall, was that, unless we appeal to transcendent standards of rationality which are immune to revision, we cannot have any way of concluding that scientific change is rational. My response has been that the reasons we have available (in a clear sense of 'reason') are used in making such judgments, and that no other considerations need or could be appealed to. Much else remains to be said in order to explain and defend the view that scientific change is in principle pervasive and void of necessary claims; but those further points must be examined on another occasion.

## NOTES AND REFERENCES

* Research for this paper was completed during a sabbatical year at the Institute for Advanced Study, Princeton, New Jersey. I am indebted to a great many people for discussions leading to the ideas expressed here, particularly John Bahcall, Hannah Hardgrave, Richard Rorty, Neal Snyderman, and Morton White.

[1] D. Shapere, "Scientific Theories and Their Domains," in F. Suppe (ed.), *The Structure of Scientific Theories* (Urbana, University of Illinois Press, 1974), pp. 518–565; also "The Influence of Knowledge on the Description of Facts," in F. Suppe and P. Asquith (eds.), *PSA 1976* (East Lansing, Philosophy of Science Association, 1977), pp. 281–298. Many of the specific cases discussed in the present paper have been developed more fully in other papers; for example, "Reason, Reference, and the Quest for Knowledge," *Philosophy of Science*, March, 1982; "The Concept of Observation in Science and Philosophy," *Philosophy of Science*, December, 1982.

[2] It is also true, however, that the gauge-theoretic approach *might* ultimately be discarded in favor of some other which might (for example) be more conducive to a unification of weak, electromagnetic, and strong interactions with gravitation. It has been suggested by Misner, for example, that formulations in terms of harmonic maps should be explored in this connection. (C. Misner, "Harmonic Maps as Models for Physical Theories," *Physical Review D*, Vol. **181** (1978), 4510–4524.)

[3] One might wish to argue that, even though people were not aware of adhering to the principle that considerations must be relevant to the subject-matter at hand in order to count as reasons, they nevertheless always adhered to it, and must have done so, because the principle is an "analytic truth." Such a claim is not, however, cogent: notions of relevance, subject-matter, and argument were far too vague in earlier times to support it without distorting the thought of those times. Nor should one claim that, even though the principle may have been discovered, it is nevertheless a necessary truth, not subject henceforth to revision or rejection. For what are counted as (or the "criteria" for identifying) a "subject-matter" and a "relevant consideration" are clearly subject to revision in the light of further scientific discovery. And I have argued elsewhere ("The Character of Scientific Change," in T. Nickles (ed.), *Scientific Discovery, Logic, and Rationality* (Dordrecht, D. Reidel, 1980), pp. 61–116); "Reason, Reference, and the Quest for Knowledge," *loc. cit.*) that the "meaning" of the "principle" in question is exhausted by such criteria (or, for certain purposes in certain contexts, by the family of such criteria that have been developed so far in history), and that even the "principle" itself might be rejected under certain scientifically-determined circumstances.

[4] D. Shapere, "Scientific Theories and Their Domains," *loc. cit.*, Part IV; "Notes Toward a Post-Positivistic Interpretation of Science," Part II, in P. Achinstein and S. Barker (eds.), *The Legacy of Logical Positivism* (Baltimore, Johns Hopkins University Press, 1969), pp. 115–160); "Natural Science and the Future of Metaphysics," in R. Cohen and M. Wartofsky (eds.), *Methodological and Historical Essays in the Nature and Social Sciences* (Dordrecht, D. Reidel, 1974), pp. 161–171.

[5] More accurately, "any specific and compelling reason." In "Scientific Theories and Their Domains" (*loc. cit.*), I distinguish between various types of scientific problems, some of which ("problems of theoretical inadequacy") provide compelling grounds for considering a theory to be false, while others ("problems of theoretical incompleteness (with respect to a given domain)) do not provide such grounds.

[6] The occurrence of the word 'reason' in (b) does not mean that this analysis is viciously circular, since the development of the concept of a "reason" is a "bootstrap" process of finding – in effect, hypothesizing – that certain considerations can be counted as reasons, using those hypothesized reasons as bases for finding further relevance-relations in the light of which the original "reasons" can be critically evaluated, and so forth. Thus at any given stage, what is to count as a reason presupposes prior "reasons," and, specifically, reasons for doubt. But such presupposition does not imply that the prior "reasons" cannot be criticized and rejected as reasons.

CHAPTER 13

# SCIENTIFIC THEORIES AND THEIR DOMAINS

## I. FRAMEWORK OF THE PRESENT ANALYSIS

If we examine some relatively sophisticated area of science at a particular stage of its development, we find that a certain body of information is, at that stage, taken to be an object for investigation. On a general level, one need only think of the subject matters called "electricity," "magnetism," "light," or "chemistry"; but both within and outside such standard fields, there are more specific examples, such as what are taken to be subfields of the preceding subjects. Further, those general subjects themselves are, in many cases, considered to be related in certain ways. I will refer to such bodies of related items as *domains*, though we will find that, in the sense in which this concept will prove helpful in understanding science, and in particular in understanding the concept of a scientific "theory," more is involved than the mere relatedness of certain items.

The preceding examples are familiar ones, and there might appear to be no problem about considering them to be "fields," or, as I have called them, domains. However, this is far from being the case. In order to bring out some of the complexities involved, let us briefly review certain aspects of the history of these examples.

It is by no means obvious that all the phenomena which we today unhesitatingly group together as forming a unified subject matter or domain under the heading of "electricity" really do form such a unity. Earlier investigators had indeed associated the known phenomena: William Gilbert discovered that some twenty substances besides the previously known amber would, when appropriately rubbed, attract light bodies. His investigations led him to conclude that electric phenomena are due to something of a material nature which is liberated by the rubbing of "electricks."[1] Succeeding workers agreed, holding that electricity consists of one (or two) fluids. However, the nineteenth century, with a larger body of apparently related information available, saw reasons to question the unity of the subject: was the "electricity" associated with physiological phenomena of the same sort (and to be explained in the same way) as that associated with inanimate objects, or was it peculiar to, and perhaps even the distinguishing characteristic of, living things? Was the "electricity" produced on the *surfaces* of certain objects by rubbing them identical with the "electricity" produced in the *interiors* of certain bodies by a voltaic cell? Even by the time of Faraday, according to Whittaker,

> the connection of the different branches of electric science with each other was still not altogether clear. Although Wollaston's experiments of 1801 had in effect proved the identity in kind of the currents derived from frictional and voltaic sources, the question was still regarded as open thirty years afterwards, no satisfactory explanation being forthcoming of the fact that frictional electricity appeared to be a surface-phenomenon, whereas voltaic electricity was conducted within the interior substance of bodies. To this question Faraday now applied himself; and in 1833 he succeeded in showing that every known effect of electricity – physiological, magnetic, luminous, calorific, chemical and mechanical – may be obtained indifferently either with the electricity which is obtained by friction or with that obtained from a voltaic battery. Henceforth the identity of the two was beyond dispute.[2]

In *De Magnete,*[3] Gilbert had called attention to a number of differences between electricity and magnetism, which Whittaker summarizes succinctly as follows:

> Between the magnetic and electric forces Gilbert remarked many distinctions. The lodestone requires no stimulus of friction such as is needed to stir glass and sulphur into activity. The lodestone attracts only magnetisable substances, whereas electrified bodies attract everything. The magnetic attraction between two bodies is not affected by interposing a sheet of paper or a linen cloth, or by immersing the bodies in water; whereas the electric attraction is readily destroyed by screens. Lastly, the magnetic force tends to arrange bodies in definite orientations; while the electric force merely tends to heap them together in shapeless clusters.[4]

Thus there were strong grounds for believing that electricity and magnetism constituted distinct subjects for investigation, for which different explanations were to be given. Nevertheless, in spite of these clear differences, reasons accumulated over the succeeding two or three centuries for *suspecting* that these differences might prove to be superficial, and that there was some deep relationship to be found between electrical and magnetic phenomena. These considerations were very varied in kind; among them were ones like the following:

> The suspicion [in the eighteenth century] was based in part on some curious effects produced by lightning, of a kind which may be illustrated by a paper published in the *Philosophical Transactions* in 1735. A tradesman of Wakefield, we are told, "having put up a great number of knives and forks in a large box, and having placed the box in the corner of a large room, there happen'd in July, 1731, a sudden storm of thunder,

lightning, etc., by which the corner of the room was damaged, the Box split, and a good many knives and forks melted, the sheaths being untouched. The owner emptying the box upon a Counter where some Nails lay, the Persons who took up the knives, that lay upon the Nails, observed that the knives took up the Nails."[5]

Subsequent investigations by Franklin, Oersted, Ampère, and Faraday (among others) of the relationships between electricity and magnetism culminated in the synthetic theories of the latter half of the nineteenth century, among which Maxwell's theory was triumphant. It is clear that in some sense Maxwell's theory provided an "explanation" of (a theory explaining) the phenomena of electricity and magnetism; nevertheless, as we shall see (and also for reasons whose character we shall see), that "theory" became part of a larger body of information which called for a further, deep explanation.

The phenomena of electricity, particularly in the nineteenth century, also came more and more to be connected with chemical phenomena; and through this association of electricity with chemistry, the suspicion – growing gradually into an expectation or even a demand – arose that a unified theory of electricity and matter should, in some form, be sought. Indeed, as time progressed, this form itself began to be clearer: the unity should be sought in the structure of the atom. Thus the areas which seemed to offer the possibility of precise clues to that structure – particularly the spectra of the chemical elements and the periodic table – became crucial areas of investigation. Lines of potentially fruitful research in the quest for the expected unifying theory began to be generated.

The relations between electricity and light underwent a similar development. Faraday's demonstration of the effect of a magnetic field on the plane of polarization of a light ray provided one sort of consideration leading to the belief that there was a deeper relationship to be sought between magnetism and hence, because of the developments summarized above, electricity and light.

Even in such a cursory survey, dealing with such familiar areas of scientific investigation as electricity, magnetism, chemistry, and light, it becomes clear that the grounds for considering the elements of each such domain as constituting a *unified* subject-matter, and as exhibiting relations with other domains which lead to the formation of larger domains, are highly complex. The general situation, in these cases at least, may be summarized as follows. Although in more primitive stages of science (or, perhaps better, of what will become a science), obvious sensory similarities or general presuppositions usually determine whether certain items of experience will be considered as forming a body or domain, this is less and less true as science progresses (or, one might say, as it becomes more unambiguously scientific). As part of the growing sophistication of science, such associations of items are subjected to criticism, and often are revised on the basis of considerations that are far from obvious and naïve. Differences which seemed to distinguish items from one another are concluded to be superficial; similarities which were previously unrecognized or, if recognized, considered superficial, become fundamental. Conversely, similarities which formerly served as bases for association of items come to be considered superficial, and the items formerly associated are no longer, and form independent groupings or come to be associated with other groups. The items themselves often, in the process, come to be redescribed, often, for scientific purposes, in very unfamiliar ways. Even where the earlier or more obvious associations are ultimately retained, they are retained only after criticism, and on grounds that go beyond the mere perceptual similarities or primitive uncritical presuppositions which formed the more obvious bases of their original association.

An important part of gaining an understanding of science, and of the nature of scientific theories in particular, must be to examine the character of such grounds for the establishment of domains. But this question immediately leads to another: for in order that the area constituted by the related items be an area *for investigation*, there must be something problematic about it, something inadequate in our understanding of it. A domain, in the sense in which that term will be used here, is not *merely* a body of related information; it is a body of related information about which there is a problem, well defined usually and raised on the basis of specific considerations ("good reasons"). In addition, that problem must be considered important (also on reasonable grounds, not on the basis of some "subjective value judgment"); it must be worth making the effort to resolve. Further, as we shall see, it must – in general, though in certain rather well-circumscribed sorts of cases not necessarily – be capable of being "handled" at the current stage of science.

In earlier stages of the development of a field, curiosity, general puzzlement, or general uncritical presuppositions undoubtedly play a predominant role in generating such problems – in determining whether an area is problematic and worth investigating. With the advance of science, however, and indeed constituting in part the very notion of scientific progress, the considerations leading to certain questions being raised about a domain, and to considering them important, become more specific, precise, and subject to constraints: the generation of problems and priorities becomes more a matter of reasoning. Investigation of the character of this reasoning constitutes another aspect of the attempt to gain an understanding of science. More specifically, and more directly relevant to our present concerns, such analysis will help us to understand the nature and function of scientific theories. For as we shall see, different sorts of questions, reflecting different sorts of inadequacies, can arise, even regarding the same domain; and we shall find that only certain of these sorts of questions are answered in terms of what are commonly and appropriately referred to as "theories."

A further aspect of scientific reasoning which arises at this point concerns the ways in which, at least often when problems arise at sophisticated stages of science, it is possible to formulate promising lines of research to pursue in attempting to answer those problems, and sometimes, to assign at least rough rankings of degree of promise to such lines of research, thereby establishing research priorities. Further, expectations – sometimes amounting to demands – arise regarding the character of any satisfactory solution to those problems, even in advance of having any such solution (and therefore certain proposed solutions are judged as more or less "attractive"). And in many cases these expectations or demands do succeed in anticipating the actual solution ultimately arrived at; but whether they do or do not, reasons are given supporting the expectations, and the character of those reasons must be analyzed. (In addition, of course, the patterns of reasoning in cases in which the actual solution is *not* in agreement with those expectations or demands must also be analyzed.) Again, in primitive stages of a science, such conclusions are no doubt based primarily on general and uncritical presuppositions.[6] But it is characteristic of sophisticated science that it tends more and more to depend on reasoned arguments for the generation of lines of research, the judgment of them as more or less "promising," and the generation of expectations regarding the character of possible answers to problems. An understanding of such reasoning is essential if we are to understand the nature of science, and in particular of scientific theories.

We may summarize the preceding discussion in terms of five major questions.

(1) *What considerations (or, better, types of considerations, if such types can be found) lead scientists to regard a certain body of information as a body of information – that is, as constituting a unified subject matter or domain to be examined or dealt with?*

(2) *How is description of the items of the domain achieved and modified at sophisticated stages of scientific development?*

(3) *What sorts of inadequacies, leading to the need for further work, are found in such bodies of information, and what are the grounds for considering these to be inadequacies or problems requiring further research?* (Included

here are questions not only regarding the generation of scientific problems about domains, but also of scientific priorities – the questions of importance of the problems and of the "readiness" of science to deal with them.)

(4) *What considerations lead to the generation of specific lines of research, and what are the reasons (or types of reasons) for considering some lines of research to be more promising than others in the attempt to resolve problems about the domain*?

(5) *What are the reasons for expecting (sometimes to the extent of demanding) that answers of certain sorts, having certain characteristics, be sought for those problems*?

Clearly, a further question, not explicitly emerging from our earlier discussion, also needs to be raised:

(6) *What are the reasons (or types of reasons) for accepting a certain solution of a scientific problem regarding a domain as adequate*?

Of these six questions, only the last has been seriously and carefully examined by philosophers of science (in connection, largely, with discussions of "inductive logic"). In the present paper, however, I will focus on aspects of certain of the other five which are relevant to the understanding of the nature and function of scientific theories.

It must not be supposed, however, that complete answers to one of these questions will be wholly independent of answers to the others, so that they can be dealt with in isolation from one another in a piecemeal attack; we will, indeed, discover interdependencies. Nor should it be supposed that there are not other questions which must be dealt with in the attempt to understand the nature of science – though I would argue that the above are very central ones. But in any case, a complete understanding of scientific reasoning would include a treatment of all these questions.[7] And a complete and systematic investigation of all of them would serve as a test (and at the same time, as seems to be the case in so many scientific investigations, a clarification) of three general assumptions about science, each successive one of which constitutes a stronger claim than, and presupposes the truth of, its predecessors. Traditionally, these three assumptions have been widely accepted (though not without occasional and recurrent opposition) as true of science; today, however, they are being subjected to powerful attack, in one form or another, from a number of quarters. The assumptions may be stated as follows:

I. *Scientific development and innovation are often appropriately describable as rational*. (Obviously, there is a question here of the appropriate precise sense or senses of 'rational'; however, as was indicated above, the proposed investigation itself would provide, in part at least, a specification of such a sense or senses, and thus be simultaneously a test and a clarification of the assumption. Such a procedure seems to be common in scientific inquiry, and itself requires analysis which it has never received, the classical philosophical attitude having been that "meanings" must be specified completely and precisely before any test can be undertaken – a procedure which, even if it is clear itself, is rarely if ever realized in any inquiry whatever.) I will call this the *postulate* (or *hypothesis*) *of scientific rationality*.

II. *The rationality involved in specific cases is often generalizable as principles applicable in many other cases*. (It might be supposed that the truth of II – and also perhaps of III – is a necessary condition for calling a subject or method "rational." However, at least one writer has in effect denied II while apparently maintaining I: "The coherence displayed by the research tradition ... may not imply even the existence of an underlying body of rules and assumptions."[8] That is, science, in its "normal" stages, operates according to an underlying rationale, even though any attempt to express that rationale in terms of explicit *rules* is, the author claims, usually [and perhaps always and necessarily] doomed to failure. Because of this possibility, it seems advisable to keep these postulates separated.) I will call this the *postulate* (or *hypothesis*) *of the generalizability of scientific reasoning*.

III. *These principles can in some sense be systematized*. This may be called the *postulate* (or *hypothesis*) *of the systematizability of scientific reasoning principles*.

The present essay can deal only with some aspects of the six questions listed above, and, in fact, will deal only with those aspects which are relevant to the concept of a scientific *theory*. It cannot pretend to deal with all aspects of that concept (or with all uses of the term 'theory'). Even within these limitations, it cannot claim completeness: it is no more than a preliminary and, in many cases, only a suggestive rather than thoroughly detailed examination of the issues.

## II. ASPECTS OF THE CONCEPT OF A DOMAIN

We have seen that, in science, items of information come to be associated together as bodies of information having the following characteristics:

(1) The association is based on some relationship between the items.
(2) There is something problematic about the body so related.
(3) That problem is an important one.
(4) Science is "ready" to deal with the problem.

Earlier, I called bodies of information satisfying these conditions *domains*. We will find that, in science, such bodies of information have other characteristics besides these four. On the other hand, we will find that, although it is generally considered desirable for (4) to be satisfied (to an appropriate degree) in order that an area can count as fully "scientific" (that is, that it can count as a "domain" in the present sense), nevertheless areas which satisfy conditions having to do with (1), (2), and (3) are often counted as "scientific" (domains) in a somewhat borderline sense even if they fail to satisfy (4). Similar qualifications will be found necessary in regard to (3).

Needless to say, these four characteristics of domains require further clarification. The remainder of this section will be devoted to discussion of aspects of (1), (3), and (4) – the concepts of "domain," "importance," and "readiness" – which are directly relevant to analysis of the concept of "theory" in science. The character of problems regarding domains (2), and in particular of problems whose answers are theories, will be discussed in Section III together with related issues concerning the generation of expectations about those answers and of lines of research which indicate promise of producing such answers.

The present essay will not provide a systematic examination of reasons for associating items into domains, or for associating domains into still larger domains. Our purpose here being the analysis of the concept of "theory," we will take for granted the *existence* of such bodies of information, and concern ourselves primarily with the generation of their problematic character, with a view to understanding how problems arise which require "theories" as their answers. A few comments about the generation of domains are necessary here. First, the mere existence of *some* relationship between items or types of items is not itself sufficient to make that body of items an object for scientific investigation. After all, not much ingenuity is required to find some relationship between any set of items. But in the case of those associations in science for which I will reserve the name "domains" the associations are (for reasons which will not be examined here) well grounded, though "well groundedness" is certainly a matter of degree.

Not only are the relationships well grounded; they are also distinguished, among the total set of relationships discoverable or constructible between any body of items, by being what we may call (in keeping, I think, with the frequent usage of scientists) "significant." That is, relative to the state of science at a given time, the relationships of importance in science – those which are the bases of domain-generation – serve as good reasons for suspecting the existence of further relationships, and also of more comprehensive or deeper relationships.[9] (The sense of "deeper" here will be examined in Section III below.) And the more relationships are found between the items of the domains, or between two domains between which some relations have been found, the stronger the suspicion will be that there is a more comprehensive or deeper relation to be found. The significance of the relations, like their well groundedness, is a matter of degree: just as the relationships which form the basis of association of the domain may be only suspected, so also their suggestion of a deeper or more comprehensive relationship may be. Or the indication may, in either case, be stronger than a mere suspicion.

This makes it clear that the suspicion of further or deeper unity involved in considering a body of information to be a domain is itself a hypothesis that may turn out to have been mistaken. In other words, that a body of information constitutes a domain is itself a hypothesis that may ultimately be rejected: either what were originally supposed to be adequate grounds (relative to the state of science at the time) for considering the information to constitute such a unified domain should not have been as compelling as they were taken to be, or else new information, uncovered later, so altered the situation that, relative to that new state of science, the grounds for suspecting disunity (for disregarding the earlier grounds) come to outweigh those for suspecting unity. In either case, it would be found that the hope for a unifying account must be abandoned, and the previous domain, so to speak, split. Such cases are common in the history of science: although we find a general trend toward unification, the history of science does not consist of a steady, unwavering march toward greater and greater unity.[10]

A further aspect of the concept of a domain may be obtained by looking at domains from a somewhat different point of view. So far we have seen the domain as a body of information concerning which a problem has arisen. However, this description may be inverted: *the domain is the total body of information for which, ideally, an answer to that problem is expected to account*. In particular, if the problem is one requiring a "theory" as answer, the domain constitutes the total body of information which must, ideally, be accounted for by a theory which resolves that problem. (The notion of "accounting for," and the force of the qualification implied by the word "ideally," will be discussed later.)

The preceding point makes it clear that the concept of a domain is intended to replace the old "observational-theoretical" distinction as a fundamental conceptual tool for illuminating the nature of science. "That which is to be accounted for" includes, especially at sophisticated stages of scientific development (and thus at those stages where characteristically *scientific* reasoning is most apparent and predominant), elements from both the traditional "observational" and "theoretical" categories. As conceived by philosophers of science, the distinction between "observation" and "theory" has proved to be unclear, partly because what they considered "observational" is found, in actual scientific usage, to be "theory laden," but also – a point not usually emphasized by critics of that tradition – because what they considered to be "theories" are often treated in science the way "facts" or "observations" ("the given") are. But furthermore, not only did the distinction, as conceived by philosophers of science, prove unclear, but also, even to the extent to which it was clear, it provided little or no insight into scientific reasoning, but even, in its philosophical form, obscured the character of that reasoning.[11] (None of this is to say that there are not important uses of the terms "observation" and "theory" in reference to science, but only that the analyses given by philosophers of science have in general failed to capture those uses.) I have argued elsewhere[12] that the distinction may have marked, rather confusedly, another problem of philosophical interest, the "epistemological" problem of the "ultimate foundations of knowledge." The solution (not yet forthcoming) of that problem would be at best only marginally relevant to the understanding of science, and would, indeed, itself be facilitated by solutions to the problems stated in Section I above.

The present analysis will recognize fully the *mutual* interdependence of "observation" and "theory," rather than considering observational and theoretical terms or propositions to be of radically distinct types. For us, what was, at a certain stage of science, a "theory" answering a problem about a domain can, at a later stage, itself become a domain (or, more usually, a part of a domain) to be investigated and accounted for. Thus Maxwell's theory, having achieved a successful unification of electricity, magnetism, and light, itself became, for reasons which will be examined later, a part of a larger domain. To put the point in another way which is natural to scientists but not to philosophers, what was a "theory" becomes a "fact" – but a "fact" concerning which there is a problem.[13] Conversely, "observations" in science are "theory laden" in certain clear ways, though, as I have argued elsewhere,[14] the objectivity of science is not thereby endangered.

Sometimes, indeed, the items of a domain are purely "theoretically determined" (which is not to say that they are, in such cases, "theories," at least in an unambiguous sense), in that no instances of them have actually been found. (They are "hypothetical entities" or "theoretical entities," but not in any of the usual senses of the positivistic tradition.)[15] Such was the case with neutrinos in the 1930s, neutron stars until the end of the 1960s (when they were found), tachyons (putative particles traveling faster than light), "superstars" (putative "stars" with masses in the range of $10^5-10^8$ suns), and intermediate bosons.[16]

It is because theories and theory-determined entities, for example, can be or are parts of domains that I have preferred to speak of "items" or "elements" of domains rather than in more traditional terms like "facts," which have associations that make them unsuitable for covering the sorts of things that go together to form objects of scientific investigation. My talk of items of "information" (rather than, say, of "knowledge") is also partly motivated by the occurrence of theory-determined entities whose existence is merely proposed as a possibility; further reasons for that choice of terminology, as well as some dangers inherent in it, will appear later.

The account given thus far emphasizes that the items of a domain are generally problematic not in isolation but rather through being *related* to other such items in a body about which the problem exists: scientific research is in such cases generated by items in association with one another, rather than by "facts" in isolation from one another. Philosophers of science, working within a tradition according to which all "facts" are "atomic," have been blinded to this primary source of scientific problems.

Nevertheless, though such cases are extremely common, they are not the only sources of research-generating problems in science. However, a great many other types of such problems can be treated against the background of an understanding of these. For example, there are cases which are naturally described as cases involving "*a* (single) puzzling fact or observation," in which one isolated "fact," or a single class of such facts, unrelated to any others in a domain-like class, are problematic. A paradigm case is that of the peculiar radio source discovered in the constellation Vulpecula in 1968, which was found to emit radio signals with an extremely regular and short repetition period (later, when more such objects were discovered, they were christened "pulsars"). Such cases, when they concern single, isolated facts or observations, are clearly not problematic by virtue of being members of a class of related objects. However, their problematic character arises in the same way that, we will find, many problems about domains become specific and research becomes directed: by virtue of what we will call "background information," in the light of which such isolated facts as those presently being considered appear puzzling.[17] Still other types of research-generating problems, having to do with inadequacies of theories (rather than of domains), will also be found to be interpretable *via* the concepts of "domain" and "background information."

There are certain sorts of problems, and associated research, which arise regarding domains themselves, in the sense that the solutions to those problems are not "theoretical." These have to do with, for example, determination of the extent of the domain (extending it by incorporating further items, or establishing relationships with other domains, or narrowing it in corresponding ways), or refinement of the precision of definition or measurement of the properties and relationships (particularly the domain-generating relationship) of its items. Generally, however, such investigations are not undertaken for their own sakes: to do so is to engage in purely "hack" work. They are considered appropriate to undertake only when other circumstances create the inadequacy (as when new technological developments make possible a significantly greater precision in measurement of the items or extension of the domain), or when a *theoretical* problem regarding the domain has come to be regarded as important. Recognition of the importance of the domain almost invariably leads to intensive efforts to make the domain more precise, and to determine its extent more accurately. Such activity often eventuates not only in greater precision with regard to the known elements, but also in the discovery of new ones, and even, on occasion, in a shifting of the characterization of the domain itself. Such shifting may, in turn, lead to a restatement (instead of a mere refinement) of the problem which initially gave importance to the investigation of the domain. Again, however, such investigations are not undertaken (or, more accurately, are not considered important research) unless there is reason to suppose that a theoretical problem regarding the domain is important, and, further, that science is "ready" for the investigation (at least in certain respects, to be discussed in the following section, not related to the clarity and precision of the domain itself). To clarify these points, let us turn to the notion of "importance" and "readiness."

The fact that there is reason to suspect a deeper relationship between the elements of a body of information is not in itself sufficient ground for undertaking a systematic investigation of it; the domain, and the theoretical problem regarding it, must be judged as to "readiness" and "importance." Such judgments are based on a number of factors. For one thing, the grounds for suspicion of the existence of a domain-generating relationship may be more or less strong; and one factor determining whether an investigation of the domain is, at some stage of scientific development, feasible or worthwhile is the strength of these suspicions, and therefore the clarity, precision, and significance of the formulation of the domain and the theoretical problem regarding it. "Readiness" for investigation is determined partly by the precision and clarity of the domain, or at least by the precision and clarity achievable with current instrumentation, as well as by the precision with which the problem regarding the domain is stated. As was indicated at the end of the preceding section, there are also factors having to do with "readiness" which are not related to the clarity and precision of the formulation of the domain and its theoretical problem. For example, the existence of promising lines of research enters into judgment of "readiness": the domain is more ready for investigation if ways can be seen which promise solution of the problem. Mathematical techniques for dealing with the problem should also be available. As to "importance," this is largely a matter of the relation of the domain and its (theoretical) problem(s) to other domains which there is reason to believe are related to the one under consideration in such a way that there is a higher domain of which this and the others are parts. If there is reason to suppose that there is such a "higher" domain, so that a unitary theoretical explanation of the whole body appears called for, and if, for example, the domain under consideration is the "readiest" for investigation in the preceding senses, or if there is reason to think that the domain under consideration offers the clearest clues to such a theoretical explanation, then it is considered "important."

At the beginning of Section II above, it was mentioned that in some cases domains which are "important" enough need not be "ready" in order that their investigation be considered reasonable and appropriate at a certain stage, and conversely. Briefly, if the problem is deemed sufficiently important, and has achieved a certain level of precise formulation, then investigation of the area in question is often considered at least marginally appropriate despite the "unreadiness" of current science in other respects to deal with the problem. Such conditions may, indeed, as was noted previously, serve as incentives to try to *make* the state of science ready to deal with the problem: research will be generated to increase the precision of data about the domain, to develop technology for doing so, and so on. (However, in some circumstances, especially when such technological developments do not seem feasible, investigation even of domains which are recognized as important will often be looked down on as excessively "speculative," sometimes to the point of being "unscientific.") On the other hand, if the domain-generating relationship and the problem are sufficiently precise, and if current science is in other respects ready to deal with the problem, then, even though the problem is not, relative to the state of science at the time, a very important one, or not clearly important, its investigation may be considered appropriate – though such investigations may be looked on as uninteresting or even as "hack" work. Such research can, of course, *eventuate* in important results.

## III. THEORETICAL PROBLEMS, LINES OF RESEARCH, AND SCIENTIFIC THEORIES

We have examined, to the extent that is relevant to present concerns, the concept of a domain. In doing so, we have made a distinction between those problems which are concerned with the clarification of the domain itself, and other problems calling for a "deeper" account of the domain. The former type of problems will henceforth be referred to as *domain problems*; the latter I will call *theoretical problems*, inasmuch as answers to them are called "theories." Later, a third general class of scientific problems will be distinguished, which are problems regarding theories themselves; these I will call *theoretical inadequacies*. In the present section, I will consider theoretical problems and the characteristics of answers to them.

It will not, however, be possible to give a complete classification and analysis of types of such problems, and therefore of theories; I am not sure that such complete classification can or should be given. I will focus on types of theories that are, in some natural sense, paradigmatically *explanatory*, and, yet more specifically, on two types of such theories, which I will call (a) *compositional* and (b) *evolutionary*. The types of problems calling for such answers will be similarly designated. To anticipate briefly, a compositional problem is one which calls for an answer in terms of constituent parts of the individuals making up the domain and the laws governing the behavior of those parts. The parts sought need not be "elementary," though in the cases to be considered here, they are.[18] Evolutionary problems, on the other hand, call for answers in terms of the development of the individuals making up the domain; paradigm examples are the Darwinian theory of biological evolution, theories of stellar evolution, and theories of the evolution of the chemical elements. There are types of theories which do not fit into these two categories, at least not easily: the "theory" constructed on the basis of the fundamental postulates of thermodynamics is an example, as is the special theory of relativity. But the two types to be considered are centrally important, and I believe that an understanding of them is a prerequisite to an understanding of other types represented by these and other examples. At any rate, they will illustrate certain general features of "theoretical problems" and the "theories" which are answers to those problems. I will begin with three case studies, from which such general features, as well as specific features, of these two types of theories will emerge.

## 1. *The Periodic Table of Chemical Elements*

A type of domain whose analysis will prove particularly fruitful are *ordered* domains: that is, domains in which types of items are classified, and those classes themselves arranged in some pattern, for example, a series, according to some rule or ordering principle. The series may (as in the case of the periodic table of chemical elements) or may not (as in the case of spectral classifications of stars, for example, the Morgan-Keenan-Kellman system) be periodic (repeating). Orderings of domains are themselves suggestive of several different sorts of lines of further research. As we would expect from our earlier discussion, some such problems (and associated lines of research) have to do with clarification and extension of the domain: for example, refinements of measurements of the property or properties on the basis of which the ordering is made, with a view to refining the ordering; or the search for other properties which vary concomitantly with those properties.[19] Answers to such problems are not what one would naturally call "theories." (However, one might "hypothesize" that a certain property varies concomitantly with the ordering property, even though the former may not be directly measurable with current techniques, and even though there is no very clear theoretical basis for the hypothesis, though the hypothesis seems needed on other grounds. In such case, we might find the word "theory" being used.) Nor does the fact that the ordering sometimes allows predictions to be made (for example, predictions of new elements and their properties on the basis of the periodic table) turn such ordered domains into "theories."[20] (In particular, the periodic table is not "explanatory," even though predictions can be made on its basis alone.)[21]

On the other hand, there are other kinds of problems and lines of research suggested by ordered domains which are concerned with attempts to construct "theories." The mere existence of an ordering relation, and still more, of a periodicity in that ordering, raises the question of accounting for that order; and the more extensive, detailed, and precise the ordering, the more strongly the existence of such an account is indicated.[22] But further, the existence of such an ordering tends to make some properties appear more significant than others (for example, atomic weight), and those properties are then looked upon as furnishing "clues" to the discovery of the presumed deeper account.

These and other features of ordered and periodic domains are illustrated by the case of the periodic table.

By the early 1870s, it had become finally clear that if the chemical elements were arranged in a table ordered according to atomic weights, and if due allowance were made for undiscovered elements, then certain periodicities in the properties of the elements (as well as certain "horizontal relationships") would be revealed.[23] Many investigators refused to believe that any further problem was raised by these relationships; to them, the "elements" were truly fundamental, not composed of anything more elementary. This was the case particularly among working chemists, to whom the "atom" of the physicists had never been at all useful, and indeed appeared to be mere speculation.[24] However, the relationships between the elements, reflecting both order and periodicity, indicated to many that there was some more fundamental composition of the elements.[25] So extensive, detailed, and precise were those relationships that even the existence of exceptions, in which the order or the periodicity, one or the other, had to be violated if the other were satisfied, could not shake the conviction of an underlying composition which would, ultimately, remove even those anomalies. (It is perhaps significant that, for a long period, only one such anomaly was known which appeared anywhere near troublesome.[26]) Hence the conviction grew that the periodic table was to be given some deeper explanation; and in particular, since the fundamental ordering factor, the atomic weight, increased by discrete "jumps" (which were, in most cases, rather close to integral values) rather than by continuous gradations, that deeper explanation was expected to be in terms of discrete components. Thus that composition was to be understood in terms of constituent *massive* particles (whose step-by-step increase in numbers was reflected in the increases of atomic weights which furnished the ordering principle of the table), and the structure of which involved repetitions at various intervals (reflected in the periodicity of *other* properties of the domain which were "significant," that is, which were related in the periodicity of the table).

This expectation increased in strength until it reached the status of a *demand*, reinforced by the following considerations: (a) more and more other areas revealed themselves as domains in which an atomistic explanation was expected (for example, the case of chemical spectroscopy, to be discussed below); (b) atomistic explanations became more and more successful (statistical mechanics) or at least more and more promising (Kelvin's vortex atom was applied in a great many areas) in other domains; and (c) reasons accumulated, as shown in Section I, for suspecting the domain under consideration (the chemical elements related through the periodic table) to be itself related, as part of a larger domain, to others, including ones in which atomistic explanations were either expected, demanded, or actually provided. All three of these sorts of considerations were, of course, open to question: (a′) the expectations of atomistic explanations in other areas could be criticized as not being sufficiently well founded, and in any case as only expectations; (b′) the atomistic explanations advanced in other areas were perhaps not as completely successful (statistical mechanics still had its difficulties, and the vortex atom was in no case applied in precise mathematical detail[27]) or necessary (perhaps "phenomenological thermodynamics" was all that was needed in the domain of heat) as was claimed; and (c′) although some relationships between the domain under consideration and others did exist, there were also differences which might obviate the suspicion that those domains should be looked upon as parts of a higher domain requiring a unitary and, in particular, compositional explanation. But the considerations leading to the expectation of a compositional theory to account for the periodic table were strong enough to constitute reasons in favor of a search for such a theory; that expectation, and the research which it guided, were not shaken substantially by such considerations as (a′) to (c′), any more than they were disturbed by the tellurium-iodine anomaly. Nor were they shaken by the failure of successive atomistic theories to account successfully for the features of the periodic table, and still less by the opposition of the energeticists and the positivistically minded philosophers, to whom *any* atomistic explanation was unscientific. Considerations (a) to (c), when added to the indications of the periodic table itself (the domain constituted by the periodic table, that is), were strong enough to constitute reasons in favor of a search for such a theory. That the reasons were not logically conclusive did not make them any the less reasons, and good ones, relative to the state of science at the time, nor did it make action in accordance with them any the less rational.

This discussion is clearly generalizable as a principle of reasonable scientific research:

*To the extent that a domain D satisfies the following conditions or some subset thereof, it is reasonable to expect* (*or demand*) *that a compositional theory be sought for D*:

(Ci) *D is ordered*;
(Cii) *the order is periodic*;
(Ciii) *the order is discrete* (that is, based on a property which "jumps"

in value from one item to the succeeding one), *the items having values which are* (with the limits of experimental error) *integral multiples of a fundamental value*;

(Civ) *the order and periodicity are extensive, detailed, and precise*;[28]

(Cv) *compositional explanatory theories are expected for other domains*;

(Cvi) *compositional theories have been successful or promising in other domains*;

(Cvii) *there is reason to suppose that the domain under consideration is related to such other domains so as to form part of a larger domain.*[29]

Although this principle can and will be generalized still further in a number of respects, I will refer to it (or to the more generalized version of it) as the *principle of compositional reasoning*.

In the light of this principle, one might even speak of the "degree of rationality" involved in pursuing a search for a compositional theory: the pursuit of a certain line of research in the expectation of finding a compositional theory for a given domain is more rational, the more of points (Ci) to (Cvii) are satisfied, and the more each of them is satisfied. And the demand that any acceptable theory accounting for the periodic table be a compositional theory is also the more rational, the more of (Ci) to Cvii) are satisfied, and the more each of them is satisfied. (Such demands are therefore not "dogmatic" in the sense of being irrational, even though the explanation ultimately accepted might not be in accord with those demands. Of course, the extent to which the demands are justified should always be appreciated.) This notion of "degree of rationality" will be found also applicable to other kinds of scientific research and expectations, theoretical or otherwise.[30]

The *problem* regarding the domain, in the case of the periodic table, then becomes not merely to give an account of the domain, but more specifically, to give a compositional theory for it. Lines of research which it is reasonable to pursue are indicated by the character of the theory expected on the basis of the characteristics of the domain, and are made still more specific by other considerations to be discussed later ("background information").

## 2. *Spectroscopy*

Some of the complexity of reasoning in science, however, begins to reveal itself when we investigate another area of late nineteenth-century science which was widely expected to eventuate in a compositional theory. Almost from its inception, spectroscopy was aimed not only at the identification of chemical elements through their characteristic spectra, but also at the development of an explanatory theory having to do with atoms or molecules. As early as 1836, the British spectroscopist Talbot had written: "The definite rays emitted by certain substances, as, for example, the yellow rays of the salts of soda, possess a fixed and inviolable character, which is analogous in some measures to the fixed proportion in which all bodies combine according to the atomic theory. It may be expected, therefore, that optical researches, carefully conducted, may throw some additional light upon chemistry."[31] And half a century later, Rydberg, one of the leading figures in the field, maintained that spectral data "relate to the motions of the least parts of matter, the atoms themselves, in a way that we can expect . . . to find the most simple functions to express the relations between the form of moving bodies, their dimensions, and the active forces."[32] Hence studies of those data could be expected to lead to "a more exact knowledge of the nature of the constitution of atoms."[33]

Thus, throughout the history of the subject, an intimate connection was presumed to obtain between the characteristic lines of the spectra of chemical elements and the characteristics (usually associated in this case, especially in the nineteenth century, with vibrations)[34] of ultimate constituents of chemical substances. So widespread was this belief that one authority on the history of spectroscopy declares that "I have found no spectroscopist who did not admit an atomic theory. As we have seen, 'understanding spectra' became almost synonymous with 'understanding atoms and molecules,' and to spectroscopists atoms and molecules were as real as spectra themselves. Thus the end of the century's anti-atomistic movement, which had its origin in the application of thermodynamics to chemical phenomena, could only have met with resistance from spectroscopists. It had apparently no influence on spectroscopy."[35]

And yet it was not until 1885 that Balmer discovered the first clear and unambiguous ordering relationship between any spectral lines. Thus, *contrary to the case of the periodic table, the conviction that a compositional theory would be found for the spectral domain was not based on ordering relationships, but preceded their discovery*. On what, then, was the expectation based that spectra would be accounted for in terms of a compositional theory? Undoubtedly it had its roots partly in considerations (Cv) to (Cvii) above (having to do with expectation or achievement of success by compositional theories in related domains), together with analogues of (Ciii) and (Civ).

Although there was no *ordering* of lines in the sense of there being a known formula relating them (or even a qualitative expression of reasonable generality and precision, of their arrangements) and measurements of their positions were not related to one another by integral multiples of some fundamental value, nevertheless the lines of elements are in general discrete (though not always sharp), and furthermore maintain their relative positions, as well as a number of other characteristics (for example, under similar conditions, intensity and degree of sharpness), with great preciseness. Thus Rydberg could adduce as a reason for investigating spectra not only that they related to atomic motions, but also that spectral data were "without comparison the richest and most uniform of all relating to all of the known elements."[36]

But a further consideration, which did not emerge from our study of the case of the periodic table, enters into this case: namely, the existence, in another domain, of a way of approaching problems which, by analogy, could offer promise here. Almost from the outset, it was felt by many that the key to success in this domain lay in constructing a theory of atomistic (or molecular) vibrations on the analogy of sound: the various spectral lines of an element would prove understandable as harmonics of a fundamental vibration and could be revealed by an appropriate Fourier analysis. (Indeed, rather than believing that the discovery of an ordering formula would provide a clue to construction of an atomic or molecular theory, many spectroscopists before Balmer believed that the key to discovery of that ordering formula lay in a consideration of the harmonics of fundamental atomic vibrations.) This case differs from ones in which a theory is *related*, or for which there is good reason to suspect is related, as part of a larger domain to the one under consideration.[37] In the present case, the domain from which the analogue approach is taken need not be so related, that is, what relations there are between it and the domain under consideration are either very general, very tenuous, or not very "significant." Furthermore, what is borrowed is not necessarily the theory of the original domain, but rather an analogue thereof: at best, it is possible to speak only of *adapting* that theory, not of *applying* it without alteration (as one would attempt to do if expecting to unify two related domains). (Of course, such adaptation often involves *interpreting* the current domain so as to make adaptation or application of the imported approach possible; in the present case, the lines had to be looked at – and there was good ground for such interpretation – as records of wavelengths of vibrations, and the explanatory entities, therefore, as vibrators.) Indeed, the adaptation may not, in its original application, even have been compositional. Nor, in fact, need it have been a "theory"; it may have been merely a mathematical technique or a way of approaching a problem. Needless to say, such analogical adaptations are available in the case of ordered domains also.

Thus the present case allows us to give a more general formulation of conditions (Ciii) and (Civ) for compositional theories, and to add a further condition (Cviii). In the present case, of course, with the qualification to be mentioned below, conditions (Ci) and (Cii), having to do specifically with ordered and periodic domains, are inapplicable:[38]

(Ciii′) *the items of the domain have discrete values which are preserved (at least under similar conditions) from situation to situation (or, their relations to one another are preserved from situation to situation), even though no general formula (or qualitative principle) expressing the relations of those values is known.* (The existence of the preserved relations is, of course, a rational incentive to suspect the existence of such a general formula and to search for it. In this sense, (Ciii′) may be seen as implying an analogue of (Ci), namely: (Ci′) *there is reason* (in this case, in the preservation of relations of lines) *for supposing an ordering formula to exist*. If such a formula is found, as in the case of Balmer's discovery, the domain becomes, to the extent that the formula deals with a proportion of the totality of items (lines) of the domain, ordered. If the formula does not yield values of the items which are integral multiples of a fundamental value, then the rationale for expecting a compositional theory will be more complex, depending in general on (Cviii), below.

(Civ′) *a number of features of an extensive range of items of the domain are open, with techniques available, to detailed and precise description or measurement.*

(Cviii) *a theory (or, more generally, a technique or method) which has been successful or promising in another domain (even though, unlike the case of* (Cvii), *that domain is not related or suspected of being related as part of a larger domain to the domain D under consideration) shows promise of being adaptable, with an appropriate interpretation of the items of D, to D.*

In spite of the fact that this and other forms of reasoning take place in science, there are often alternative lines of reasoning available which lead to different conclusions; and the issue of which line has the strongest arguments in its favor is not always clear-cut. Thus, whether a certain body of information constitutes a domain or not; whether a certain item is or is not a part of a domain (that is, whether or not a theory for that domain is responsible for accounting for that item); the extent to which a certain problem is important; the extent to which the state of science is "ready" to investigate a certain problem; the degree of promise of a certain proposed line of research; whether a certain specific sort of answer to a problem is reasonably to be expected; whether a certain proposed answer to a problem is adequate – all these can be, and in any given situation in the history of science are apt to be, debated. This is not to say that such issues are *never* unambiguously clear, and still less to say that they do not, in general, become more so as science progresses: for substantive scientific knowledge, as it accumulates, imposes more and more stringent conditions on the character and interrelations of domains, the kinds of questions that can reasonably be asked regarding them, the reasons for considering those problems important and ready for investigation, the moves that it is reasonable to make in trying to answer the problems, the kinds of answers to those problems that can be expected to be found, and the conditions an answer must satisfy if it is to be acceptable. Nevertheless, in all these respects, what is maintained by scientists is of a hypothetical nature, and there is generally room for disagreement, without it being perfectly clear which side has the strongest reasons in its favor.

Thus, for example, although most, if not all, prominent spectroscopists of the nineteenth century believed with Talbot, Mitscherlich, and Rydberg that spectral lines are produced by atoms, there were some who did not believe that a study of those lines would lead, at least easily, to knowledge of the constitution of atoms. There were two distinct lines of argument by which this conclusion was arrived at. On the one hand, Kayser and Runge, two of the leading figures in the history of spectroscopy, held that, although spectral lines are ultimately produced by atoms, their study could provide us only with knowledge of *molecules*, not of atoms. Their reasons for this attitude were complex, having to do partly with experimental results and partly with conceptions of how those results could have been produced.[39] On the other hand, yet another major figure, Arthur Schuster, was pessimistic about the possibility, or at least the ease, of gaining insight through spectra even into the nature and structure of molecules, much less of atoms – not because spectra do not have their origin in the vibrations of atoms or molecules, but because spectra are too complex.

> . . . we must not too soon expect the discovery of any grand and very general law [from investigation of spectra], for the constitution of what we call a molecule is no doubt a very complicated one, and the difficulty of the problem is so great that were it not for the primary importance of the result which we may finally hope to obtain, all but the

most sanguine might well be discouraged to engage in an enquiry which, even after many years of work, may turn out to have been fruitless. We know a great deal more about the forces which produce the vibrations of sound than about those which produce the vibrations of light. To find out the different tunes sent out by a vibrating system is a problem which may or may not be solvable in certain special cases, but it would baffle the most skilful mathematician to solve the inverse problem and to find out the shape of a bell by means of the sounds which it is capable of sending out. And this is the problem which ultimately spectroscopy hopes to solve in the case of light.[40]

These examples only illustrate some of the kinds of disagreements to which science, for all its accumulated knowledge and constraints, is subject.[41] Nevertheless, the existence of such disagreements, and the frequent unclarity as to which view is correct, do not mean that no rationale exists in science and its development: the viewpoints *often*, if not always, have reasoned arguments in their favor, even if those arguments are not always telling or accepted. And the situation may, and often does, become more clear-cut with further research. The possibility of rationally based disagreements, in fact, plays an important role in science: the possibility helps to ensure that reasonable alternatives will be explored.

### 3. *Stellar Spectral Classification and Stellar Evolution*

We saw that, in the case of the periodic table, the expectation of a compositional theory arose almost immediately; few workers, however, were interested in questions of the evolution of the elements.[42] On the other hand, it was not long after the first spectral classifications of stars were published that those classifications were associated with expectations (and presentations) of an evolutionary theory explaining the classification. In fact, some of the pioneers of stellar classification were among those who presented such theories. So strong and persistent was this tendency that by the end of the century, the historian of astronomy Agnes Clerke could write that "Modes of classifying the stars have come to be equivalent to theories of their evolution."[43] It will prove illuminating to examine the roots of this difference between the case of the periodic table and that of astronomical spectral classifications.

By 1863, the work of Huggins and others had established, on the basis of spectral analysis, that the stars are composed of the same elements as are found on earth. Classification of stars on the basis of spectral features began at about the same time with the work of Secchi, Vogel, and others. The resulting classifications correlated well with the colors of the stars.

At this point, there was no clear order or sequence among the different classes.[44] However, some astronomers (notably Vogel), on the analogy of changes in cooling materials in terrestrial cases, proposed that the different colors (and therefore the correlated spectral classes) were indications of an *evolutionary sequence*. The hottest (and, on this view, the youngest) stars would be blue or white, while red stars were in their old age.[45]

Further, some writers, noting the presence of strong hydrogen lines in the spectra of the white (and therefore, on the theory under consideration, young) stars, conjoined to that theory an hypothesis about the composition of the elements: generally, either that of Prout, according to which all elements are composed of hydrogen (and whose atomic weights would therefore be expected to be integral multiples of that of hydrogen), or else some other view of the fundamental composition of the elements (for example, a modification of Prout's hypothesis, to the effect that the ultimate constituents had an atomic weight of one-half that of hydrogen, thus removing what was to the Prout hypothesis an embarrassing anomaly, namely that of chlorine, with a well-documented atomic weight of 35.5).[46] And on the basis of that combination of a theory of the composition of the elements and a theory of stellar evolution, they proceeded to develop an evolutionary theory of the chemical elements (Prout's own view having been, at least in its usually understood form, purely compositional): as stars age, higher elements are built up out of hydrogen, so that older stars are composed of a larger proportion of heavier elements.[47] This view, though it was not without its immediately obvious difficulties, found some support in the fact that, as a general trend, there are an increasing number of higher element lines as we proceed through the spectral classes, ordered according to color, from white to red.[48] Thus, in summary, the older stars were held to be red and composed of heavier elements – on the assumption, of course, that the *total* composition of a star was accurately reflected in the spectral observations of its surface. (It was not until 1921 that Saha demonstrated that the differences of spectra did not reflect even a difference of *atmospheric* composition, but merely one of temperature. The differences in chemical composition of most stars, though highly significant for interpretation of their energy production, internal structure, and evolution, are small. The assumption made by the early astrophysicists, however, was certainly the reasonable one to make at the time.)[49]

The situation seems to have been this: while the spectral classes did admit of at least a rough ordering[50] in terms of increasing numbers and intensities of heavy element lines (coupled with decreasing intensities of hydrogen lines), such orderings were not in general made by pre-twentieth-century pioneers[51] *except* in conjunction with a theory (or analogy) imported into the domain, namely, the color changes that take place with cooling.[52] Nevertheless, a sequential ordering of spectral classes on the basis of decreasing hydrogen and increasing heavier element lines could have been achieved, at least in rough fashion, even without importation of the "background information" (as I will call it) concerning colors of cooling bodies. And if it had been so achieved, its existence would of itself have served as good ground for suspecting that an evolutionary process might be involved, simply on the basis of the fact that the ordering was based on (depending on which end one cared to look from) increase or decrease of certain factors.

However, importation of the background information concerning color changes of cooling bodies made three further contributions to the interpretation of the sequential ordering as an indicator of a possible evolutionary process. First, it increased the strength of the suggestion that an evolutionary sequence might be involved by showing that the sequential ordering of the domain on the basis of lines could be correlated with a *temporal* order having to do with the cooling process. Second, it suggested a *direction* of the evolution – a direction which could not have been extracted from the ordering on the basis of lines alone, unless one made arbitrary assumptions (for example, that the evolution is from "simple" to "complex"). For, the final death of a star being a cold, burned-out state, the red stars should be the oldest. (Unfortunately, the suggestion was not clear with regard to the beginning of the sequence: were the white stars like Sirius the youngest, blazing forth suddenly at their birth, and gradually burning themselves down to a red old age? Or as Lockyer maintained, were certain red stars young, gradually heating up by some process, if not Lockyer's meteoritic one, to white maturity, after which they declined to a second red stage just preceding death?)[53] And third, by itself constituting the outline, at least, of a theory, an answer to the evolutionary problem regarding the domain, it suggested directions which research could take: directions in which the theory needed to be laid out in detail.[54]

In spite of the residual ambiguity, in this case regarding the beginning of the temporal process of stellar evolution, we can see clearly at work here two more principles of reasonable scientific research, this time applying to the (or a) way in which a problem arises, with regard to a domain, for which an *evolutionary* answer (explanatory theory) can reasonably be expected and sought:

(Ei) *If a domain is ordered, and if that ordering is one which can be viewed as the increase or decrease of the factor(s) on the basis of which the ordering is made, then it is reasonable to suspect that the ordering may be the result of an evolutionary process, and it is reasonable to undertake research to find such an answer (which we have called an evolutionary theory).*

(Eii) *The reasonableness of such expectation is increased if there is a way (for example, by application or adaptation of some background information such as a theory from another domain, whether unrelated or [preferably] related) of viewing that sequential ordering as a temporal one, and still more if a way is provided of viewing that ordering as having a temporal direction.*

Clearly, there are also analogues for evolutionary theories of conditions (Civ) to (Cviii) which were stated earlier for the case of compositional theories. I will call (Ei) and (Eii) together the *principle of evolutionary reasoning*. It should be added that (Ei) *alone* constitutes only a weak reason for undertaking research in quest of an evolutionary theory; for, without (Eii), little or no specific direction is provided for research.

Condition (Ei) was not applied in science before the second half of the nineteenth century;[55] its acceptance as a new general reasoning principle was due in no small measure to the success of Darwin's evolutionary account of biological species. The principle was, however, only gradually accepted, and this perhaps explains (together with the very real difficulties involved in the attempt) the failure, as it seems to us, of pioneers in stellar spectroscopy to try to order the classes unless they did so with an evolutionary idea already in mind. In any case, this example shows that new reasoning principles, as well as new substantive information, can be introduced into science as part of its maturation. Today the principle seems a natural one to use – so much so, in fact, that we often have difficulty in understanding why it was not applied by earlier thinkers.

Why was it, then, that condition (Ei) was applied by a considerable number of workers in the domain of astronomical spectral classification, but only rarely in that of the periodic table, where one also had a sequential ordering, and even theories, like Prout's, of the composition of the elements? (Recall that Prout's view that elements are built of hydrogen was seen by most of its adherents only as a compositional theory, not an evolutionary one.) Perhaps chemists were so used to thinking of "elements" as "always having been there" that, despite the view of some of them that the elements were *composed* of something still more elementary, they found it difficult to think of them as having been built up in a historical sense. Astronomers, on the other hand, were more used to thinking in terms of origins and development,[56] and many of them evinced a lively interest in the new biological ideas. But there is also a less speculative and sociological answer to the question: for *astronomical evolutionary theories played the same role vis-à-vis the periodic table that the "background information" regarding cooling of hot bodies played vis-à-vis spectral classification*. (The role, namely, summarized in condition (Eii), above.) That is, having interpreted spectral sequences as temporal, theories of stellar evolution could now be applied to interpret the increasing sequence of atomic weights in turn as a temporal, and as a temporally directed, one. Without this application of (Eii), the sequence of chemical elements, even seen as composed of increasing numbers of like parts, could *at best* suggest, in accordance with (Ei), that there might be something evolutionary involved. But that alone would have provided only weak incentive and, even more important, guidance in the search for such a theory. Thus the comparative absence, among chemists, of theories of the evolution of the elements is quite understandable in terms of the need of some way of satisfying (Eii) before the search for an evolutionary theory of the elements could seem attractive.[57] And if chemists also had an occupational block against seeing the use of the "Darwinian" principle (Ei), the comparative absence among them of interest in an evolutionary theory of the elements would be still more understandable.

I would not want to claim that (Ei) and (Eii) cover all the kinds of reasoning involved whenever an evolutionary theory is suspected and sought. In particular, the case of biological evolutionary theories is undoubtedly too complex to be dealt with adequately in terms of them. But the present case does illustrate one sort of reasoning pattern involved in the expectation of and search for such theories, and a very fundamental and important one at that. (The same qualifications should be understood in regard to the reasoning principles extracted from the other two cases dealt with above, dealing with compositional theories.)

We have seen, then, that certain bodies of information, related in certain ways, raise problems of various sorts, to some of which the answers that can reasonably be expected are what are called "theories." Conversely, a "theory" is what counts as a possible answer to any one of those types of questions.

It might be objected that, although the analyses given here may be relevant to a great many questions *about* theories, they tell us nothing about the "nature" of theories, which after all is the subject of the present conference. And one might suggest an analogy: if we want to know what babies are, it is irrelevant to inquire about how they arise, how they behave, what sorts of problems arise concerning them and how to deal with those problems. (The latter three questions – with reference to theories, of course, not to babies – will be discussed in the remainder of this paper.) Similarly, it might be argued that, if we want to "understand what a scientific theory is," it is irrelevant to inquire about the rationale underlying how theories (or the need or expectation of them) arise, or how they function in science, or what sorts of problems arise concerning them, or how those problems may be dealt with. The problem about the "nature" of theories – so this objection might continue – concerns the *definition* of the term 'theory' (or alternatively, perhaps, the criteria for identifying them in the first place), and so is *presupposed* by any attempt to deal with the questions raised in the present paper.

Now, I do not wish to deny that there is a problem about formulating a definition of 'theory,' or a general set of criteria for identifying theories (if indeed either of these is what philosophers who concern themselves with the problem of the "nature" of scientific theories are after). But there are a number of important points to make about that concern. In the first place, as the present paper has tried to demonstrate, there are important, different *kinds* of theories in science, so that a selection of their common features may have to be so general as to become unenlightening. (And in any case, how are we to get at such common features if we do not first recognize and try to come to grips with the different kinds of theories actually found in science?) Furthermore, it is notorious that philosophers of science have provided no generally acceptable definition of 'theory' (or set of criteria for identifying theories), and that the usages to which philosophers, especially recently, have put the term 'theory' are often so vague and ambiguous as to be scandalous.[58] Far from *presupposing* a solution to the problem of the definition of 'theory' (or of criteria of identification of theories), a more detailed study of the roots, roles, and problems of theories might in fact *help* resolve that problem. For certainly it is easier to identify paradigm cases of theories in science than – if the experience of philosophers is any evidence – it is to formulate a general definition of 'theory' or a set of general criteria for identification of theories.

But far more important, the investigation of the problem of 'theory' definition (or of criteria for identifying theories) can provide almost nothing in the way of understanding scientific theories. Here the analogy with babies may be turned against our hypothetical objectors: for what understanding of babies would we get if we arrived at a general definition of 'baby' or a set of general criteria for identifying babies? (Who has a problem about identifying babies anyway?) On the other hand, understanding of babies *is* gained through studying their origins and development, their behavior, the problems that can arise concerning them, and the ways in which those problems can be met.[59] The case is similar with scientific theories: the problem of the "nature" of scientific theories – the problem of "understanding what a theory is" – is interpreted in a far more illuminating and fruitful way in terms of an investigation of the questions indicated here than in terms of the problem of "definition" or "criteria of identification." It is no wonder that the work of philosophers of science has seemed so barren and irrelevant to actual science. For they have, too often, concerned themselves with relatively barren and irrelevant problems.

## IV. THEORETICAL INADEQUACIES AND THEIR TREATMENT

Earlier, two general classes of scientific problems were distinguished: domain problems, having to do with, for example, clarification and extension of the domain, and theoretical problems, calling for a "deeper account," a "theory," of the domain. But once a theory has been presented in answer to a theoretical problem, a further type of problem can arise, namely, problems concerning the adequacy of the theory itself. I will refer to this class of problems as *theoretical inadequacies*. These will be surveyed in the present part through an examination of Bohr's theory of the hydrogen atom (his quantum theory).[60] That theory showed the way toward resolving the problems discussed earlier in the present paper, concerning the composition of the chemical elements, the characteristic features of the periodic table, and the characteristics of the spectra of the chemical elements. However, the theory was seen very clearly to be inadequate in certain respects; and it is those early inadequacies, rather than the later ones, which led to the need for developing a new theory, with which we shall be concerned here. Our examination will focus on the following aspects of the case: what were considered to be inadequacies in the theory in its early phase, and the reasons for considering it to be inadequate in those respects; what were considered to be reasonable directions to move in trying to overcome those inadequacies; the reasons why those lines of research suggested themselves in the first place, and why they appeared plausible; and what were considered to be adequate solutions of those problems or removal of those inadequacies. Needless to say, only a few selected aspects of these questions, even in reference to this case, can be considered in this paper; other aspects of them, in reference to different cases, have been studied in Part II of my "Notes toward a Post-Positivistic Interpretation of Science," which should be considered to supplement, and in some respects to detail further, the present discussion.

In the course of examining inadequacies and their treatment in the Bohr case, we will also be led to touch on the question of the "function" of theories: the question, that is, of whether scientific theories are to be interpreted "realistically" (as claiming that atoms or electrons, for example, really exist), or as "instruments" for calculation or correlation of data (a view sometimes expressed by saying that theories, or at least the "theoretical terms" occurring in them, are "idealizations," "simplifications," "models," "abstractions," "logical constructs," and so on). However, this question, even though it is highly relevant to the understanding of the nature of scientific theories, will not be examined directly here, inasmuch as I have dealt with it in considerable detail elsewhere.[61]

## 1. *Inadequacies of the Bohr Theory*

A. *Incompleteness*. Although Bohr's theory won immediate acceptance largely through its deduction of the Balmer formula and the Rydberg constant, and its "physical interpretation" (explanation) of the hitherto unexplained ("empirical") denominators in the Balmer formula, it also gave or promised to give (by being extended to other elements besides hydrogen) explanations of a large number of other features of the more general body of information already strongly suspected to be deeply related, for example, the periodic table, and, in particular, chemical valency.[62]

Nevertheless, it was realized quite early that, even for the spectral domain, the Bohr theory offered no way to account for the intensities and polarizations of the spectral lines.[63] And thus, with respect to *this* domain (which by this time was recognized as being a subdomain of a larger body of information for which a unified account was expected), the theory can be said to have been *incomplete*, in the following sense: (1) there was no reason at this stage to suppose that the theory was not fundamentally correct; and (2) there was no reason to suppose that the theory as stated could not be *supplemented* so as to account for the intensities and polarizations.

This example, when coupled with the analysis given earlier in the present paper, suggests a general notion of the completeness or incompleteness of a theory. A domain, it will be remembered, is the total body of information for which a theory is expected to account. *A theory can, then, be judged complete or incomplete relative to that body of information*. Of course, as has been pointed out, the limits of the domain are not always well defined, and it is not always generally agreed whether a certain item is part of the domain or not. Hence debate about the completeness or incompleteness of a theory is always possible. But this is not to say that the issue is never clear.

Although the theory may be complete or incomplete with regard to some given domain, it may not even be clearly applicable to some larger domain in which its original domain has come to be included. Nevertheless, despite the fact that a domain actually undergoing investigation at a given time is usually a subdomain (on more or less strong grounds) of some broader body of related information, judgments of completeness or incompleteness of a theory are ordinarily made with respect to the specific (sub)domain for which it was designed. Nevertheless, in certain contexts, it may be judged as to completeness with respect to the total inclusive domain or some portion thereof larger than the original domain of the theory. And it is judged more highly the more features of the total domain (above and beyond those of its original subdomain) it accounts for or shows promise of accounting for (as Bohr's theory showed initial promise of being extendible to other elements than hydrogen). The stronger the grounds for supposing the original domain of the theory to be related to another domain as part of a yet larger domain, of course, the more strongly the theory is expected to account for the items of that other domain, and the more it will be likely to be judged as to completeness with respect to the more inclusive domain.

That a theory is incomplete in this sense is of course an "inadequacy"; nevertheless, it is no ground for *rejecting* the theory as false; a theory can be *incomplete*, with respect to a given domain or subdomain, without being fundamentally *incorrect*; judgments of incompleteness are not tantamount to rejection, "falsification." On the other hand, a theory can be known to be fundamentally incorrect, and yet still be useful, especially if there is no better alternative theory available; for often the limits of applicability of the theory – the location, so to speak, of the incorrectness – is known.[64] It is thus misleading and ambiguous to say that a theory is not "rejected" until a better theory is available, for the theory can continue to be *used* (and thus not "rejected" in this sense) and yet known to be incorrect (and thus "rejected" in this sense).

B. *Simplification*. We will consider two major respects in which the Bohr theory as originally presented can be said to have been a "simplification."

(1) *Bohr ignored the motion of the atomic nucleus*. On what ground could this have been alleged to be a "simplification"? The supposition was based on two considerations, deriving from two previous bodies of knowledge:

(a) Classical electricity, according to which opposite charges attract each other (and the electron is negatively charged, while the nucleus around which, in Bohr's theory, it orbits, is positively charged);

(b) Classical mechanics, according to which, if a smaller body is moving, under an attractive force, in an orbit around a larger one, the latter also will move in response to attraction by the smaller, and both bodies will describe orbits about a common center of force.

In general, such "background information" – information which is neither part of the immediately relevant domain of the theory, nor part of the theory – serves (along with a number of other functions) as a basis for the distinction between a "simplified" and a "realistic" treatment of a subject or domain. This role of background information which is itself subject to revision makes it clear that the claim that a certain idea is a "simplification" is, in science, a hypothesis, which may turn out to have been mistaken (for example, Planck's attitude toward his quantum hypothesis).[65]

(2) *Bohr ignored relativistic effects of the high velocities required of his orbital electrons*. In this case, the allegation of "simplification" is again based on "background information," this time the special theory of relativity.

C. *Structure. The structure of the nucleus*, and any influence that structure might have on the processes dealt with explicitly by Bohr, were largely ignored. (There is no name for this type of "inadequacy"; I will propose one shortly.)

In one sense, this inadequacy might be thought of as an "incompleteness" of the theory; however, there are reasons for distinguishing it from the kind of "incompleteness' dealt with in A, above. In A, we know precisely what is wrong: incompleteness is there judged relative to a fairly well-defined domain; and we know, in A, that what has been omitted from account makes a difference. In short, we know what must be taken into account in order to obtain a better (more complete) theory. But in the present case, we do not know *how* or even *whether* any internal structure the nucleus might have will have any effect on the domain at hand, even though it is necessary, in the theory, to make reference to the nucleus and to attribute some general properties to it (for example, possession of a net positive charge balancing the charges of the orbital electrons). I propose that this type of inadequacy be called *black-box incompleteness.*[66]

## 2. *Treatment of Inadequacies in the Bohr Theory*

The present paper will deal with only one of the types of inadequacies discussed above, namely, simplifications. Problems of incompleteness (and also of fundamental correctness) have been examined in "Notes toward a Post-Positivistic Interpretation of Science."

In the case of simplifications, two situations of special interest, for our purposes, arose soon after the presentation of Bohr's theory. These two situations are related to the two respects discussed above in which Bohr's theory was considered to be a simplification.

(1) *The spectrum of Zeta Puppis and the motion of the nucleus.* This was a problem with which Bohr's theory was confronted almost immediately; its nature and resolution are summarized well by Rosenfeld and Rüdinger:

> Bohr had come to the conclusion that certain spectral lines (originally discovered in the spectrum of the star Zeta Puppis), which up to then had been believed to belong to hydrogen, must really be ascribed to helium. . . . The outstanding English spectroscopist A. Fowler, who had himself discovered some of the spectral lines in question, was, however, not yet convinced: he pointed to a small but real discrepancy between the experimental results and the values found by a simple application of Bohr's formula. Now, since Bohr's assertion was based on a correspondence argument, such a disagreement, however small, would in fact mean the breakdown of the whole basis for his theory. In the subsequent discussion in "Nature" Bohr showed, however, that Fowler's objection could be answered by taking into account the motion of the nucleus around the center of gravity of the atom. By this more exact calculation Bohr was not only able to demonstrate the finest agreement between the calculated and the observed spectral lines, but he could also predict that a series of other spectral lines from helium, so far unobserved, which according to the simple theory would coincide with some of the Balmer lines, should in fact appear very slightly displaced in relation to these. The lines were discovered the following year by Evans at the places predicted.[67]

Thus the situation was as follows: a problem arose for the theory (Fowler's objection that certain lines did not appear at the positions calculated on the basis of Bohr's theory); and a reasonable line of research in attempting to answer the problem was to look at areas in which simplifications had been made. One might even speak here of a general *principle of nonrejection of theories*: when a discrepancy is found between the predictions of a theory and the results of observation or experiment, do not reject the theory as fundamentally incorrect before examining areas of the theory in which simplifications have been made which might be responsible for the discrepancy.

(2) *Sommerfeld and the fine structure of spectral lines*. In the preceding case, a problem arose which was answered by giving a more realistic treatment instead of, as earlier, a simplified one. The reverse order of events, however, might have taken place: the more realistic treatment might have been given without the problem ever having come up, and the altered prediction of the positions of the lines made. This was, in fact, the order of events in the present case: Sommerfeld, realizing the simplification involved in ignoring the relativistic effects of the high velocities of the orbital electrons, took into account those effects, giving a more realistic treatment of the situation, and in the process arriving at the prediction of a fine structure of spectral lines. Again we see how "background information," in making the distinction between simplifications and realistic treatments, also determines what counts as a reasonable line of research in the face of a problem confronting the theory (Zeta Puppis case), or even in the absence of one (Sommerfeld case). Of course, this is only one sort of case in which certain moves in science having to do with theoretical inadequacy acquire reasonableness; but it should be sufficient to show that (and even how), in dealing with such problems just as with domain problems and theoretical problems, certain moves in science sometimes, at least, are reasonable, plausible, even if they should ultimately turn out not to provide an answer to the problem at hand.

The present paper has examined a number of cases from the history of science, with a view to exposing some of the reasoning patterns existing there with regard to the generation of scientific areas, problems, and attempts to deal with those problems – all with the hope of shedding some light on the nature and function of scientific theories.[68] The patterns discerned are justly labeled "reasoning-patterns": in particular, their formulation requires no reference to psychological or sociological factors, and their operation in concrete cases is, or at least can be, independent of such factors.

Nevertheless, one of the main points of this paper has been that, although scientific development is often reasonable, there is no question here of anything that could appropriately be called a "*logic of discovery*," for there is no *guarantee* that a certain line, or even *any* reasonable line, of research *will* eventuate in a solution of the problem. Rather than speaking of a logic of discovery, it is less misleading, and more faithful to the spirit of science, to describe the analyses given here as having been concerned with the rationale of scientific development.

## NOTES

[1] E. T. Whittaker, *A History of the Theories of Aether and Electricity* (New York, Thomas Nelson, 1951), Vol. I, p. 35.
[2] *Ibid.*, p. 175.
[3] W. Gilbert, *De Magnete* (New York: Dover, 1958), Bk. II, Ch. II, esp. pp. 95–97.
[4] Whittaker, *A History of the Theories of Aether and Electricity*, p. 35.
[5] *Ibid.*, pp. 80–81.
[6] The roots and character of some of the most general of such presuppositions have been analyzed in my essay, "The Development of Scientific Categories," forthcoming.
[7] It is necessary to emphasize this point; for many, perhaps most, philosophers of science, even when they have raised such questions as 1 to 5, have denied that they were relevant to the attempt to understand science; for, according to those philosophers, such questions (with the possible exception of certain aspects of 2) have to do only with the "psychology," "sociology," or "history" of science, rather than with its "logic," which alone is relevant to the *understanding* of science.
[8] T. S. Kuhn, *The Structure of Scientific Revolutions* (Chicago: University of Chicago Press, 1970), p. 46.
[9] For example, the discovery of a concomitant variation of properties may in itself be only a weak indicator of a deeper relationship between two types of items or between two domains, *unless* there is reason to suppose that those properties are "significant." Thus Faraday rotation (rotation of the plane of polarization of a ray of light as a function of magnetic field strength), alluded to earlier, was an indication of a deeper relationship between magnetism and electricity precisely because polarization was a manifestation of what, in view of the Young-Fresnel theory of light, was a centrally important feature of light – namely, its transverse wave character. On the other hand, the differences between electricity and magnetism noted by Gilbert did not have to be taken too seriously because there was little or no ground, at the time or for long after, for distinguishing significant from nonsignificant features of electricity and magnetism. When due account is taken of this aspect of "significance" of relationships, Mill's methods are indicative that further research is in order, though they may not, in general, specify precisely what research to undertake. It is perhaps worth pointing out here that, as can be seen from the examples given in this note, "significance" is a function of what will later be called "background knowledge": information which, though not a part of the domain under consideration, is accepted (on presumptively good grounds) and utilized in interpreting and dealing with that domain and its problems.
[10] The reasons for suspecting a deeper unity among the elements of a body of information, and thus creating a domain for investigation, have nothing to do (or need have nothing to do) with a faith on the part of the scientist concerned in some "Principle of the Unity (or Uniformity or Simplicity) of Nature." The reasons adduced in particular scientific contexts are specific, not general, scientific, not metaphysical or aesthetic. A belief in such a general principle would in any case never be sufficient to provide or even suggest specific directions as to where to look in the search for unification – what lines of research are promising, and the general outlines of what a successful solution of a problem might be expected reasonably to look like. And yet, as this paper argues, sometimes such directions *are* indicated, with more or less clarity and precision, in actual scientific cases. As we shall see, the "readiness" of a scientific problem to be investigated

is, in part, a function of the specificity with which such directions are indicated; thus something so general as a "faith in unity," while as a matter of sociological fact it might happen to be shared by a majority of scientists at some time, would not constitute a reason for investigating any particular problem, or for abandoning the solid ground of "ready" problems for a flight of unifying speculation.

[11] D. Shapere, "Notes toward a Post-Positivistic Interpretation of Science," in P. Achinstein and S. Barker, *The Legacy of Logical Positivism* (Baltimore: Johns Hopkins University Press, 1969), pp. 115–160; see especially Part I, pp. 115–131.

[12] D. Shapere, "The Concept of Observation in Science and Philosophy," *Philosophy of Science*, December 1982.

[13] That "facts do not raise their own questions" has become a cliche. On the present analysis, there are good objective reasons why a certain problem regarding a domain is raised – and though it may not (always) be raised *by* the "facts" (or domain) which the question is *about*, it is nonetheless misleading to suggest that the alternative is that "it is *we* who pose questions *to* nature," suggesting that there is no objective ground for the questions. On the contrary, the alternative is that the questions are usually raised by (or in the light of) *other* "facts" (information which is taken for granted, presumably on good grounds) relative to the situation or "facts" under consideration.

[14] D. Shapere, "The Concept of Observation in Science and Philosophy," treats the character of "observations" in science. My concern here is with "theories."

[15] They are, for example, "hypothetical" – unconfirmed – to a degree to which the theory need not be (consider tachyons and the special theory of relativity). But they are also hypothesized as being *entities*, as positivistic "theoretical entities" were often not, as theoretical terms were often looked on as not *really* referring to existent things (or even as claiming to refer to such things).

[16] There are some instructive differences between the examples mentioned. Tachyons constitute a subject for investigation only because they are *not ruled out* by the theory of special relativity. (Thus, although their investigation presupposes that theory, it is not strictly speaking an investigation *of* that theory, except insofar as it may involve a study of that theory to see if it can be extended so as either to require or rule out the existence of tachyons.) In the case of neutron stars, the existence of the entities was *required* by the theories concerned – general relativistic gravitation theory, together with broader theories of stellar evolution – if there exist (as there do) stars exceeding a certain mass, and if that excess mass is not lost prior to the star's arrival at a certain late stage of its evolution. Again, "superstars," like tachyons, are not ruled out as possible (though earlier, it had been thought that stars of masses greater than about 100 suns were ruled out on the basis of stability considerations). Unlike the case with tachyons, however, there were *other*, independent reasons for asking whether there might be superstars and, when Hoyle and others showed that there could be, for investigating their properties: namely, the need to find some way of accounting for the apparent large energies expended by radio galaxies and quasars (that is, the need to resolve a problem arising in an independent domain). Finally, the intermediate boson, still undetected in 1971, though it belongs in this general class, was proposed not on the basis either of following from or not being ruled out by a theory, but rather on the basis of analogy with another theory for another (but in many respects similar – the analogy was not an arbitrary one) domain: the intermediate boson was to play the role for weak interactions that is played by the photon for electromagnetic interactions and by the pi-meson for strong

interactions. It should be added that "investigation" of such theoretically determined entities includes not only the attempt to find them, but also to investigate their properties from a purely theoretical point of view. This is particularly obvious in the case of superstars, where the motivation is to see if they could, if they existed, be the source of the energy required to account for quasars.

[17] As we have already seen, background information also contributes to the *raising* of problems regarding domains, by determining their "significance."

[18] Such theories are related to, and indeed are a subclass of, the "existence theories" discussed in "Notes toward a Post-Positivistic Interpretation of Science" – theories, briefly, which make existence claims. Field theories are examples of existence theories which are not readily subsumable under the present heading of "compositional theories."

[19] In the case of ordered domains, one can speak of domain problems having to do with the *incompleteness* of the domain, because of unoccupied places in the ordering; for example, Mendeleev and other pioneers of the periodic table gave reasons for considering there to be "gaps" in their orderings – for considering certain elements to exist which had not yet been discovered. Discovery of such elements, however – though filling in those gaps – while it did increase current knowledge, did not, in any usual sense of the word, increase "understanding" or provide "explanation" of the system of elements.

[20] It is true that, especially in the decade before 1870, several alternative orderings of the chemical elements had been suggested. (See F. P. Venable, *The Development of the Periodic Law* (Easton, Pa.: Chemical Publishing, 1896); J. W. van Spronsen, *The Periodic System of Chemical Elements: A History of the First Hundred Years* (New York: Elsevier, 1969); for a collection of primary sources, see D. M. Knight, ed., *Classical Scientific Papers: Chemistry, Second Series* (New York: American Elsevier, 1970). Could those proposed "orderings" be called "theories"? The question is an empirical one of usage, and what we find is that terms like "classification," "system," "table," or "law" are used; but the term "theory" is used also, but only very occasionally. This is true not only of present usage, but also, as far as I have been able to determine, in regard to the proposals made during the decade 1860 to 1870. The rarity with which the term "theory" is applied to such cases is, I think, significant. However, even to the extent to which one might be tempted to use it in reference to the ordering of the domain of chemical elements, none of the analysis given here would be vitiated. For our purposes, all that is relevant is that *once the character of the ordering had been determined, that ordering itself became part of the domain*. That is, even if the ordering is considered to have been "theoretical" at a certain stage, nevertheless once it had been settled on through the work of Mendeleev and Meyer, it became an integral aspect of the body of information which was to be accounted for – that is, it became part of the domain. (And in any case, an "ordering" would be a very different *kind* of theory from a compositional one.)

As to the use of the term "law" in reference to the periodic table, it is important to remember the historical background of that term in this connection. In his paper of 1871, Mendeleev stated the "periodic law" as follows: "The properties of simple bodies, the constitution of their compounds, as well as the properties of the last, are periodic functions of the atomic weights of the elements." (D. Mendeleev, "The Periodic Law of the Chemical Elements," *The Chemical News*, Vols. XL (1879) and XLI (1880) (translation of the 1871 article); reprinted in Knight, *Classical Scientific Papers*, pp. 273–309; quotation is on p. 267. Interestingly, Freund refers to this article as "The Periodical Regularities of the Chemical Elements": I. Freund, *The Study of Chemical*

*Composition: An Account of Its Method and Historical Development* (New York: Dover, 1968, reprint of the 1904 edition), p. 500. But "the true function, expressing how the properties depend on the atomic weights, is unknown to us." (Mendeleev, in Knight, *Classical Scientific Papers*, p. 288.) That is, Mendeleev, like nearly all other workers in the late nineteenth century, conceived this functional relationship to be a mathematical one *whose precise form remained to be discovered*; and although he was not averse to calling his periodic table a "law," it is clear that he (like others) considered the true expression of this law to be a mathematical one, and that his statement of it was only a vague one which however was clear enough to allow rough results to be achieved. Again in his Faraday lecture of 1889, Mendeleev expressed the same view: "The Periodic Law has shown that our chemical individuals display a harmonic periodicity of properties, dependent on their masses. Now, natural science has long been accustomed to deal with periodicities observed in nature, to seize them with the vise of mathematical analysis, to submit them to the rasp of experiment." (Mendeleev, "The Periodic Law of the Chemical Elements," Faraday Lecture, June 4, 1889; reprinted in Knight, *Classical Scientific Papers*, pp. 322–344; quotation on page 328.) But after discussing the inadequacies of attempted formulations of this function, and listing what he took to be requirements of an adequate expression of it, he found it necessary to conclude that "although greatly enlarging our vision, even now the periodic law needs further improvements in order that it may become a trustworthy instrument in further discoveries" (*ibid.*, p. 337). This was the universal view: that the true "law" was yet to be found (earlier, Mendeleev had explicitly declared that "I designate by the name of *Periodic Law* the mutual relations between the properties of the elements and their atomic weights, relations which are applicable to all the elements, and which are of the nature of a periodic function"; see Freund, *The Study of Chemical Composition*, p. 469). As late as 1900, this was still the view: "We have not been able to predict *accurately* any one of the properties of one of these [noble] gases from a knowledge of those of the others; an approximate guess is all that can be made. The conundrum of the periodic table has yet to be solved." (Ramsay and Travers, *Argon and Its Companions*, 1900; quoted in Freund, *The Study of Chemical Composition*, p. 500.) The relevance of this disheartened conclusion to the present point is made clear by Ida Freund's comment: "The special feature of the conundrum thus referred to by Professor Ramsay is how to find the formula for the function which would correlate the numerics of the atomic weight with the properties susceptible of quantitative measurement. Another aspect of it is that of the expression of the atomic weights themselves by means of a general algebraic formula. This problem is an attractive one, and several attempts have been made to solve it, in spite of the fact that the only indications of the direction in which success may be expected are of negative nature" (Freund, *The Study of Chemical Composition*, p. 500).

Thus, although the periodic table was widely referred to as a law, the general opinion of the time was that it could be called a "law" only in a rather loose sense, the true law being the precise mathematical expression of the "function" relating the atomic weights and the other properties of the elements and, presumably, their compounds. Views claiming that the table should be called a law because it permits prediction must take into account this feature of the historical situation. (Van Spronsen, *The Periodic System of Chemical Elements*, makes a deliberate decision to refer to it as the periodic "system" rather than "law," and so forth.)

[21] In some respects, similar remarks hold for the multitude of conclusions about stars (for example, regarding distances, masses, intrinsic luminosity, ages, chemical composition, internal structure – few of which are "directly observable" in anything like a positivistic sense, if in any natural sense at all) which can be drawn on the basis of the Hertzsprung-Russell diagram, the astrophysical analogue of the periodic table. In this case, however, more of a "theoretical" character has become embedded in the use of the H-R diagram than in the corresponding use of the periodic table for prediction of missing elements. Indeed, a close comparison of these two cases would demonstrate clearly the absurdity of claims that scientific predictions and other conclusions always, and in some sense that is both normal and univocal, involve or presuppose the use of "theories." Such an investigation would also reveal much of importance about the interactions and interpenetrations of "theory" and "classification." The latter is, in many cases, far from being unintellectual drudge work.

[22] This is *one* motivation of the attempt, mentioned in the preceding paragraph, to refine the measurement of the ordering properties: to test the strength of the indication that there is a deeper account to be found.

[23] Relations between isolated groups of elements had been noted much earlier and held to be indications of some common composition at least of the elements so related. As early as the second decade of the century, Döbereiner had constructed "triads" of related elements: for example, he found the (then supposed) atomic weight of strontium (50) to be the mean of those of calcium (27.5) and barium (72.5), and this was taken as grounds for questioning the independent existence of strontium. Later, groups of more than three elements were found to be related, and again compositional theories of various sorts were suggested for those elements. Such views, of course, were based only on what are now called "vertical" rather than on "horizontal" relationships, which were disclosed fully only by the periodic table. In general, nineteenth-century science found many reasons besides the relations embodied in the periodic table for supposing the chemical elements, or at least some of them, to have common constituents; certainly there is no ground for the allegation that the view of the transmutability of the elements had died with the alchemists and was not revived until the work of Rutherford and Soddy; it is one of the liveliest strands in nineteenth-century science. A good survey is W. V. Farrar, "Nineteenth-Century Speculations on the Complexity of the Chemical Elements," *British Journal for the History of Science*, Vol. II (1965), pp. 297–323, though he neglects to include explicit discussion of the important isomeric theories of the elements.

[24] For the opposition of chemists to physical atoms, see D. M. Knight, *Atoms and Elements: A Study of Theories of Matter in England in the Nineteenth Century* (London: Hutchinson, 1967); W. H. Brock, *The Atomic Debates* (Leicester: Leicester University Press, 1967); and W. McGucken, *Nineteenth-Century Spectroscopy: Development of the Understanding of Spectra 1802–1897* (Baltimore: Johns Hopkins University Press, 1969). In this connection, perhaps Lavoisier's conception of elements as "the substances into which we are capable, by any means, to reduce bodies by decomposition" (A. L. Lavoisier, *Elements of Chemistry* [New York: Dover, 1965], p. xxiv) joined with Dalton's views and a too-rigid empiricism to discourage many chemists from "unscientific," "metaphysical" speculations about the composition of the elements. Mendeleev himself, though on other occasions he was not at all averse to such speculations, remarked in his Faraday lecture that "the periodic law, based as it is on the solid and wholesome

ground of experimental research, has been evolved independently of any conception as to the nature of the elements; it does not in the least originate in the idea of an unique matter; and it has no historical connection with that relic of the torments of classical thought, and therefore it affords no more indication of the unity of matter or of the compound character of our elements, than the law of Avogadro, or the law of specific heats, or even the conclusions of spectrum analysis" (Knight, *Classical Scientific Papers: Chemistry, Second Series*, p. 332). The statement following the word "therefore," of course, is a complete *non sequitur.*

[25] The situation was, however, complicated (a) by the fact that what were and were not elements was not always clear, and (b) by the fact that atomic weights were not well known and were difficult to measure, especially without making certain assumptions. With these and other difficulties to overcome, much "theory" was used in the attempt to construct orderings of the elements before Mendeleev and Meyer. See the references in note 20.

[26] The one case which approached being a clear one, because of the large difference in atomic weight involved, was that of tellurium and iodine – though, as we shall see momentarily, even this case was not as clear as van Spronsen makes out. Although the ordering of cobalt and nickel was also anomalous, "The atomic weights of the latter pair of elements differed so slightly that at first the extent of the problem was not appreciated, the more so because their properties differed so little" (van Spronsen, *The Periodic System of Chemical Elements*, p. 236). The third anomaly, with regard to the order of argon and potassium, was not known until 1894, and even then there were difficulties about determining the atomic weight of argon. It was with the discovery of this anomaly, according to van Spronsen, that "the problem became extremely disturbing" (*ibid.*). Even the case of tellurium and iodine was not admitted as clear by many authorities, including Mendeleev himself, who in 1889 declared triumphantly that "the periodic law enabled us also to detect errors in the determination of the atomic weight of several elements . . . Berzelius had determined the atomic weight of *tellurium* to be 128, while the periodic law claimed for it an atomic weight below that of idoine, which had been fixed by Stas at 126.5, and which was certainly not higher than 127. Brauner then undertook the investigation, and he has shown that the true atomic weight of tellurium is lower than that of iodine, being near to 125" (Faraday Lecture, in Knight, *Classical Scientific Papers: Chemistry, Second Series*, p. 339; for an amusing if vitriolic response to Brauner's work, see Freund, *The Study of Chemical Composition*, p. 505). As to the fourth anomaly, that of the pair protoactinium-thorium, it never presented a problem, as protoactinium was discovered in 1918, five years after Moseley, Rydberg, and Van den Broek demonstrated that atomic number, rather than atomic weight, provided the fundamental ordering principle.

[27] For the vortex atom, see McGucken, *Nineteenth-Century Spectroscopy*, esp. pp. 165–175.

[28] Thus, because it fails to satisfy this condition to a very high degree, Bode's "law," which gives a simple mathematical ordering relationship approximating the distances of the planets out to Uranus reasonably well, cannot be *clearly* included as part of a domain concerning the planets. Ordering only seven items, it is not very extensive; nor is it very detailed, not holding for Neptune or Pluto at all, and not relating to any other planetary characteristics besides distance; nor is it very precise, since it holds only approximately for the planets to which it does apply. In other words, the failure of a

theory of (say) the origin of the solar system to account for Bode's "law" cannot be considered a very great weakness of that theory. For such a theory is only weakly (that is, only to the rather low degree that Bode's "law" is extensive, detailed, and precise) required to account for that "law." It is not clear, in other words, that Bode's ordering really sets a *problem* regarding the planets. To put the point more generally, membership in a domain is also a "matter of degree," and is thus subject to debate. That is, whether a theory is reasonably to be expected to account for a certain item is itself a matter which can be questioned. This kind of move in defense of a theory has often been made in the history of science.

29 (Cv) to (Cvii) clearly have to do with determinations of the "importance" of the problem.

30 The whole tenor of the present analysis has been in the direction of making the line between "science" and "nonscience" a matter of degree rather than a sharp distinction in terms of some "line of demarcation."

31 McGucken, *Nineteenth-Century Spectroscopy*, p. 8. Jammer is thus in error in stating that "Mitscherlich [1864] was the first to point out that spectroscopy should be regarded not only as a method of chemical analysis ... but also as a clue to the secrets of the inner structure of the atom and the molecule"; M. Jammer, *The Conceptual Development of Quantum Mechanics* (New York: McGraw-Hill, 1966), p. 63. Mitscherlich had declared that "[The difference in the spectra of the elements and compounds] appeared to me of great importance, because by the observation of the spectra a new method is found of recognizing the internal structure of the hitherto unknown elements, and of chemical compounds." A. Mitscherlich, "On the Spectra of Compounds and of Simple Substances," *Philosophical Magazine*, 4, 28 (1864), p. 169; quoted in C. L. Maier, "The Role of Spectroscopy in the Acceptance of an Internally Structured Atom, 1860–1920," Ph.D. dissertation, University of Wisconsin (1964), p. 38. Maier's comment on this passage is: "To suggest that Mitscherlich is here implying an internal structure of the elemental atoms would be a distortion in terms of the context of the times. He refers to the internal structure of elements and compounds, not of atoms and molecules. It is far more probable that Mitscherlich had reference to the utility of this new spectral distinction in deciding whether elements and compounds were actually structured into entities such as the atom and molecule than that the atoms themselves were internally structured" (pp. 38–39). Maier could not, of course, have had Jammer's later remark in mind here, but his point certainly holds against the latter. However, perhaps Jammer was thinking only of Daltonian "atoms," in which case his remark would be perfectly in order. In this connection, Maier's own reference to "elemental atoms" is confusing: as he himself notes, it is perfectly correct to interpret Mitscherlich's statement as referring to a possible internal structure of the elements (Daltonian atoms).

32 McGucken, *Nineteenth-Century Spectroscopy*, p. 155.

33 *Ibid.*, p. 155. It is certainly necessary to reject Dingle's assertion that "Rydberg's work, fundamentally important though it is, was purely empirical" (H. Dingle, "A Hundred Years of Spectroscopy," *British Journal for the History of Science*, Vol. I (1963), p. 209). Maier argues effectively against Dingle's interpretation (Maier, "The Role of Spectroscopy," pp. 102ff.).

34 See Maier's discussion of what he calls "the acoustical analogy" and its guidance of the search for mathematical formulas relating the spectral lines (Maier, "The Role of Spectroscopy," Ch. III). McGucken's discussion is also highly illuminating in this

connection and complements that of Maier very well (McGucken, *Nineteenth-Century Spectroscopy*).

[35] *Ibid.*, p. 204.

[36] *Ibid.*, p. 155. In our terms, Rydberg here adduces the high degree of "readiness" in addition to "importance" as an argument for investigating spectra.

[37] In this kind of case, where there is not good reason to suspect that the two domains are related as parts of a larger domain, it is appropriate to speak of the importation of ideas from one domain into another as based on "analogy." Where there are such good reasons, it is more appropriate to speak of such importation as being based on "evidence" (or, more generally, on "reasons"). No doubt there are always *some* grounds, at least very weak ones, for suspecting that two domains may be related as parts of a larger domain; and to this extent, the difference between introducing new ideas on the basis of "analogy" and on the basis of "evidence" is a matter of degree – as is the distinction between principles (Cvii) and (Cviii), which will be discussed below. But this fact does not sanction the obliteration of the distinction by those who maintain that all hypotheses in science are introduced on the basis of "analogy."

[38] It should be recalled that (Ci) and (Cii) did not *have* to be fulfilled in order that a compositional theory could be reasonably expected; their fulfillment, above and beyond the fulfillment of some of the other conditions, only provided increased grounds for expecting such a theory. Condition (Ciii), concerned with discreteness and integral multiplicity, while even it is not a *necessary* condition for expecting a compositional theory, does perhaps by itself provide stronger grounds for such expectation, other things being equal, than condition (Ci) alone would. It should be noted that condition (Ci) is undoubtedly susceptible to considerable generalization.

[39] McGucken, *Nineteenth-Century Spectroscopy*, p. 156; the reasons behind this attitude are surveyed by that author on pp. 73–83.

[40] *Ibid.*, pp. 125–126. Maier conveys a picture of mass desertions of the field of spectroscopy: "As the complexity of spectra became apparent, more and more workers turned away from it as a method of practical analysis. . . . The field of chemical analysis was left to a few stalwarts . . . " (Maier, "The Role of Spectroscopy," p. 40). Though this is perhaps something of an exaggeration, considering the widespread employment of spectral analysis by astronomers, it was undoubtedly a very common attitude. Even Kayser, in 1910, had come to the point of declaring that "there is little prospect that in the future qualitative analysis will apply spectroscopic methods to a large extent . . . I come to the conclusion that quantitative spectroscopic analysis has shown itself as impractical" (quoted in *ibid.*, p. 41). It is interesting to note that this attitude was repeated by Niels Bohr, who finally gave a successful explanation of spectra in atomistic terms: "The spectra was a very difficult problem. . . . One thought that this is marvelous, but it is not possible to make progress there. Just as if you have the wing of a butterfly, then certainly it is very regular with the colors and so on, but nobody thought that one could get the basis of biology from the coloring of the wing of a butterfly" (quoted by J. Heilbron and T. Kuhn, "The Genesis of the Bohr Atom," in R. McCormmach, ed., *Historical Studies in the Physical Sciences* [Philadelphia: University of Pennsylvania Press, 1969], Vol. I, p. 257).

[41] A splendid example of several different kinds of disagreement of the sorts discussed here is found in the debates regarding interpretation of the photoelectric effect in the years immediately preceding Einstein's "Concerning a Heuristic Point of View about the

Creation and Transformation of Light" (*Annalen der Physik*, 17 (1905), pp. 132–148; translated in H. Boorse and L. Motz, eds., *The World of the Atom* [New York: Basic Books, 1966], Vol. I, pp. 545–557) and in the succeeding years up to Compton's interpretation of his x-ray scattering experiments in 1924. Historical aspects of the case are well presented in M. Klein, "Einstein's First Paper on Quanta," *The Natural Philosopher*, No. 2 (New York: Blaisdell, 1963), pp. 57–86, and R. Stuewer, "Non-Einsteinian Interpretations of the Photoelectric Effect," in R. Stuewer, ed., *Historical and Philosophical Perspectives of Science* (Minneapolis: University of Minnesota Press, 1970), pp. 246–263.

[42] For descriptions of such theories, see Venable, *The Development of the Periodic Law*, and van Spronsen, *The Periodic System of Chemical Elements*. It is highly significant that a large proportion of these evolutionary theories of the chemical elements were proposed by men whose primary work, or a considerable part of whose work, lay in fields outside chemistry. An explanation of this phenomenon will be offered shortly.

[43] A. Clerke, *Problems of Astrophysics* (London: Black, 1903), pp. 179–180. One is reminded here of McGucken's statement, quoted earlier, that in the nineteenth century, "'understanding spectra' became almost synonymous with 'understanding atoms and molecules'"(McGucken, *Nineteenth Century Spectroscopy*, p. 204). There were, as usual, dissenters from the prevailing view: the British astronomer Maunder wrote in 1892 that "spectrum type does not primarily or usually denote epoch of stellar life, but rather a fundamental difference of chemical constitution" (O. Struve and V. Zebergs, *Astronomy of the Twentieth Century* [New York: Macmillan, 1962], p. 187). Once again we are reminded of the attitudes of some chemists toward compositional theories of the elements.

[44] Even the *Henry Draper Catalogue of Stellar Spectra*, which became the basis of modern classifications, in its first volume of 1890, divided the stars into sixteen classes denoted by the letters A through Q (with J omitted), the alphabetical order not corresponding to any ordering among the classes. It was only early in the twentieth century that some of these groups were omitted, or combined and relettered, and the order changed, ultimately becoming the present O-B-A-F-G-K-M-R-N-S ("Oh, be a fine girl, kiss me right now, sweet"), an arrangement which does provide a sequential ordering.

[45] Other theories, for example, that of Lockyer, held that some red stars are young and heating up – according to Lockyer through meteoritic impacts – to the white stage, while others had cooled from the hotter stage. It is worth noting that an alternative view of the colors of stars had been proposed earlier: Doppler had suggested that blue stars are moving toward us, while red stars are moving away. Such an interpretation, however, would imply corresponding shifts of spectral lines, which are not observed.

[46] Such views of element composition were combined with a variety of theories of stellar evolution other than the one under consideration.

[47] There were also theories according to which greater age saw a greater *breakdown* of heavier elements, *ending* with the ultimate constituents.

[48] The existence of this trend does not itself seem to have been taken as a basis for class orderings in the early stages of astronomical spectroscopy: criteria of ordering would have had to be complex and quite beyond the knowledge of the times. It was, however, clear enough, once the classes had been ordered according to color, to provide additional support for that proposed ordering.

49 The history of vicissitudes in the presumed relation between spectral classification and evolution of stars is a fascinating one, worth examining for the insight it would provide regarding the rationale of scientific change. By 1928, one of the major theoreticians of astrophysics, James Jeans, could write that, although "The early spectroscopists believed that the spectrum of a star provided a sure indication of the star's age," Saha's ionization and excitation theory had shown that "The linear sequence into which the spectra of stars fall is merely one of varying surface temperature. Clearly this circumstance robs stellar spectra of all direct evolutionary significance" (J. H. Jeans, *Astronomy and Cosmogony* [New York: Dover, 1961], p. 166). "The problem of stellar evolution is now seen to be quite distinct from that of explaining the distribution of stars in the [Hertzsprung-] Russell diagram, and, furthermore, the problem can expect no assistance from the observed distribution of stars in this diagram" (*ibid*., p. 172). Only a few more years were to pass before the pendulum began to swing back in the direction of a connection between spectral classification and stellar evolution – though that connection came to be seen as far subtler and more complex than anything envisioned by the early pioneers like Vogel and Lockyer. Those later views of a connection between spectral classification and stellar evolution, however, involve a deep penetration of "theory" into the domain (as summarized in the classification and the H-R diagram). (For example, Chalonge's system, by relying heavily on the "Balmer jump" in hydrogen spectra as a basis for spectral classification, ties itself closely to the theory of the hydrogen atom; by this means, it becomes highly useful and precise, even though it is limited in applicability to early-type stars [approximately G0 and earlier] which show hydrogen lines and the Balmer jump with sufficient clarity in their spectra.) Analysis of the modern conceptual situation in this area, though it would be very instructive as to the interrelations of "theory" and "observation," and as to the ways in which a new "theory" accounts for such a domain (consisting of an intimate mixture of older classification and later theory), and therefore as to the nature of theories, is too involved to be discussed here.

50 As was mentioned above, the attempt to lay out such an ordering would not have been free of difficulties. For example, many spectral lines had not been identified, so that the correlation of "lateness" of spectral type with "heaviness" of composition, rather tenuous at best, could not always be assured. Again, what are now called O- and B-type stars have weaker hydrogen lines than A-type (white) ones, even though they are bluer and hence, on the theory under consideration (as well as according to modern astronomy), hotter.

51 Recall the case of the *Draper Catalogue*, discussed in note 44.

52 This importation was also supported by the obvious explanation of starlight in terms of the stars being hot and radiating. The resultant loss of energy by radiation would also be clearly suggestive of an evolutionary process. Again, an importation into the stellar domain is at work – in this case, supported by the newly verified view that the stars have the same composition as the earth, so that the same processes and laws can be expected to be at work in both (that is, reasons have been found for supposing the terrestrial and stellar domains to be, in this respect, parts of a larger domain).

53 Lockyer's reasoning was apparently vindicated later by Miss Maury's discovery that, despite the fact that red stars had the same lines in their spectra, in some such stars the lines were more strongly defined than in others. Hertzsprung, in 1905, established by statistical arguments the validity of Miss Maury's conjecture that this indicated that there were two radically different types of red stars. H. N. Russell, a few years later, used this

fact in his construction of a new version of Lockyer's general view of stellar evolution. See Struve and Zebergs, *Astronomy of the Twentieth Century*, pp. 195–200.

[54] This third point leads to a suggestion which I am not prepared to develop fully in the present paper. Thus far, I have been speaking of theories as answers to questions. While there is a point to this, it should be remembered that those questions themselves, in the cases considered, involve a general idea of what their answer will be like. In this sense – and it is a sense which seems *prima facie* to fit a great many cases in the history of science – a theory is *gradually developed* by a process of increasingly precise and detailed statement of the initial vague idea; there is then no single point in time at which one can say unambiguously that the theory has been *arrived at*. It would then be misleading to speak, in all cases, as if there were a single event of proposing an answer to the theoretical problem. If this suggestion is borne out, as I strongly suspect it will be, one source of philosophers' difficulties with the notion of "theory" will have been exposed: for when does an idea become precise enough to be called a "theory"? The difference would seem to be more a matter of shading than of sharpness. Theory development would then be more appropriately describable, in some cases at least, as *a process of convergence from generality to (relative) precision* than as a precisely datable event like answering a question.

[55] The few exceptions – most notably, the Kant-Laplace "nebular hypothesis" of the origin of the solar system – are not, however, paradigm examples of evolutionary theories. They have to do with the *origin* of a system and its development only to a certain stage, after which, as far as the theory is concerned, it ceases to develop further. Such theories, while they do have much in common with paradigmatically evolutionary theories, should perhaps be distinguished from them as a separate category of "genetic theories."

[56] At least in the sense of "genetic theories," if not strictly speaking of evolutionary ones; see preceding note. Also, the view of stars as radiating bodies called for an evolutionary theory, as was pointed out in note 52.

[57] In the light of this need, it is no wonder that theories of the evolution of the chemical elements tended to be proposed by men who worked in areas other than chemistry – and, in particular, in areas in which evolutionary theories were being developed. It is almost as if a theory of chemical evolution *needed* to come from another (appropriate) area (see note 42). Note, too, that the application of theories from those domains to the chemical one was based on good reasons, in particular, on the similarity of composition of earth and stars, as established by spectral analysis.

[58] See, for example, D. Shapere, "The Structure of Scientific Revolutions," *Philosophical Review*, 73 (1964), pp. 383–394; "Meaning and Scientific Change," in R. Colodny, ed., *Mind and Cosmos* (Pittsburgh: University of Pittsburgh Press, 1966), pp. 41–85; D. Shapere, "The Paradigm Concept," *Science*, Vol. 172 (1971), pp. 706–709. On the other hand, the older "positivistic" approach to the analysis of 'theory' (in terms of an interpreted axiomatic system) was not only very unilluminating or inadequate, but also was positively misleading, and thus positively interfered with the attempt to understand scientific theories (see my "Notes toward a Post-Positivistic Interpretation of Science," Pt. I.)

[59] This is perhaps a linguistic remark about the term 'understanding,' rather than an empirical one about babies.

[60] N. Bohr, "On the Constitution of Atoms and Molecules," *Philosophical Magazine*, Vol. 26 (1913), pp. 1–25, 476–502, 857–875.

[61] D. Shapere, "Notes Toward a Post-Positivistic Interpretation of Science," Pt. II. The analysis given there is developed further in "Natural Science and the Future of Metaphysics," in R. Cohen and M. Wartofsky (eds.), *Methodological and Historical Essays in the Natural and Social Sciences*, Dordrecht, Reidel, 1974, pp. 161–171.
[62] Bohr himself did not originally approach the development of his theory with the intention of accounting for the features of spectra. In fact, "In a letter to Rutherford dated 31 January 1913, he had, in fact, explicitly excluded the 'calculation of frequencies corresponding to the lines of the visible spectrum' from the subject matter he took as his own. His program for model building [?], like that of Thomson which it closely followed, relied mainly on chemical, scarcely on optical, evidence" (Heilbron and Kuhn, in McCormmach, *Historical Studies in the Physical Sciences*, p. 257). It was not until very late in the preparation of his theory that Bohr realized the possibility of giving an account of spectral lines in terms of his theory. Nevertheless, it was the success of his theory in regard to spectra that won immediate acceptance for his theory. Heilbron and Kuhn also argue convincingly that Bohr did not approach his theory via an attempt to patch up Rutherford's theory of the atom either, entering the investigation of that theory also only relatively late in his pre-1913 work. His theoretical interests had centered on the electron theory of metals, the topic of his doctoral dissertation, and it was concern with these problems that culminated in the 1913 trilogy – leading him into the question of the stability of the Rutherford atom along the way.
[63] Use of the correspondence principle as a basis for calculating the polarizations of the lines is not considered here as a "part of the theory." The principle was not, in any case, very successful with regard to the intensities.
[64] "Fundamental incorrectness" is, of course, another type of "inadequacy," the one most emphasized by philosophers of science, even to the exclusion of considering other types of inadequacies. This type of inadequacy, important though it is, will not be discussed further in the present paper; it has been dealt with, although, I would now say, not thoroughly, in "Notes toward a Post-Positivistic Interpretation of Science." In particular, an example is given there in which "the limits of applicability of the theory – the location, so to speak, of the incorrectness – is known."
[65] In general, the background information is accepted as true, or at least as the best hypothesis available. In the present case, however, confusion might result from the fact that the background information leading to the distinction between a "simplified" and a "realistic" treatment of atomic motions had classically been associated with two theories that were now contradicted by the Bohr theory – the very theory to which they are here being applied in making this distinction. How is this possible? Several different sorts of considerations play a role. Often the items of information relevant to making the distinction in particular cases can be separated from the general context of the theory which is being contradicted; and even where this separation cannot be clearly accomplished, the area in which the theory is known to be incorrect may not be the area relevant to making the distinction in the case at hand.

Simplifications have much in common with what were called "idealizations" in "Notes toward a Post-Positivistic Interpretation of Science." In that article, typical cases of idealizations were analyzed in terms of three features of their use: (1) there exist certain problems to be solved; (2) mathematical techniques exist for dealing with those problems if the entities dealt with (or their properties) are considered in a certain way, even if it is known, on the basis of the theory, that those entities could not really be that way (in some cases, the allegations that the entities could not really be that way are based not on the theory at hand, but on other grounds); and (3) in many cases, it is provable that the difference between a realistic and an idealized treatment will be insignificant (for example, below the limits of accuracy required by the problem at hand; or below the limits of experimental accuracy) relative to the problem at hand.

There are, however, some differences between what were in that paper analyzed under the name "idealizations" and what are being considered here as "simplifications," though the differences are not necessarily sharp ones, and there are borderline cases. For example, the conclusion that certain concepts are "idealizations" is, in the most typical cases, drawn from the theory at hand (as in the case of the point-charge electron considered in "Notes toward . . . "), whereas what we have here, in a perfectly natural way, called "simplifications" are employed in the original *construction* of theories (rather than in the *application* of an already constructed theory to a specific problem), and so are considered to be "simplifications" on the basis of *previous* theories or *other* knowledge, rather than, as in the case of "idealizations," on the basis of the theory itself. Again, in the case of the most typical "idealizations," we usually do not have available the mathematical techniques required for a (direct) realistic treatment (though approximation techniques may be resorted to), whereas in the case of "simplifications," as will be seen shortly, we often do.

66 Of course, there were indications of an internal structure of the nucleus available at the time; but since those indications were independent of Bohr's theory, they are irrelevant to the present point.

67 L. Rosenfeld and E. Rüdinger, "The Decisive Years, 1911–1918," in S. Rozental, *Niels Bohr* (New York: Wiley, 1967), pp. 59–60. For further details, see Whittaker, *A History of the Theories of Aether and Electricity*, Vol. II, pp. 113–115; Jammer, *The Conceptual Development of Quantum Mechanics*, pp. 82–85.

68 It should be noted that the cases selected are all related to one another, and that their discussion leads naturally into that of a number of crucial developments in twentieth-century science. Their analysis can thus provide a background against which those more modern developments can be profitably examined.

CHAPTER 14

# REMARKS ON THE CONCEPTS OF DOMAIN AND FIELD*

Every inquiry, or at least every inquiry that aims at knowledge, has a subject-matter – that which the inquiry is about, which is the object or set of objects studied. Let us, for the sake of brevity, and without further analysis for the present, call the set of things studied in an investigation the *domain* of the inquiry, and the particular things that make up the domain the *items* of the domain. (These "items" might, in particular inquiries, be spoken of as "objects," "processes," "behavior," "facts," or in still other ways ["fields," "virtual bosons"]; but such ways of speaking involve problems which, being irrelevant to present purposes, I wish to avoid by speaking more generally and neutrally of "items." Sometimes it is convenient to speak of classes of items as themselves items which are objects of investigation.) The domain of a scientific inquiry generally includes many items, the items studied being considered as examples of types or classes or sets of items.

An inquiry is also characterized by certain problems about its domain, and by certain techniques for studying the domain in order to answer those problems; those problems and techniques, as well as the description of the items and the items considered to belong to the domain, can undergo change as a result of the inquiry. It is often convenient to talk of a scientific inquiry or class of inquiries (and in particular of an evolutionary series of domain items, problems, and techniques) as a scientific *field*[1] or area of science, in order to refer not only to the domain, but also to the problems concerning it and the techniques brought to bear on the domain in the attempt to resolve those problems. (A scientific field is often also characterized by a theory or set of theories of its domain.) With regard to all such terminology, however, we must keep three points constantly in mind.

First: What might most immediately be thought of as scientific fields – physics, astronomy, chemistry, biology, geology, *etc.* – are too broadly conceived to capture much that goes on in scientific research. The real give-and-take of science, the real wrestling with concrete problems, takes place at levels of *subfields* of these – fields that go by such names as "high-energy physics," "solid-state physics," "rare-earth chemistry," "galactic astronomy," "radio astronomy," "genetics," "plate tectonics" – and still more at levels of "fields of specialization" within these subfields – such as "hadron jets," "stellar rotation," "quasi-stellar radio sources," "transfer RNA," and in many cases more specific subareas of these. (In these examples, the subfield or field of specialization is frequently named in terms of the domain, which is a subdomain of a larger domain; in other cases, the name is given in terms of a certain technique employed. The latter sort of names, however, tend to be abandoned in favor of domain-based names when the domain becomes better understood. This is not universally the case, however: some physicists continue to prefer the name "high-energy physics" to "particle physics" on the ground that the former name avoids questionable assumptions implicit in the latter.) Fields of specialization are sometimes demarcated by the interest or approach of the individual scientist and sometimes by considerations intrinisic to a more general field or subfield (*e.g.*, specialized study of stellar rotation might gain importance, and therefore the status of a specialized subfield, due to its relevance to the more general subfield of evolution of planetary systems, or for more general problems concerning stellar evolution, or in still other ways connected with more general subfields of astronomy and their problems).

Second: Scientific fields are very fluid sorts of things, and shift more rapidly the more specific their concerns (the more "specialized" they are). Because fields of specialization are in many cases defined by the importance of their problems in a larger context, and because problems often are resolved or abandoned and others become important, what is considered a "field of specialization" at one time may be rather transitory; the same may be true of subfields on a more general level, though their lifetimes are generally longer. The techniques, problems, and, as we shall see more fully below, domains shift from period to period in the development of science, and even the most general levels of scientific fields may ultimately be changed in profound ways – we see an example in the essay, "Alteration of Goals and Language in the Development of Science." (The fundamental problems, and the domain itself, may change in such drastic ways that one sometimes says that an old field has been abandoned and a new one created. Less spectacularly, what was considered to have been part of the domain of one field may be shifted to the domain of another – *e.g.*, the transfer of the study of heat from chemistry to physics with the abandonment of the caloric theory and its replacement by the kinetic theory in the nineteenth century.)

Third: Especially at the level of very specific subfields or areas of specialization, different researchers bring different approaches to the study of the domain, not only in the sense that they bring different techniques to bear in attempting to deal with problems, but also in the sense that they conceive the problems, and sometimes even the domain itself in certain respects, in different ways. Such differences in conception need not be arbitrary and idiosyncratic: they may be, and indeed in the cases of most importance philosophically are, the results of importing into the study of a particular domain or subdomain techniques and concepts of another field, often on the basis of some new hypothesis about the domain in question (*e.g.*, the application of chemical analysis and radioactive dating to the study of meteorites by Urey and his school, on the hypothesis that such techniques would illuminate the problem of planetary origins). In some cases, the domain itself may be reconceived and redescribed, so that one might say not merely that a domain has been approached in a new way, but that the very subfield itself has been created by the approach. In general, especially on the most concrete levels of research, the concepts of a "field of scientific research or specialization" and of the "domain" with which such a field is concerned, must be understood with the flexibility that recognizes the transitory, fluctuating, and sometimes individualistic ways in which science is structured. Nevertheless, such designations do reflect features of the structure and activity of science, and therefore, with the qualifications I have mentioned, can be utilized both as tools and as objects of study in the analysis of scientific reasoning and change.

A multitude of problems of importance for the understanding of scientific reasoning arise in connection with scientific domains, problems, and, more generally, fields. Among the most important are the following: How are domains marked off as distinct areas for investigation in science? How are the items of the domain distinguished and described? How does it come about that something regarding the items in the domain is considered to *need* investigating (*i.e.*, what is the source of the problems that make the domain an area of inquiry)? What determines (or at least suggests) the types of answers to those problems that will be considered possible or satisfactory? These questions are not necessarily independent: the answer to any one of them may well involve answers or partial answers to some or all of the rest. For example, the fact that a problem exists might itself delimit an area of inquiry. On the other hand, the answers to these questions may be different in different cases, and may in some cases be independent of one another.

My primary concern in "Alteration of Goals and Language in the Development of Science" is with certain aspects of the second of these questions – the question of the ways in which items of a domain are described – and to a lesser extent with the first – the question of the ways in which scientific domains are marked out. In particular, what I wish to do here is to fill in some details of the following picture of scientific domain-development which I suggested in "Scientific Theories and Their Domains":

> Although in more primitive stages of science (or, perhaps better, of what will become a science), obvious sensory similarities or general presuppositions usually determine whether certain items of experience will be considered as forming a body or domain, this is less and less true as science progresses (or, as one might say, as it becomes more unambiguously scientific). As part of the growing sophistication of science, such associations of items are subjected to criticism, and are often revised on the basis of considerations which are far from obvious and naïve. Differences which seemed to distinguish items from one another are concluded to be superficial; similarities which were previously unrecognized, or, if recognized, considered superficial, become fundamental. Conversely, similarities which formerly served as bases for association of items come to be considered superficial, and the items formerly associated are no longer, and form independent groupings or come to be associated with other groups. The items themselves often, in the process, come to be redescribed, often, for scientific purposes, in very unfamiliar ways.

The specific points in this picture that are expanded and elaborated in "Alteration of Goals and Language in the Development of Science" are these:

(1) The groupings of items into domains is not, in sophisticated science, generally based on obvious sensory similarities between items; nor are the items described in such terms. Nature does not happen to come once and for all divided, on the basis of anything immediately given in experience, into "areas" or "fields" for investigation. (That the items may be identified, or their presence inferred, on the basis of some "observable characteristics" is not in question here. The presence of an electron may be inferred from the curvature-characteristics of a line in a photograph of a bubble chamber in an electromagnetic field under certain experimental conditions; an A2 star may be identified by the Balmer lines in its spectrum. These are certainly "observable" features; but that *those* features should be considered as identifying or indicating the sorts of entities to be studied is not a matter of anything that could be called immediate or obvious sensory characteristics. That is something that has been arrived at as a product of a long and complex investigation.)

(2) The grouping of items into domains – and consequently the structuring of science into fields and subfields – changes with the development of science. (This is not to say that it always *must* change.) Such regrouping is brought about as a result of new knowledge, or knowledge-claims. (Not all new knowledge-claims result in regrouping of domains.)

(3) Circumstances arise in the development of science in which, with

the realignment of fields and their domains in the light of new knowledge-claims, redescription of the items of the domains may also occur. In general, as science proceeds, the connections between knowledge-claims, domain groupings, and descriptions (and often naming) tend to become tighter and tighter.

In "Alteration of Goals and Language in the Development of Science," I will not, except briefly and where relevant, attempt to discuss the sorts of reasoning by which such changes in domain groupings and description take place; nor will I go into the ways in which the dangers of such procedures are guarded against in science – for example, how scientific objectivity is protected and in general preserved despite the fact that prior knowledge-claims come to be "built into" the descriptive language and associations of the objects of scientific investigation (the "theory-ladenness" problem, which will be discussed in detail in Chapter IV of *The Concept of Observation in Science and Philosophy*). What I will be concerned with in "Alteration of Goals and Language in the Development of Science" will be to point out *that* such changes take place, some of the circumstances in which they take place, and some of the advantages that accrue to the scientific enterprise through such changes.

## NOTES

* This paper is an adaptation, for present purposes, of the opening portions of an earlier paper, "The Influence of Knowledge on the Description of Facts" (in F. Suppe and P. Asquith, eds., *PSA 1976*, East Lansing, Philosophy of Science Association, 1977, Volume II, pp. 281–298. In the present volume, it should be viewed as an introduction to the paper that follows, "Alteration of Goals and Language in the Development of Science."

[1] Further aspects of the concept of a "field" have been examined by my former students, L. Darden and N. Maull, in their "Interfield Theories," *Philosophy of Science*, XLIV (1977), pp. 43–64.

CHAPTER 15

# ALTERATION OF GOALS AND LANGUAGE IN THE DEVELOPMENT OF SCIENCE*

## I. NAMING AND DESCRIBING: THE CASE OF CHEMISTRY

In the Near East and Europe in the ancient and medieval periods, the language in which material objects were described and named was based primarily on the sensory appearances of substances, usually their color, but sometimes their smell, fusibility, solubility, consistency, geometrical form, or some other particularly striking sensory property. Occasionally a geographical location which was the original or major source of the substance provided the basis of the name; in other cases, the use (*e.g.*, medicinal) did so. Such characteristics, generally, remained the bases of naming and classification until well into modern times.

Perhaps the most prominent general idea of matter in pre-modern times was that substances grow to maturity, like vegetables, in the earth. In the middle ages, this view was given philosophical credence in terms of Greek philosophical ideas. There is, it was widely claimed, only one kind of "earth," one element earth, which exists in various degrees of potentiality and actuality, the latter state being one of greater perfection than the former. Earth might exist in the state of powdery, crumbly dirt, or in various higher degrees of actualization or perfection, such as stones or metals. Lead, for instance, is a very imperfect "fulfillment" of earth, only slightly above dirt – as can easily be "seen" by the ease with which lead tends to decay into "dirt" on the outside. It was an easy step from such views to the claim that there is a highest degree of actualization or perfection of the earthy element. The alchemical tradition, though it had arisen, presumably, independently of the philosophical tradition, came rather generally to adopt this viewpoint. There were, of course, variations on the general theme. Some of these had to do with the precise rank-order of perfection or actualization of the variety of forms in which earth manifests itself: is gold really the most perfect form? Is there one hierarchy, or are there two or more branches, each with its highest form? Again, while it was assumed that the alchemist could reproduce, on the miniaturized space- and time-scale of his laboratory, the grand and leisurely processes of nature, conceptions of exactly what those procedures were varied widely in their details. What, precisely, are the activities the alchemist must engage in in his laboratory in order to bring earth to perfection? In what sequence should he perform those activities? Variations of detail in respect to such questions often fed back to variations in interpretation of the philosophical underpinnings of matter-theory. It is important to point out, however, that detailed alchemical recipes focussed on altering the sensory properties of substances to convert them into the sensory properties which were considered to be characteristic of more perfect states of the earthy element. (And thus those properties, too, were rated in hierarchies of perfection.)

For present purposes, however, I want to emphasize a third type of disagreement, having to do with the causes or sources of the various degrees of perfection. We can find – usually implicit, and often not clearly distinguished from one another – three major types of views on this topic: (1) views according to which there is some single "essential" form or potentiality of earth which is to be brought to perfection; (2) views according to which the different degrees of perfection are due to the balance or imbalance of opposing "forms" (hot-cold, wet-dry); and (3) views according to which the different degrees of perfection are due to different admixtures of the other three elements (fire, air, water) with earth. On such views, it is clear that certain kinds of aims and activities that we today associate with laboratory handling of substances were not appropriate or at least not central. For example, on the first two, the "actualization-of-essential form" and "balance-of-opposites" variations of perfectionist views, the idea of separating an earthy substance from other earthy substances was at best peripheral. Even where alchemical recipes called for such "removal of impurities" (as we would say), it was at best a preliminary procedure. For what would remain would be not what we today would think of as a "purified" substance, but rather an imperfect substance that still had to be brought to perfection by the various procedures that constituted the real work of alchemy. Even the third view, that the earthy substances we find in our experience are really combinations of earth with other elements, differed fundamentally from the modern view which it seems superficially to resemble. For even where lip service was paid to the theory that earthy substances are composed of more elementary stuff, alchemical procedures were rarely conceived of as processes of putting together elemental substances to produce a more perfect compound. Neither "analysis" nor "synthesis" in the modern chemical senses was of central concern in the alchemical enterprise. And in any case, no alchemical views that we know of aimed at anything like what we would call *understanding* of material substances in terms of their constituents; rather, they aimed at bringing them to perfection.

How, then, did the modern approach to the study of matter arise? How did we come to the view that – very briefly and roughly – material substances can be understood in terms of their constituents, the arrangement of those constituents, and the forces holding them together, and that, furthermore, such understanding is what we should aim at in our laboratory dealings with material substances? For brevity, I will call this "modern" view the *compositional approach* to material substances and the understanding of them; and I will call beliefs of these sorts *compositional theories*. The present paper is not the place to give a detailed and documented account of that transition from "perfectionist" to "compositionalist" views. My purposes lie elsewhere, and can be served by calling attention to some major aspects of that transition. In particular, I will sketch the origins only of the first aspect of that approach – the view that material substances can be understood in terms of their constituents –omitting any discussion of the second and third aspects of the total view – that also involved in understanding material substances are the arrangement of the constituents and the means (forces) by which they are held together.

The later middle ages and early modern period witnessed the discovery of vast numbers of new substances. Procedures for producing, operating on, and identifying substances became increasingly more sophisticated; and with this sophistication came an increased emphasis on compositional theories as a means of bringing order to the bewildering variety of material substances. Three- and five-element theories became widespread, and the older four-element theory received a new boost, being given a truly compositional twist. But in the sixteenth and seventeenth centuries, all such theories were confronted by their opponents with a crucial objection: what, precisely, does fire (the primary means of operating on material substances) do to the substance on which it acts? One view was that fire merely separates substances into their constituent parts, while not altering those parts in any way. This, of course, was an essential assumption of the compositional views. But how could we know that the fire did not act on the substance so as to *alter* it, so as to *produce* the breakdown products which did not exist as constituents of the substance prior to the application of the fire? Robert Boyle, opposing the current element theories, so argued, as did his great contemporary, Van Helmont.

There was of course no way to establish that breakdown products *were* pre-existing constituents except by trying – by seeing whether a set of substances could be found which would be breakdown products of a great many reactions, which themselves could not be broken down by any means at the disposal of the "chymists" of the time, and by seeing whether those products could be reassembled to form the original substance. In other words, they had to discover whether "elements" exist; and in the process, they had to clarify what they intended in speaking of elements. As opposed to the older tradition which maintained that something could be spoken of as an element only if it *actually* was ultimate, the newer "chymists" introduced the concept of an element as a product actually arrived at in the the laboratory, and which could not be further broken down by any means *at their disposal*. Further, as opposed to the older element theories, and indeed to the three-, four-, and five-element theories of the sixteenth, seventeenth, and much of the eighteenth centuries, the newer chymists maintained that not every element need appear as a breakdown product in every reaction; only some of the total set of elements would be constituents of any given substance. And finally, the candidates for elementhood would have to prove themselves true building-blocks by being actually put together to form the original substance of which they were the alleged constituents.

Vindication of this viewpoint was a long and complex process, extending well into the nineteenth century; indeed, as we shall see, it has faced a succession of perils and challenges in the twentieth century, and is even today experiencing another threat. But by the third quarter of the eighteenth century its promise was beginning to show. The ground had by then been well prepared for Lavoisier to bring together a list of elements in the new sense which by and large – *i.e.*, with some important exceptions and qualifications – would have a long and successful subsequent history. Even so, the successful defense of the breakdown-products view was not by itself sufficient to bring the new compositional view to full fruition. There remained the question of how to interpret specific chemical changes. In the light of the long philosophical and alchemical tradition, the powdery, crumbly calx of mercury would have seemed simpler (formerly interpreted as less perfect) than the metal; something, it seemed, should have to be *added* to the calx to produce the metal. The fact that the calx weighed more than the metal was not necessarily, or even, in the light of those same traditions, probably of any significance. For the idea of substances having a negative weight was an old and respectable one; in particular, a long and honorable tradition maintained that fire had a negative weight or intrinsic "lightness." If the metal were to be taken as a compound of calx and the matter of fire (phlogiston), the weight of the calx plus the negative weight of the phlogiston would add up to the lesser weight of the metal. On the other hand, if the calx, on heating, *released* a part of its substance, another breakdown product "oxygen," under the action of heat, the metal would, other conditions being satisfied, be the element and the calx the compound. (This question in turn depended on the hypothesis that "airs" could combine chemically with earth, a hypothesis that was strongly opposed, and with quite powerful reasons, by many.) In other words, if and only if all elemental substances had non-negative (*i.e.*, positive or zero) weight would the interpretation of this and a multitude of other reactions be made unambiguously in the way they have been since Lavoisier. Again, this latter view turned out to be be successful. Further details of the new theory were drawn in by the specifics of Lavoisier's theory, and, though most of those were shown in the following decades to be wrong, the compositional view was well on its way to two centuries of elaboration and development.

But the chemical revolutionaries did not stop here. For in the process of discovering new substances and new techniques for handling them, they and their predecessors had come more and more, in the eighteenth century, to the feeling that sensory qualities were inadequate as bases of naming and classifying material substances. On such bases, substances by then known to be different had been confused; and the same substance, if produced by two different methods, was often considered to be two different substances. Sensory qualities – what we today have come, on the basis of deepened understanding, to call "physical" as opposed to "chemical" properties – had proved themselves to be superficial, and a systematic method of naming which would reflect the deeper relations between substances came to be seen as desirable. Thus, with the rise of compositional views, demand arose for a revision of the names of substances, a revision which would reflect composition. "Denominations should be as much as possible in conformity with the nature of things"; "the denomination of a chemical compound is only clear and exact in so far as it recalls its constituent parts by names in conformity with their nature." So spoke Guyton de Morveau, perhaps the leading advocate of such reform; and the revision of chemical terminology was carried into effect by him in collaboration with Lavoisier, Fourcroy, and Berthollet.

There were thus five major ingredients in the "chemical revolution" as I have portrayed it: (1) the very possibility of compositional analysis, and the idea that knowledge of composition gives understanding of matter; (2) the concept of an element as a breakdown product, and (in support of (1)) the existence of elements; (3) the concept of non-negative weight as being centrally relevant chemically; (4) Lavoisier's oxygen theory of acids, metals, and calxes (which I have not discussed, as it provides only specific details of what I take to be the most fundamental aspects of the shift); and (5) the reform of chemical nomenclature to reflect composition (as conceived in Lavoisier's list of elements). As I remarked above, vindication of the compositionalist approach did not take place at one blow; it was a long time in coming, and indeed the viewpoint has had its ups and downs in the succeeding two centuries. We will encounter some of them later.

What I want to emphasize about this case are the following three points.

(1) The transition from the perfectionist to the compositionalist approach to matter-theory constituted a profound alteration of the *goals* of matter-theory and laboratory dealings with matter. In connection with the alchemical tradition, it is difficult to feel comfortable in saying that that tradition was even engaged in "investigation of matter," or in "trying to understand material substances" – so deeply are such expressions enmeshed in our compositionalist modes of thought.

(2) The shift also brought with it a revision of the vocabulary in which material substances, the objects of the matter-theorist's "investigations," are named and described, and that renaming and redescription involved the incorporation of the compositionalist view of the nature of matter into the naming and description of the objects of inquiry.

(3) Contrary to some widespread ideas in current philosophy of science, the new viewpoint can be understood as having been arrived at on the basis of what fully deserve to be called "reasons." The compositional view gradually gained credence through the discovery of a multitude of new substances; through the strong indications that old practices of naming and classification in terms of sensory qualities were unsuitable, superficial, non-fundamental, and productive of error – that a new, systematic, and more fundamental idea of *appropriateness* of names and descriptions was needed; through the discovery that indeed elements could be found, and that therefore the major uncertainty of compositional theories – whether chemical analysis alters a substance to produce the breakdown products – could be answered; through the gradual vindication of the view that all matter is characterized by non-negative weight, by demonstrating that such a view could be applied systematically and consistently in the interpretation of chemical reactions; and by a host of other considerations also. The considerations I have tried to bring out in this brief account should, however, be sufficient to show that the abandonment of the "perfectionist" viewpoint and adoption of the "compositionalist" one was a development which was fundamentally rational, despite the fact that it involved a radical change of the goals, methods, and conceptions of the subject-matter of thought about and laboratory handling of material substances. The world *could* have been of a "perfectionist" sort,

or, more generally, of a sort that is, at a fundamental level, non-compositional; the objection that chemical analysis alters substances might have proved valid. Nevertheless, there were considerations having to do with what was found out about material substances that indicated the acceptability of the compositionalist viewpoint. And though the notion of a "reason" (rational consideration) here has been left on an intuitive level, surely any account of scientific change, and any account of what a "reason" is, should be able to provide a basis for the intuition that emerges from this case.

If I am right about this, then the goals of inquiry can be altered in the course of inquiry, and for reasons having to do in a centrally important way with what we find out in the course of inquiry. We might even say that what counts as an "inquiry" may be altered, though the intuitive inapplicability of the concept of inquiry to what the alchemists were doing might lead us to search for a more general term to talk about whatever the activity was that *both* the alchemists and compositionalists were engaged in. This is why, for example, I have consistently retreated from talking about the alchemists' "inquiry into (or study of) material substances" to a more neutral, more general, but correspondingly vaguer description in terms of "laboratory handling of material substances." The continuities between what they were doing and what post-Lavoisier chemists do are real, and must not be submerged by focussing solely on their different goals and consequently taking the alchemical and modern chemical traditions to be "incommensurable." But though those continuities can be fairly clear, a common term for the activities of both traditions can be adequate only at the price of generality and vagueness.

So also, if I am right, the way we name and describe substances may *on some occasions* be altered in the light of what we learn about nature. More generally, our naming and description of the objects (or processes, *etc.*) of our investigation, of the subject-matter of our inquiry, may be altered in the light of what we learn about how those objects are to be understood, and of how they are interrelated. Information we have gained about those entities or processes, including information about how they are to be dealt with (understood), is incorporated into the naming and descriptive vocabulary we use in talking about them as objects of inquiry. (To use a favorite but highly misleading metaphor of philosophers of science in the past two or three decades, our "observations" become "theory-laden.") And those names or descriptions are subject to further alteration in the light of further findings.

Even if, rather than replacing an old set of names as occurred in the chemical revolution, we retain the same vocabulary, the "meanings" – as yet I suppose no technical analysis of this term – associated with the words are subject to revision. Consider, for example, the history of the term 'electron,' and, more generally, 'particle.' The term 'electron' was introduced by Stoney in 1894 to refer to the discrete unit of charge which seemed indicated by a proper interpretation of Faraday's earlier experiments on electrolysis. But with the discovery that that charge was associated with a material particle, the term came to be applied to the particle rather than to the charge. This was a natural shift of usage, because scientists since Lavoisier and in some areas since Newton had agreed, on the basis of abundant reason, that mass was the fundamental property of an entity, all other properties being possessed ("carried") by the massy entity but not being essential characteristics of it. So the material mass was really the object of concern, and it just happened to carry an indivisible unit of charge also.

In succeeding years, what counts as an electron has altered drastically. Quantum mechanics relativized the concept of a particle in general to specific types of experimental situation, and quantum electrodynamics blurred the distinction between particle and field still further. The assignment of quantum numbers to "particles" grants no special status to mass as opposed to charge or any other of the numerous properties, such as spin, parity, lepton number, *etc*., that have been discovered and assigned to the electron in those years. The behavior of this "particle" is in a great many respects unlike that attributed to it originally; in particular, its half-integral spin requires that it obey a special kind of statistics (Fermi–Dirac) as opposed to the Maxwell–Boltzmann statistics of classical particles and also to the Bose–Einstein statistics of particles with integral spin. In the meantime, during its long career since the end of the nineteenth century, the electron has been hypothesized to have many different properties, some of which were finally accepted as belonging to it while others were rejected. Some of the properties so hypothesized were in contradiction to others: for example, according to Lorentz, the electron could not have a zero radius; but the theory of relativity required that it have that very characteristic. Later accounts of the electron (and other elementary particles) managed to escape contradictions only by what many consider to be a makeshift expedient, renormalization, which, according to some scientists, provides no insight into nature but only allows us to "get the right numbers." On the other hand, one could hope that the success of renormalization is an indication of something deeper which we do not yet understand but which we may in time.

It seems natural to characterize this tale by saying that the electron is, or at least is treated as being, something *about which* we can have competing theories; that we may at certain stages assign certain properties to it, but at later stages we may assign others; that the electron is something which may have properties that we have not yet discovered, and to which we may assign properties which it does not have; that there may be many things about the electron that we do not yet understand. Indeed, it is easy to suppose further that this is the case for *all* substantive scientific terms. But what we are to make of this characterization – what brings it about that we treat the electron as thus "transtheoretical" (or perhaps "transdescriptional") – is something we shall have yet to investigate in this paper.

In the second part of this essay, I want to elaborate on these cases and the conclusions to which they (and of course many others) seem to lead regarding scientific change, generalizing those conclusions, extending them, and carrying their analysis and implications further. As I have tried to make a case for these points in other papers, I will not present or argue for them in detail here, but will discuss them only to the extent that they provide needed background for the treatment of issues about "meaning," "reference," and "necessary truth."

## II. THE INTERNALIZATION OF REASONING IN SCIENCE

In previous papers I have used the term "domain" to refer to those bodies of information which constitute the subject-matter of a given field or area of science: that is, which are considered to be related, the relation being conceived in terms of characteristics of the items, and regarding which items there is a common problem or set of problems and a reasonable expectation of a common sort of solution to those problems. (This is oversimplified; as one example of a complication that would have to be taken into account in a full treatment of domains, sometimes items of information are considered together on the basis of a common method of approach to them, rather than on the basis of common characteristics of the items. But such complications will not be important here.) The case of the development of chemistry discussed in Section I illustrates some central features of such domains and the dynamics of their change.

(1) That a given body of information (or what is taken to be information) *does* constitute a domain, to be studied as a related body, is itself a hypothesis – a hypothesis for which reasons may be given, those reasons themselves being in principle open to debate. (I will explain later what I mean by "in

principle.") Thus we have seen that earthy substances (roughly, solids) were in earlier times grouped together. The reasons for the grouping consisted of whatever considerations led to distinguishing between earth, water, air, and fire – for example, the rather primitive and naïve observations of their distinct properties and behavior. Not seeing compelling reasons for fundamental distinctions between types of earth, and on the basis of an analogy with vegetable growth and decay, the alchemists, at least, saw the "problem" with regard to this domain as one of bringing earths, or certain of them, to perfection. (The analogy was, to be sure, supported by observations allegedly showing such things as the decay of lead.)

(2) The division of experience into subjects for investigation (domains) is not something that is given once and for all, in any simplistic way (*e.g.*, by naïve sensory experience), but is something that can in principle evolve and change; and that restructuring of the fields of scientific inquiry can be instituted for reasons (rather than just on the basis of our "interests"). Such restructuring may be of any of the following sorts.

(a) The central problem or problems of the domain – the goals sought – may be reconceived. For the newer chemistry of the eighteenth century, it was the composition, not the perfectability, of material substances that was the problem about material substances.

(b) Distinctions may be made between items previously identified, and identifications made of items previously distinguished. We have seen this in the case of the development of chemistry, and have also seen the connection between such needs for identifications and distinctions and the rise of dissatisfaction with previous bases of classification and practices of naming and describing.

(c) New items of information can be added to the domain, and items previously included can come to be excluded. Airs and waters were clearly in the domain of Lavoisier's chemistry, whereas their role in chemical combination, and therefore their relevance to the study of such combination, had been rejected (on the basis of arguments) by many leading chemists before. New substances were constantly being discovered which, for the compositionalist, required chemical "analysis." On the other hand, Lavoisier included light and a new, weightless matter of heat, caloric, in his table of elements; later, both were separated from the subject-matter of chemistry. Again, with the clarification of the distinction between mixtures and compounds toward the end of the eighteenth century, a whole body of material "substances," mixtures, were removed from the concerns of the chemist. In general, whether any given item of (putative) information belongs to a given domain, in the sense (for example) that the same account can be expected of it as of other items in the domain, is in principle debatable. Whether a given claimed item of information even *is* an item of information can also in principle be questioned. That is, the responsibility of a given field of science for investigating and accounting for a given item or claimed item is in principle subject to debate. Indeed, entire domains can be split, what was a unified subject-matter coming to be viewed as the realms of two separate fields; conversely, separate domains can come to be fused, as were the subjects of light and electricity with the work of Maxwell. Such unification may be a drawn-out process: with the discovery of the electric battery, not only did a new technique for separating substances become available, but chemical combination was shown to be affected by electricity; this led to theories according to which electrical forces are the causes of chemical union. With the joining of electricity and light in Maxwell's theory, broader unity yet was suggested; but that fuller unification was achieved only with a clarification of the distinction between the fundamental forces of nature and the development of a fundamental theory of the electromagnetic interaction, quantum electrodynamics. And that distinction in turn has, over the past decade or so, begun to be broken down.

(d) The properties which are held to be relevant to the investigation of the items of the domain can undergo revision. At earlier stages, colors and other sensory properties were the relevant properties to be altered by alchemical processes. Weight was often not considered relevant, even by the "chymists" of the sixteenth through eighteenth centuries. It became a prime consideration to Lavoisier and his successors. Lavoisier, however, admitted also weightless substances; but the triumph of "mode-of-motion" theories of heat and light removed those items from the category of substances and transferred them from the domain of chemistry to other fields. Central in that development was the abandonment of the idea that any material substance could be weightless: all such substances were thenceforth held to be characterized by positive (non-zero, non-negative) weight, or rather, mass – thenceforth, that is, until the reintroduction of massless material entities with the photon, the neutrino, and other massless denizens of the modern elementary particle zoo.

(3) As we have seen, revisions of domains can sometimes take the form of renaming and redescribing the items therein; in the case of the chemical reform of nomenclature in the eighteenth century, the revisions incorporated the belief that knowledge of composition gave understanding of (compound) material substances; and a haphazard language and procedure of naming in terms of sensory qualities was replaced by a systematic language and

procedure of naming new substances in terms of their constituents. But even where renaming and redescription do not take place, profound alterations can and generally do occur in what is said about the entities in a domain. Indeed, as in the shift from Stoney's use of 'electron' to Thomson's, change can take place in whether an object of investigation is considered to be a property or an entity. Even such categorial distinctions as that between "entities" and "interactions" can be blurred, as that one has been in quantum field theories, where (for example) bosons are particle manifestations of interaction fields.

(4) In some cases it becomes possible in science to explain how it was that we were able to group items into a certain domain, or why a certain approach to explanation of information of a certain kind proved feasible – and even why that information was information of a certain *kind*, why we were right in grouping certain things together as being of the same kind. Thus one may ask why the compositional approach, in the form in which it was practiced in chemistry after Lavoisier, was successful. The answer is readily forthcoming in terms of our present knowledge of the fundamental interactions of nature. The binding energies responsible for chemical combination are electromagnetic in nature, and are of roughly the same order of magnitude as the energies available to us through the application of heat and electricity to substances. That fact made it possible to break down "compounds" (now understood as molecules). On the other hand, electromagnetic forces are separated from the strong forces, which hold atomic nuclei (protons and neutrons) together, by an enormous gap: those latter forces are far beyond the power of fire and batteries and the like to affect. Compounds can be broken down into their elements (atoms) by everyday-sized forces; but a huge leap in energy technology was required to enable us to break nuclei apart. Thus it is that, given a narrow, well-defined range of energies, ordinary materials can be broken into "elemental" constituents. We can even understand why, from our everyday perspective at least, there *are* elemental constituents, why there *are* certain fundamental kinds of substances: they are the last substances we can break down until we have vastly more energy available to us; and each of the distinct kinds is characterized by its own unique makeup of nuclear constituents and orbital electrons.

It is important to note that things might well have been otherwise. For example, the range of chemical binding energies might have been far greater than it is: some molecules might have resisted being split by any means at the disposal of nineteenth century scientists. They would then not have been able, in the case of some and perhaps even many substances, to defend the view that those substances were compounds of elements. They might not have been able to discover certain elements because of their tenacity in holding on to other elements. This situation might have been widespread; and whether the compositional approach would then have been maintained, the exceptions being explained in terms of the inadequacies of currently available energy sources to split them, or whether the compositional approach would have been at least temporarily abandoned, would have depended on many considerations; it certainly would not have deserved the support it had in the actual event. Or alternatively, instead of there being a wide gap in strength between the electromagnetic and strong (nuclear) forces, there might have been more of a continuum or overlap, and what we consider elements would have been as easily splittable as compounds (no doubt with unfortunate consequences for us). Breakdown products of a substance would then have depended on the amount of energy applied in a particular situation: for example, the breakdown products arrived at by subjecting a substance to a certain temperature would have been different from those obtained at some higher temperature. We would thus not have found elements. (This was in fact the very objection Robert Boyle raised against the element theories of his time: soap, for instance, acted on in one way, gave certain breakdown products; acted on in another way, it yielded others. Hence breakdown products were not elements.) In more extreme versions of such universes, we would indeed find approximations to a Heraclitan flux.

There are, of course, more radical ways in which things might have been, or may yet be found to be, other than what we now suppose. The present state of elementary particle theory is a case in point. To date, quarks have not been observed except within the confines of the individual particles (hadrons) of which they are said to be constituents: there, within the hadronic context, we find particles moving freely and having the characteristics quarks are supposed to have; but they have not been found outside this context. In the face of this situation, three alternatives suggest themselves. First: perhaps there are free quarks, but higher energies or more careful experiments are required to observe them. Second: perhaps quarks *are* constituents of hadrons, but are confined within the latter by a force which, unlike other forces of nature that we have known, increases with distance – so that, when we try to pull two quarks (for example) apart, we must apply more and more energy, until enough energy will have been fed in to produce a new quark-antiquark pair, which in turn will couple with the old quarks to entrap each of them once again in a hadron. It must be said of this alternative that, while such quark confinement is compatible with most versions of the latest theory of the strong interaction, quantum chromodynamics, it has not been shown to follow therefrom. Continued failure to observe free quarks, and to deduce quark confinement from quantum chromodynamics, could well prove disturbing enough to lead to serious consideration of more radical alternatives. Third: as one of these, perhaps quark confinement is an indication that, at a truly fundamental level, the very notion of a particle is only an approximation, valid only under certain restricted conditions; the concept of a particle – of an independent entity – may yet have to be viewed as appropriate only in an "asymptotically free" limit, and our fundamental theories will no longer be characterized by their talk about particles. Perhaps we would then return to a "geometric" theory of matter, of a type that has more in common with views of Descartes, Riemann, Clifford, Eddington, Einstein, and Wheeler than with the compositionalist approach. (Perhaps such a theory might be developed through the mathematics of harmonic mappings, as already tentatively suggested by Misner.) Even the compositionalist approach to the understanding of matter, despite its enormous successes, is not in principle immune to doubt and eventual rejection. Not only might things have been other than what we suppose them to be; they may yet be so. Indeed, our past experience with the development of science suggests that new theories may turn out to be wholly unlike anything we have so far been able to imagine. That is a sense of possibility that we must make room for in our philosophical theories!

(5) Finally, what was an explanatory theory for a domain at one stage of science may at a later stage (or even, in special circumstances, for a contemporary field) be taken as a domain, the subject-matter of a field of investigation itself. Although in the case of chemistry the elements themselves were part of the subject-matter of post-Lavoisier studies, nevertheless the central problem at that stage was to explain compounds as being composed of the elements: the latter, though their properties too were to be studied, initially functioned primarily as explanatory entities. Gradually, however, their properties came more and more to be primary objects of investigation – though such studies were long hampered by ambiguities about the properties to be studied (specifically, the atomic weights) which were interestingly analogous to the problem of interpreting chemical reactions faced earlier (the problem of whether negative weights were to be admitted). Much later in the nineteenth century, with problems of determining atomic weights resolved, and with the consequent ordering of the elements according to atomic weight in the periodic table, some scientists began to view the elements themselves, so related, as indicating the possibility of a deeper explanatory

theory, and thus as themselves (with that ordering) constituting a domain.

The picture we thus get is a highly dynamic one. First, it is of a kaleidoscopic shifting of associations of items into areas for study, and of a constant re-describing and often a renaming of the objects themselves. But it is not only that we redescribe and rename objects (or properties or processes or whatever the "items" with which we deal may be), as though those "objects" (*etc.*) were given once and for all, from our first encounter with them, though we may name and describe them in different ways. On the contrary, what we take to *be* the objects (*etc.*) is itself often in flux. (I will speak of this as a process of "reconceptualization," though that term is not completely satisfactory.) Nor are the possible changes restricted to the foregoing: in addition, what is once counted as a "property" can be transferred to the category of "objects" (electron); and even distinctions between categories can be revised (quantum field theory). And too, the very goals of inquiry, the problems to be dealt with and the kinds of answers to look for and the methods by which to seek them – even, that is, what "inquiry" itself consists in – can alter (perfectionist versus compositionalist approaches to "dealing with" matter).

But secondly, however apt it may otherwise be to talk about the shifts as kaleidoscopic, they are not arbitrary or capricious: the picture we get is one of the shifts as being brought about for reasons. This paper is not the place to seek an analysis of what a "reason" is in science; here all that is required is the intuitive notion that that is how the changes in question come about. For *whatever* view we may come to about what "reasons" are, that view will have to account for the shifts that take place in science, at least in cases that are most characteristically scientific, as shifts that take place for reasons. Judging from the cases considered above, we can say at least that a full view would include success and freedom from doubt as requisites for something's being usable in science as a "reason."

Thirdly, the picture we get from the cases examined in Section I and the generalizations extracted in Section II is one of a process of building information into our ways of thinking and talking, a building in of beliefs which have (ideally) proved successful, free from doubt, and perhaps otherwise worthy. Not only do the shifts take place *as a result of* beliefs that have been found successful and free from doubt (and perhaps satisfy other conditions that make them "reasons"); we also *incorporate* such beliefs into our conception of the subject-matter of our inquiry, by "reconceptualizing" the objects (*etc.*) of our inquiry and redescribing them (electron), and in some cases even renaming them (incorporation of compositionalist views into the nomenclature of chemical substances). Such incorporation of beliefs is an instance of a more general phenomenon in the development of science, which I have elsewhere called the *internalization of relevant considerations* in science: a process of gradual reformulation, in the light of what we have learned, of the scientific enterprise – its goals, problems, patterns of explanation, and indeed all its aspects, in addition to its subject-matter – so as to make all these so tightly bound by relevance considerations that any problem we confront can be dealt with solely in terms of the subject-matter which that problem is about. (Indeed, the very problems come to be formulable in terms of the way we conceive, name, and describe our subject-matter.) That process, which is essentially one of gradually distinguishing the scientifically relevant from the irrelevant and the unscientific, of gradually demarcating science from non-science, is an ideal we have learned to seek, but which is far from fully achieved. It exemplifies a third intuition (in addition to "success" and "freedom from doubt") that must go into any analysis of what a "reason" is: namely, that, in any argument concerning a subject-matter, those considerations will be relevant as reasons that have to do with that specific subject-matter. (The question, "What does *that* have to do with the subject we're discussing?" is a challenge that the consideration adduced by our opponent is irrelevant, that it does not constitute a reason either way in our dispute.)

Finally, the picture at which we have arrived is one of constant openness to the possibility that doubt *may* (though it need not) arise, that our present views, including the ways we conceptualize "objects" (*etc.*) and name and describe them, may have to be revised or rejected and replaced. Despite the fact that compositionalist views were built into the very language in terms of which we name and describe material substances, that view may even yet have to be rejected. Indeed, science is *never* irrevocably committed to *any* view which it holds, no matter how confidently it may build on that view and build it into our very modes of speech and thought. Our reasons for belief are never conclusive; it is always possible that doubt may arise. This is what I have had in mind in speaking, throughout this Section, of something being "in principle" subject to doubt or debate. (But of course the bare possibility of doubt is not a reason for doubting, in the sense that it picks out any specific belief as the one to worry about. The mere possibility of doubt does not preclude our building on those beliefs about which doubt has not, at least significantly, arisen yet.)

## NOTE

* Section I of this paper is a presentation, in quite different form, of the major case dealt with in an earlier article, "The Influence of Knowledge on the Description of Facts" (in F. Suppe and P. Asquith, eds., *PSA 1976*, East Lansing, Philosophy of Science

Association, 1977, Vol II, pp. 281–298. References to primary and secondary sources will be found in that paper; two extremely valuable secondary works are: M. Crosland, *Historical Studies in the Language of Chemistry* (Cambridge, Cambridge University Press, 1962), and R. Multhauf, *The Origins of Chemistry* (New York, Franklin Watts, 1967). Section II of the present paper goes well beyond the earlier one in discussing this case. The essay immediately preceding this one in the present volume, "Remarks on the Concepts of Domain and Field," should be viewed as an introduction to this one.

CHAPTER 16

# THE CONCEPT OF OBSERVATION IN SCIENCE AND PHILOSOPHY*

*(Summary Version)*

The solar neutrino experiment, designed to test our fundamental theories of the source of energy which sustains the stars and life, is one of the most sophisticated and important experiments of the past two decades. But it leads to some philosophical puzzles. For although the central core of the sun lies buried under 400,000 miles of dense, hot, opaque material, astrophysicists nevertheless universally speak of the experiment as providing "direct observation" of that central core. What can be meant by such talk? Is the puzzled philosopher simply ignorant of the ingenuity of modern science? Or are the astrophysicists using the term loosely or incorrectly or misleadingly, in a kind of sociological aberration that philosophers must gently tolerate while realizing that that usage has nothing to do with "real" observation? Or are the philosopher and astrophysicist perhaps interested in entirely different and unrelated problems, their respective usages being, from their respective points of view, equally legitimate? Or are the usages perhaps related, but in ways more complex than might be supposed from these or other usual alternatives?

The key to understanding the astrophysicists' use of 'observation' and related terms in their talk about neutrinos coming from the center of the sun is to be found in the contrast between the information so received and that based on the alternative source of information about the solar core, the reception of electromagnetic information (photons). In the latter case, our knowledge of electromagnetic processes, and of conditions inside the sun, tells us that, under the conditions of temperature and pressure existing there, the distance a photon can be expected to travel without interacting with some other particle is well under one centimeter. Therefore a packet of electromagnetic energy produced in the central core will take over 100,000 years to reach the surface. In that journey, it will be absorbed and re-radiated or scattered many times, and the original character of the radiation, and therefore of the information carried by it, will be drastically altered (it won't

* This article is a summary of part of Chapter II of a forthcoming book, *The Concept of Observation in Science and Philosophy*, and is reprinted here by permission of the publishers, Oxford University Press. A much longer portion of that chapter also appeared in *Philosophy of Science*, December, 1982, under the same title.

even be the same photon). But once it reaches the solar surface and passes into interplanetary space, the radiation will proceed to us with only low probability of being interfered with and altered. It is in this sense that our "direct" electromagnetic information about the sun is only of its surface; we only "observe" the surface. All conclusions drawn from that information about the deeper regions must be "indirect," "inferential."

Contrast this with information about the central core received *via* neutrinos emitted therefrom. The extremely "weak" character of the interactions of neutrinos with other matter, and the consequent low probability of their being interfered with in their passage over long distances, even when passing through dense bodies, enables them to pass without hindrance from the center of the sun to us. Any information they carry is thus unaltered by interactions along the way.

My suggestion is that we take this contrast seriously as a basis for interpreting the expression 'directly observed (observable)' and related terms in contexts like the astrophysicists' talk about direct observation of the center of the sun. The analysis I propose is this:

> $x$ is directly observed (observable) if:
> (1) information is received (can be received) by an appropriate receptor; and
> (2) that information is (can be) transmitted without interference to the receptor from the entity $x$ which is said to be observed (observable) and which is the source of the information.

(In this summary, I will ignore the modifier 'directly'; it is discussed in the paper.)

The remainder of my paper elaborates on this analysis, explaining in detail its content and implications. In particular, a detailed study of the solar neutrino experiment shows that, and precisely how, specification of what counts as observation is a function of the current state of physical knowledge, and can change with changes in that knowledge. (I will speak later of what counts as "knowledge" in these contexts.) More explicitly, current physical knowledge specifies what counts as an "appropriate receptor," what counts as "information," the types of information there are, the ways in which information of the various types is transmitted and received, and the character and types of interference and the circumstances under which and the frequency with which it occurs. These functions of prior knowledge are brought out by examining *first*, what I call, for convenience, the "theory of the source" – the account given of the release of information by the source (the entity $x$, in this case the core of the sun); *second*, the "theory of the transmission" of that information; and *third*, the "theory of the receptor" of the information. That detailed study produces a more general account of "observation" in such cases, and also some conclusions which are central to understanding the knowledge-seeking enterprise in science. The general account proceeds as follows.

The body of physical science includes assertions about the existence of entities and processes which are not accessible to the human senses – assertions to the effect that those senses (I will speak only of vision) are receptive to only a limited range of the electromagnetic spectrum. The eye thus comes to be considered a particular sort of electromagnetic receptor, there being other receptors capable of detecting other ranges of that spectrum. In other words, the extension of knowledge, in the discovery of the electromagnetic spectrum, leads to a natural extension of what is to count as observational: the very fact that information received by the eye becomes subsumed under a more general type of information, leads to all receptors of that information, including the eye, being considered on a par as "appropriate receptors." But besides electromagnetic interactions, current physics recognizes (with qualifications noted in the paper) three other fundamental types: strong, weak, and gravitational interactions. And this leads to a further generalization: an "appropriate receptor" can be understood in terms of instruments capable of detecting the interaction concerned, and therefore the entities interacting according to the laws of current physics. This extension of knowledge also produces a clarification of the concept of 'information' relevant in examples like the neutrino case: for the four fundamental types of interaction lead to there being, as of the present epoch in physics (subject to the noted qualifications), four fundamental types of information emitted by objects; those same four types of interaction also govern the reception of that information. And the laws of current physics (the laws of the relevant type of interaction) also govern the sense in which that "information" *counts as information*: that is, how, and the extent to which, and the circumstances under which, the receptor-interaction can be used by us to draw conclusions about the source. The conditions under which such conclusions can be drawn are expressed, where observation is concerned, in the two conditions stated earlier as to when an object can be said to be observed. For these purposes, however, as the detailed study of the solar neutrino case shows, knowledge of the four fundamental types of interaction is by itself insufficient to permit the drawing of conclusions about the source – to permit us, that is, to say that an observation has taken place. In the theory of the source, we must add both general laws about the *kind* of object the source is (in our case, the general laws of stellar structure) and specific information about the particular object (in our case, ultimately the mass and distribution of chemical composition in the sun, determination of the latter in turn demanding knowledge of the age of the sun, the chemical abundances on the solar surface, and the theory of stellar evolution). A combination of fundamental theory and other knowledge, both general and specific, must also be brought to bear in the theories of the transmission and of the receptor of the information. In the latter, for example, we must employ theories of nuclear reactions, experimental determinations of reaction rates, cosmic ray physics, the chemistry of noble gases, the properties of cleaning fluid, information about the radioactive content of the rock walls of the cave in which the receptor is located, technological information as to how to air-proof the apparatus (and theoretical information as to why this must be done) and of how to clean it, technological information about the capabilities of radioactive-decay counters, and much else. In all three components of the "observation-situation," moreover, the kinds of errors and inaccuracies to which the information is or may be subject is also given by current knowledge (in our case, for example, the range of uncertainties in nuclear reaction rates).

Thus, what we have learned about the way things are has led to an extension, through a natural generalization, of what it *is* to make an observation, and furthermore, various aspects of that knowledge are *applied* in making specific observations. It would thus be a mistake to say that there is *no* connection between the astrophysicists' use of the term 'observation' in this case and uses (at least certain ones) of that term which associate it with sense-perception. Rather, the relation lies in the fact that the astrophysicists' use is a generalization of (certain) uses having to do with sense-perception, and in the fact that whatever reasoning has led to our current understanding of the electromagnetic spectrum, the fundamental interactions, and the means of receiving information conveyed by the entities and processes we have found in nature, functions also as *reasoning* leading to the generalization. The generalization, that is, is not made capriciously, arbitrarily, but rests on reasons.

But there is a still more general point – also of a rational sort – behind the scientists' extension of the concept of observation, and it is this further point that brings out the contrast between his "observation" and that with which the philosopher is usually concerned. The philosopher's use of the term 'observation' has traditionally had a double aspect. On the one hand, there is the *perceptual* aspect: "observation," as a multitude of philosophical analyses insist, is simply a special case of perception, usually interpreted as consisting in the addition to the latter of an extra ingredient of focussed attention. "The problem of observation" is thus seen as a special case of "the problem of perception," to be approached only in the light of an understanding of the latter. On the other hand, there is the *epistemic* aspect: the *evidential* role that observation is supposed to play in leading to knowledge or well-grounded belief or in supporting beliefs already attained. For the empiricist tradition in epistemology proposed that all knowledge (or well-grounded belief) "rests on experience," where "experience" was interpreted as sense-perception. In that tradition, as indeed in most other philosophy, these two roles have been identified: the question of observational basis for beliefs or knowledge was interpreted as the question of how *perception* could give rise to knowledge or support beliefs.

In science, however, these two aspects have come to be separated, *and for good reason*. Science is, after all, concerned with the role of observation as evidence, whereas sense-perception is notoriously untrustworthy (in specific and rather well-known ways; the non-specific way or alleged way that leads to philosophical skepticism is completely irrelevant here). Hence, with the recognition that information can be received which is not directly accessible to the senses, *science has come more and more to exclude sense-perception as much as possible from playing a role in the acquisition of observational evidence*; that is, it relies more and more on other appropriate, but dependable, receptors. It has broken, or at least severely attenuated, the connection between the perceptual and epistemic aspects of "observation" and focussed on the latter. And this is only reasonable in the light of the primary concerns of science: the testing of hypotheses and the acquisition of knowledge through observation of nature.

It is true that there remains a crucial role for sense-perception in the acquisition of scientific knowledge. For after all, it is *we* human beings who have set up the "appropriate receptor," *we* who will use the received interaction as information. It follows that whatever information is received through the "appropriate receptor" must be transformed, in a final segment of the apparatus, into humanly-accessible form. Thus if the information comes *via* weak interactions, it must be transformed into electromagnetic information in the visual wavelengths, or into audible clicks, or into readable printout, or the like. But we must be clear as to exactly what is involved here: the human perceiver need not be present when the information is received by the "appropriate receptor"; nor need he even be present at the time the information is converted into humanly-accessible form. It has been an unquestioned assumption of the philosophical tradition that, for an "observation" to take place, the perceiver must be present when and where the information is received, and in a state and under circumstances in which he is capable of receiving that information. But as is seen in the solar neutrino case, this assumption need not be satisfied. The counts of neutrino-reception (and a great deal of their interpretation) are made by electronic devices and recorded by computers. In principle, a human perceiver need not drop by to pick up the information for years. Yet it still counts as observational evidence. Though he has set up the receptor for the purpose of advancing his own knowledge, the human perceiver plays the role of a mere user of the information received and recorded. That is the sole remaining link between observation (in its role as evidence) and sense-perception, at least in the solar neutrino case. (I might remark that the conflation of the perceptual and epistemic aspects of observation is not peculiar to philosophers; it is also found in ordinary usage, and understandably, since ordinary people ordinarily observe (acquire evidence) by perceiving (attentively, probingly, or however).

If there is anyone who still wants to insist that we ought not refer to the activities I have been describing as "observation" (but perhaps rather as "experiment" or "detection"), perhaps the following review of my arguments will serve as a deterrent to their recommendations. First, although the astrophysicists' usage is a departure from the ordinary, it is a *reasoned* departure, characteristic, in that regard, of the departures science often leads us to make in our beliefs. Second, its being a departure does not lessen the fact of its *relation* to what is ordinarily spoken of as "observation" (when it is related to perception): it is in fact partly a generalization of that concept. And finally, this "detection" performs the very same primary epistemic roles assigned to observation by the empiricist tradition and at least some aspects of ordinary usage: of being the basis of testing beliefs and of acquiring new knowledge about nature. Indeed, it performs those roles *better* than they could have been performed without the "background knowledge" that science has accumulated and that enters into scientific observation. There is thus abundant reason for considering the word 'observation' to be appropriate in the contexts I have been discussing.

We see in the case of the solar neutrino experiment the pervasive role played by what may be called "background information." It is through those roles that science builds on what it already knows, even where its observational capabilities are concerned. It *learns how* to observe nature, and its ability to observe increases with increasing knowledge (or decreases when it learns that it was mistaken in some piece of background information it employed). In the process of acquiring knowledge, we not only learn about nature; we also learn how to learn about it, by learning what constitutes information and how to obtain it – that is, how to observe the entities we have found to exist, and the processes we have found to occur in nature.

The employment of background information in science – indeed, the necessity of employing it – has been termed by some philosophers the "theory-ladenness" of observation. Putting the matter in that way has led to a great deal of perplexity: doesn't the "loading" of observation amount to slanting the outcome of experiment? and doesn't such slanting imply that scientific testing is not objective, and indeed that what science claims is knowledge is really only fashion or prejudice? Such perplexities, and the epistemological relativism they engender, trade in part on an ambiguity in the term 'theory.' For on the one hand, that term *is* used (not always appropriately) to refer to the background information which enters into the conception of an observation-situation. But on the other hand, it is also often used in reference to what is uncertain ("That's only a theory"). Collapsing these two senses leads to thinking of background information in science as uncertain, and from there by various paths to considering it arbitrary. But though it is true that the background information in science is *not certain* (in the sense that it could be mistaken, and in the sense that it involves a range of possible error), it is not for that reason *uncertain* (in the sense of being highly shaky or arbitrary). For wherever possible in the attempt to extract new information, what science uses as background information is the *best* information it has available – speaking roughly for present purposes, but nevertheless appropriately, information which has shown itself highly successful in the past, and regarding which there exists no specific and compelling reason for doubt. (We learn what it is for beliefs to be successful, and what to count as a reason for doubt, and when doubt is compelling in the sense of being serious enough to worry about.)

Calling all background information "uncertain" – calling it "theoretical" in the *second* of the preceding senses – emphasizes that *all* our beliefs are "doubtful" in the sense that doubt *may* arise, and that that doubt *may* prove so compelling as to force rejection of the idea in question. But that mere possibility of doubt, as we have learned in science, is *no reason* not to build on those beliefs which have proved successful and free of specific doubt; or regarding which the doubts that exist are either well-founded estimates of error ranges that are narrow enough, at least in some contexts, to permit useful investigation, or else are judged, on the basis of what we know, to be insignificant, not compelling, in some other way. Thus the fact that what counts as "observational" in science is "laden" with background information does not imply that observation is "loaded" in favor of arbitrary or relative or even, in *any useful* sense, "uncertain" views. (Nor does it imply that that background information cannot itself come to be questioned and rejected.) The employment of background information, far from being a barrier to the acquisition of knowledge about nature, is the means by which such further information comes to be attained.

My final point is this. Consider the following three sequences of descriptions of marks of various kinds on a photographic plate:

| | | | |
|---|---|---|---|
| speck | dot | image | image of a star (or of a particular star) |
| smudge | streak | spectrum | spectrum of a star (or of a particular star) |
| scratch | line | track | track of an electron. |

In each of these three sequences, as we move rightward, more "background information" is required. (In the book, I go into the kinds required.) Now certain philosophers have considered "the problem of knowledge" in something like the following way: we are to take as our starting-point the perceptual analogues of these dots, streaks, or lines (or perhaps specks, smudges, and scratches, or perhaps something still more, or even absolutely, "neutral"), and try to see how we could pass from them, without use of any "background beliefs" (including claims to knowledge) whatever, in the rightward direction of the sequences. But in the first place, that procedure is impossible (whether "logically" or "historically"): considering the dot to be an image *requires* the importation of prior information or belief; dots (and sense-data) alike are too impoverished, by themselves, to serve even as potential bases for obtaining knowledge. Relevance to being information, and to serving as a basis for obtaining further information, too, is created by richness of interpretation, and scientifically *reliable* information is established by employing, as background information to establish that reliability, prior successful beliefs which we have no specific and compelling reason to doubt.

But in fact, in science, we do not "begin" (whatever that means) with the dots (or specks or sense-data); we use the vocabulary that is strongest given what we know in the sense I have detailed. It is only when specific reason for doubt arises (for example, when we find reason to think that what we have taken as an image of a star *may* be of a quasar or a galactic nucleus or a comet) that we *withdraw* our description to the more "neutral" (with respect to the alternatives) level of speaking of it *only as an image* (of something). Further doubt can lead us to "retreat" again, to calling the mark a dot. And so forth, there being no clear reason to suppose that doubt might not arise at *any* level of description, whether our language is rich enough to provide a more neutral point of retreat or not. (Thus the very problem of the sense-datum philosopher and his cousins is suspect.)

All this is only to say again that we use our best relevant prior beliefs – those which have been successful and are free from reasons for specific and compelling doubt – to build on. Only now we see a new application for that principle: for it holds as well for our descriptions, and our vocabulary in general, just as it does in the belief contexts discussed earlier. There would be *no sense* (or at least only a humorous point *a la* Calvin Coolidge) in describing a situation in a "weak" way (for example, in a way taken from the left part of one of the above sequences) when we have *no reason* to describe it in that way, and when all the reasons we do have make a stronger description appropriate. In any situation, except when making a joke or in analogous (non-epistemic) cases, we use the strongest justified description, and only withdraw to less committal, more neutral ones when specific reason for doubt arises.

These considerations apply to the observation-situation in the case of the solar neutrino experiment: the relevant background information satisfying conditions of reliability as far as we have any reason to believe, we therefore say that our observation is of the center of the sun, or of processes occurring there.

The issues raised at the beginning of the paper are thus resolved. The use of the term 'observation' by the astrophysicists is not idiosyncratic or unrelated to certain aspects of ordinary and philosophical uses. Rather, it is an extension of such uses, a generalization thereof, made on the basis of reasons, and designed to make the most of the epistemic role of observation. The philosopher of the sense-datum sort (at least) is dealing with a problem, *his* "problem of knowledge," which differs in crucial ways from that of the astrophysicists; but the former's problem is suspect, and in any case has no relation to the knowledge-seeking enterprise as we engage in it, but conceives that enterprise in a way directly opposed to the way it is actually carried on, both in everyday life and in science. Indeed, *any* formulation of the problem of knowledge which conflates the problem of observation with the problem of perception, or which fails to recognize the necessary role of background knowledge in the knowledge-seeking and -acquiring process – any formulation of it, in other words, which falls short of the "naturalized" way in which I have been dealing with it – will be a misconception which will fail to grasp important aspects of the scientific enterprise.

But in dealing with the issues which were raised at the beginning of this paper, we have gone far beyond them. For we have come to see that, and how, science builds on what it has learned, and that that process of construction consists not only in adding to our substantive knowledge, but also in increasing our ability to learn about nature, by extending our ability to observe it in new ways. And, I might suggest, we have seen a way to understand how it is, after all, that all our knowledge rests on observation.

CHAPTER 17

# NOTES TOWARD A POST-POSITIVISTIC INTERPRETATION OF SCIENCE, PART II

## EXISTENCE AND THE INTERPRETATION OF PHYSICS

### 1. *The Logic of Idealization in Physics*

Our discussion will begin with a consideration of three examples. These cases are meant to bring out the fact that, in at least some actual physical reasoning, a distinction is made between the way or ways in which entities can or cannot exist and the way or ways in which, for the sake of dealing with certain problems, it is possible and convenient to treat those entities, even though, on purely physical grounds, we know that they could not really be that way. For the sake of convenience (and not to introduce a pair of technical terms) we shall refer to the concepts so distinguished as, respectively, "existence concepts" (or "existence terms") and "idealization concepts" (or "idealization terms"). The precise characteristics of these concepts will be brought out in the examples to be discussed. It must be remembered that for some purposes it might be necessary to point out differences between the cases classified together here, and, furthermore, that not all concepts employed in physics fit appropriately into *either* of the two classes distinguished here. The present section will focus primarily on idealization concepts, discussion of existence concepts being postponed until the next section.

1. A "rigid body" is defined classically as one in which the distances between any two of its constituent parts (particles) remains invariant. (Call this Definition R1.) If a force is applied to the body at any point, then, in order for the body to remain rigid in this sense—i.e., in order for the distances between any

*I am indebted for discussions leading to many of the ideas in this paper to the students in my graduate seminar at the University of Chicago. I also wish to thank Professors Kenneth Schaffner and Sylvain Bromberger for valuable help on many points.*

two points to remain the same—that force must be transmitted instantaneously to all other parts of the body. In other words, the force must be transmitted with infinite velocity. But according to the special theory of relativity, energy and momentum (and hence forces) cannot be transmitted with a velocity greater than that of light. Therefore, with the application of a force at one of its points, all the parts of a body cannot begin moving simultaneously; the body must be deformed. It is thus impossible, according to the special theory of relativity, that there should exist any such things as rigid bodies in the classical sense.

Nevertheless, the concept of a "rigid body" is employed by writers on special relativity, including, in a way that is central to his exposition, Einstein in his original 1905 paper on the subject.[1] How is such usage to be reconciled with the contradiction described above? There appear to be three general types of attitudes taken toward the role of the concept of rigid bodies with respect to the special theory of relativity.[2]

*a.* One may argue that for certain purposes the only sense of "rigid body" needed is the notion of a body which does not change its shape or size when *free* of external forces:

R2. A rigid body is one in which, if the body is free of all external forces and in equilibrium with respect to internal forces, the distance between any two constituent parts will remain invariant.

This is apparently the, or one, sense in which Reichenbach and Grünbaum employ the expression "rigid body" in their attempt to analyze the logical status of the concepts of space and time in science, and particularly in special relativity.[3] It must be noted, however, that this is *not* the sense in which the expression is ordinarily used in physics, where, according to R1, a rigid body is one in which, even if the body is *not* free of external forces, the distance between any two constituent parts will remain the same. In the sense of R2 *all* bodies are rigid,[4] whereas, according to the physical usage R1 (e.g., in the classical theories of rigid and elastic bodies), under the action of external forces some bodies are spoken of as rigid and others as not (and this is the interesting case for those branches of physics). It is the status of *this* notion of rigid body—R1—that is in question in the argument of the second paragraph of this section, not the notion expressed in R2.

*b*. One can maintain that the classical notion, R1, is adequate as an approximation, for practical purposes: that in many cases bodies can be constructed or visualized in which, under the conditions of some problem, the deformations produced by the prevailing external forces are so small that they can be ignored for the purposes of the problem. In this sense some rods are rigid *enough* (in the *classical* sense) to permit us to speak of, for instance, rigid measuring rods or rigid reference frames. Thus, for example, the bodies whose behavior is being referred to our system of "rigid" co-ordinates may be so far away, or so small in mass compared to the masses of the rods constituting our reference frame, that the distortions produced on the reference frame by those bodies can be neglected for the purposes of the problem under consideration (and therefore for more general purposes of theoretical analysis and exposition, where these requisite conditions may be assumed to hold). Fock gives a somewhat different practical justification of the employment of the classical concept of rigid body, although his justification is still in the spirit of the present approach.

> [For the reasons discussed earlier] the notion of an absolutely rigid body may not be used in Relativity Theory.
>
> However, this does not preclude the use of the notion of a rigid measuring rod in discussions of relativity. For this notion merely presupposes the existence of rigid bodies whose shape and size remain unchanged under certain particular external conditions such as the absence of accelerations or impulses, constancy of temperature, *etc*. Such rigid bodies can be realized with sufficient accuracy by solid bodies existing in nature and they can serve as standards of length.[5]

*c*. However, many writers have considered such a situation unsatisfactory. Treating the notion of rigid body in the classical sense as an approximation has seemed to such writers insufficiently rigorous for employment in a theory in which the notions of rigid reference frames and rigid measuring rods are claimed to play so fundamental a role. In all strictness such a notion of rigid body should be applied only under special circumstances (e.g., for rigid reference frames, where the reference frame is overwhelmingly—relative to the problem under consideration—more massive than the bodies considered in reference to it, or else sufficiently distant therefrom). Furthermore, it has seemed desirable to develop the theory of elasticity within the framework of special relativity. In connection with such considerations attempts have been made to develop a relativistic analogue of the classical concept of a rigid body. The first such effort was made by Born in 1909; however, his proposed definition was shown to fail by the successive criticisms of Ehrenfest, Herglotz and Noether, and Von Laue.[6] Other efforts have been made, including the following by McCrea:

> R3. We shall . . . define a *rigid rod as one along which impulses are transmitted with speed c*. . . . Since the theory [of special relativity] permits the existence of no 'more rigid' body of this sort, there is no objection to adopting the term *rigid* in this sense.[7]

If there exist no rigid bodies in McCrea's sense, it is not because the theory prohibits their existence, but rather because there are no bodies with a dielectric constant equal to 1 (in which case, impulse signals would be transmitted through such bodies with the velocity of light *c*; otherwise, the velocity of transmission in them would be less than *c*). As a matter of fact, no such bodies are known.

In both cases (*b*) and (*c*)—the cases that are relevant to the contradiction pointed out at the beginning of this section—we see that rigid bodies do not exist. The concept of a rigid body (in either the classical or the McCrea sense) is therefore often referred to as an "abstraction" or "approximation" (Fock) or as an "idealization" (Synge—see the quotation at the end of this paragraph).

Such references must not obscure the radical difference in the status of the classical and McCrea concepts. As far as the special theory of relativity is concerned, we might discover a body with a dielectric constant equal to 1—in which case a McCrea rigid body would have been found, and the point of calling that concept an "idealization" would be lost. But, if the special theory of relativity is correct, there *cannot be* a rigid body in the classical sense at all. The classical concept of a rigid body, insofar as it is employed in the context of relativistic physics, *must* be considered an "idealization": bodies simply *cannot* be of that sort. And the reason why that concept must be so considered is not some general philosophical thesis to the effect that *all* scientific concepts are "idealizations"; on the contrary, the reasons are purely scientific ones laid down by the special theory of relativity. It is thus clear that, if we look for reasons for calling the classical concept of rigid body an "idealization," the existence of these purely scientific reasons makes the following philosophical argument—advanced by Synge—superfluous: "In such measurements the infinitesimal rigid rod plays a fundamental part. Do such rods exist in nature? In one sense, we can say at once: Certainly not! None of the sharpened idealizations of theoretical physics is to be thought of as actually existing—they are like the point of Euclidean geometry."[8]

Nevertheless, despite the fact that there could be no such things as classical rigid bodies according to the theory of relativity, it proved useful, in the context of discussions of that theory, to talk in terms of such bodies. Furthermore, as we have also seen, it is possible to do so, at least in the context of certain kinds of problems. As the above quotation from Synge continues: "But in the world of these sharpened concepts, the idea of a rigid rod is useful and kinematically admissible."[9]

Nowhere in classical physics does one find any allegation (on scientific grounds analogous to those supplied by the special theory of relativity) that the existence of absolutely rigid bodies is an impossibility. If one wanted, then, to call the concept of a rigid body an "idealization," one had to appeal either to an assertion that there are in fact no bodies satisfying the definition or to extrascientific considerations such as a Synge-type argument to the effect that no scientific concept could be other than "idealized." However, just as there could, as far as the special theory of relativity is concerned, be rigid bodies in McCrea's sense, so also there might have been, as far as classical physics was concerned, absolutely rigid bodies in the classical sense. In particular, there was no reason (of a physical kind) why elementary particles could not be rigid bodies.[10] We shall return to this point later.[11]

2. We now pass to our second example. According to the Lorentz theory of the electron[12] (and difficulties in this regard have remained in all post-Lorentzian theories of the electron[13]), that particle cannot be a geometrical point, having zero radius. This results fundamentally from the fact that the electrostatic energy of a charged sphere of radius $r$ and charge $e$ is (except for a numerical factor) equal to $e^2/r$; this formula implies that a charged sphere of zero radius would have infinite energy, or, if we apply the Einstein relation $E = mc^2$ between energy $E$ and mass $m$, infinite (rest) mass.[14] However, the electron does not have infinite energy or mass.[15] Nevertheless, for certain purposes —for the solution of certain problems—and under certain circumstances, it is convenient and possible to treat the electron *as if* it were a point particle.

> We have written the solution of the potential problem as a sum of boundary contributions and a volume integral extending over the source charges. These volume integrals will not lead to singular values of the potentials (or of the fields) if the charge density is finite. If, on the other hand, the charges are considered to be surface, line, or point charges, then singularities will result. . . . Although these singularities do not actually exist in nature, the fields that do occur are often indistinguishable, over much of the region concerned, from those of simple geometrical configurations. The idealizations of real charges as points, lines, and surfaces not only permit great mathematical simplicity, they also give rise to convenient physical concepts for the description and representation of actual fields.[16]

As in the case of classical rigid bodies, we see from the present case again that it is on *scientific* grounds that treatment of the charged particle as a dimensionless point is considered an "idealization." The conclusion that the electron *cannot really be* a dimensionless point is not, in this case either, a logical or epistemological overlay superadded to the science concerned—a conclusion drawn solely from a more general and sweeping philosophical thesis to the effect, for example, that *all* scientific concepts are idealizations, or that all bodies are (or must be) extended, or that our ordinary concept (usage) of the expression "material object" and related terms implies that talk of dimensionless material objects is absurd.

Furthermore, not only is the *impossibility* of considering electrons really to be dimensionless points based on purely scientific considerations; the *rationale* for considering them *as if they were*—the possibility of so treating them, and the reasons why it is convenient to do so—is also scientific in character. As Panofsky and Phillips note, the fields that occur when we consider the source charges to be localized in a point are "often indistinguishable, over much of the region concerned, from those of simple geometrical configurations." It is thus *possible* to treat the source charges (at least in many problems) as if they were concentrated at a point, *even though we know, on purely scientific* (i.e., not metaphysical or linguistic) *grounds, that electrons cannot really be that kind of thing*. Furthermore, it is *convenient* to treat them in that way, for "The idealizations of real charges as points, lines, and surfaces not only permit great mathematical simplicity, they also give rise to convenient physical concepts for the description and representation of actual fields."

The electron as it really is cannot, therefore, have the zero-radius characteristic which is attributed to it for the sake of dealing with certain problems; and this distinction—between the electron as it really is and the electron as idealized because it is possible and convenient to treat it in a certain way—is one which, in this case at least, is made on purely scientific grounds.

This point is not vitiated by the fact that there are also difficulties in the Lorentz theory in considering the electron to have a *non-zero* radius. However the negative charge was held to be distributed over the extended Lorentz electron (and there were, consistent with the basic theory, a number of alternatives which could be held regarding this distribution[17]), the question naturally arose why the constituent parts of this negative charge did not repel one another, causing the explosive disruption of the electron. In order to ensure the stability of the electron it is necessary, as Poincaré showed,[18] to introduce cohesive forces that counterbalance the Coulomb repulsive forces of the electron on itself and maintain it in equilibrium. Poincaré's counterpressure, however, was not electromagnetic in nature and in fact was completely mysterious.[19]

If the "Poincaré pressure" appears to be *ad hoc* and objectionable, certainly the difficulty which thus arises in treating the electron as having a non-zero radius is not on a par with the objection against its being a dimensionless point. For in the former case the required counterforce is known, calculable: it is in fact equal to the Maxwell stress exerted by the surrounding field. Only the source, the explanation, of this counterpressure is unknown, and the hope could be maintained that it would be explained by some future, more complete theory—either a wholly electromagnetic theory (in which the Maxwell and Lorentz theories would appear as "special cases") or else a theory which would introduce entirely new forces supplementing electromagnetic ones.[20] If later, with the development of quantum theory, objections to the very notion of a radius of the electron appeared, such objections do not militate against the present point that, within the framework of Lorentzian physics, the electron—despite idealized treatments thereof—could not in reality be a dimensionless point, even though it might be possible and convenient to treat it as if it were, in order to deal with certain physical problems.

This point concerning the idealizational status of the notion of a dimensionless charged particle in Lorentzian physics is not affected by the following situation either. Recall that, according to the special theory of relativity, there cannot be such things as rigid bodies in the classical sense. This fact, as Yilmaz points out, has interesting consequences for relativistic discussions of elementary particles.

An elementary particle is, by definition, a material object which takes part in physical phenomena only as a unit. In other words, from the physical point of view it should not be useful to think of any component part to an elementary particle or to analyze it further. In order to describe the state of the motion of an elementary particle, it is sufficient to know only its position, velocity, and rotation as a whole. It is clear that this would imply a rigid structure if the particle had any classically meaningful extension at all. Thus, elementary particles must be pictured as point particles in the theory of relativity.[21]

Thus, whereas Lorentz' theory *precluded* the electron from having a zero radius, the theory of relativity *requires* that it have this characteristic. But Lorentzian electrodynamics is expressible in relativistic terms, and, indeed, in contributing the Lorentz transformations, it provided an essential ingredient in the development of Einstein's theory. Classical (relativistic) electrodynamics thus appears to contain a contradiction. This state of affairs does not, however, constitute a sufficient reason for rejecting that theory as useless—a conclusion one might be led to draw if one looked upon scientific theories as interpreted axiomatic systems. (For an axiomatic system which contains a contradiction will imply as theorems all well-formed sentences of the "language" in which the system is formulated. Thus, if scientific theories are viewed as interpreted axiomatic systems, one would have difficulty understanding how relativistic electrodynamics managed to distinguish true from false propositions.) The contradiction does, however, lead to considering relativistic electrodynamics to be not a fundamental theory but a theory requiring revision, a theory incapable, in the form in which it gave rise to contradiction, of providing a full account of particles. The contradiction even locates the area of inadequacy of the classical theory: namely, in the domain of the very small; and the realization of this fact might have led scientists, even at that time, to raise a significant question:

> Since the occurrence of the physically meaningless infinite self-energy of the elementary particles is related to the fact that such a particle must [because of the impossibility of rigid bodies, according to special relativity] be considered as pointlike, we can conclude that electrodynamics as a logically closed physical system presents internal contradictions when we go to sufficiently small distances. We can pose the question as to the order of magnitude of such distances.[22]

As we have seen, of course, the Lorentz theory, even in prerelativistic form, was already known to be incomplete, in that the stability of the extended electron appeared to require the existence of a force which was non-electromagnetic. But such incompleteness alone did not *require* that the fundamental concepts and equations of the Lorentz theory be *altered* or *replaced* —only that they be *supplemented*. In the face of the difficulty about the Poincaré pressure, those concepts and equations *might* have been altered; indeed, this was the aim of Gustav Mie's highly influential work.

> The first attempt to set up a theory which could account for the existence of electrically charged elementary particles, was made by Mie. He set himself the task to generalize the field equations and the energy-momentum tensor in the Maxwell-Lorentz theory in such a way that the Coulomb repulsive forces in the interior of the electrical elementary particles are held in equilibrium by other, *equally electrical,* forces, whereas the deviations from ordinary electrodynamics remain undetectable in regions outside the particles.[23]

But Mie had an axe to grind: in line with a "purely electromagnetic world picture (or rather, with the particular electromagnetic world picture which is based on the Maxwell-Lorentz theory),"[24] he wished to develop, along such lines, a theory which would be *absolutely complete.* But the difficulty concerning the Poincaré pressure did not require such unrelenting faith: supplementation, even along non-electromagnetic lines, might suffice. On the other hand, the present difficulty, the contradiction posed by the incorporation of the Lorentz theory into the relativistic framework, shows beyond a doubt that that theory cannot be *fundamental:* it has definite and specifiable limits wherein it cannot be adequate and *must* be replaced (and not merely supplemented), although the equations of the more fundamental theory might (and, indeed, in light of the success of the Maxwell-Lorentz theory, could be expected to) approximate more and more to those of the Lorentz theory as we pass from the domain of the small to that of the larger and larger.

Special relativity might also be blamed for the inconsistency; but in view of the large measure of independence of that theory from special assumptions about the nature of elementary particles,[25] and also in view of its applicability to domains other than electrodynamics, this location of the blame surely would have appeared unlikely.

Note that one thing that is *not* done is to dismiss the problem by saying that the theory of relativity, or classical electrodynamics, is merely an "idealization" (at least with regard to the treatment of elementary particles), or that the contradiction is merely a characteristic of the "formal structure" of the system. The difficulty is taken seriously and, furthermore, is looked upon as indicating the inadequacy of the theory as an account of nature. Such seriousness surely should be connected with the fact that the theories considered distinguish between the ways certain entities—the entities concerned in the contradiction—cannot be and the ways in which they can or must be. The theories are contradicting one another, not in the idealizations they consider useful (would that be a contradiction in any case?), but in their assertions about the way things are: the idealizations made by the theories (including the idealization of electrons as dimensionless points in the Lorentz theory) are irrelevant. And, in any case, the distinction between idealizations and the way things are is one which, in the cases considered at least, is made within the theories, on scientific grounds; the notion of "idealization" has a definite use in physics. Should we not, then, be wary of extending that notion to physics (physical theories) as a whole—of trying to resolve our perplexities about physics by assimilating physics to the status of "idealizations"? For in such extended usage the definite scientific function of idealizations—the working use of the term "idealization"—is not to be found.

Study of the cases of rigid bodies and the electron has led us to consider a new problem, namely, how the non-idealized notions are connected with the notions of "completeness" and "fundamentality" of "theories." No technical meaning of the term "theory" has been presupposed here: certain propositions, ordinarily held to be "parts of" the Lorentz and Einstein theories, have been shown to legislate certain requirements as to the way things must or can or cannot be. In conjunction with other considerations, these requirements in turn lead to the exposure of deficiencies which require that the propositions (and concepts) in question be either supplemented or modified—that the theory in question is, in precisely specifiable respects, either not complete or not fundamental. And on some occasions the steps that need to be taken, or the (or some of the) alternative steps that might be taken, or at least the domain where modifications need to be made can be specified in advance, at least to some degree. (Thus Poincaré calculated the needed but mysterious counterpressure; Landau and Lifshitz—with the benefit of hindsight, to be sure, although a hindsight which *could* have been foresight—locate the limitation of classical relativistic electrodynamics in the domain of the very small.) These are questions that have received little attention in twentieth-century philosophy of science; we shall return to them later.[26]

3. A situation analogous to that regarding the possibility of point particles in the Lorentz electron theory arises within the context of classical mechanics, wherein an infinite gravitational potential would result from the localization of gravitational mass in a dimensionless point. If the occurrence of such an infinite value is ruled out, this could count as an argument to the effect that "mass points" are impossible according to classical mechanics.

Be that as it may, however, Newton himself (or any of his successors over at least the following century, as far as I am aware) did not employ such an argument, which, in the form presented here, had to await the introduction of the notion of potential. Newton's own reason for maintaining that particles are not *really* mass points (as they are considered in his mathematical theory) appears to be based on Rule III of his "Rules of Reasoning in Philosophy": "The qualities of bodies, which admit neither intensification nor remission of degrees, and which are found to belong to all bodies within reach of our experiments, are to be esteemed the universal qualities of all bodies whatsoever."[27] On the basis of this rule, Newton argues that "the extension, hardness, impenetrability, mobility, and inertia of the whole, result from the extension, hardness, impenetrability, mobility, and inertia of the parts; and hence we conclude the least particles of all bodies to be also all extended, and hard and impenetrable, and endowed with their proper inertia."[28] One is tempted here to say that Newton has appealed to "extrascientific" considerations for his ascription of non-zero extension to elementary particles. Yet the status of general principles of inductive inference, such as Newton's Rule III, seems to me to be doubtful: scientific considerations have, after all, been relevant to its rejection—namely, considerations issuing primarily from quantum theory.[29]

Whatever the rationale for supposing "mass points" to be idealizations in Newtonian mechanics, however, the rationale for supposing their employment to be both useful and possible is clear. (1) There are certain problems to be solved—problems relating to the positions, velocities, masses, and forces of bodies. (2) Mathematical techniques exist for dealing with such problems if the masses are considered to be concentrated at geometrical points (namely, the geometrical techniques of Newton and, for later scientists, the methods of the calculus). (3) It is, as Newton showed, possible to treat spherically symmetrical bodies as if their masses were concentrated at their centers (and, incidentally, Newton conceived the elementary mass-particles to be spherically symmetrical). And, as for bodies not spherically symmetrical, they could be considered in the same way, provided the distances between their centers was large in comparison to their radii—a condition fulfilled, happily, by the earth-moon system. More generally, bodies could be considered as if their masses were concentrated at their centers of gravity.

These same kinds of considerations are relevant in a wide variety of cases in science which may be (and ordinarily are) classified as "idealizations." This is not to suggest that there are not other kinds of concepts in science which may be (and are) called "idealizations," the rationale of whose use differs from the cases considered in this section. Nor is it to suggest that there is not more to say about the rationale, the logic, of idealization in the cases that have been considered here: for some purposes it might be useful, for example, to note differences between the cases dealt with above. Thus the impossibility of classical rigid bodies in the light of special relativity is a result of a direct contradiction of that classical notion with a fundamental principle of special relativity; but the impossibility of Lorentzian point particles results only from the rejection of the possibility of an infinite energy (or, if Einstein's principle $E = mc^2$ is used, perhaps from the fact that electrons are "observed" not to have infinite mass). And for certain purposes these differences—passed over in the above account—might well be highly relevant. Again, for certain purposes it might be useful to talk in terms of relative *degrees* of "idealization." Nevertheless, the cases as presented above illustrate the point of crucial importance for *present* purposes, namely, that there are often good *scientific* reasons for distinguishing between the way in which a certain entity is asserted to be (or not to be) and the way in which it is *treated* (although, again for scientific reasons, it could not really be that kind of thing) for the sake of convenience in dealing with certain scientific problems. It is this distinction and the insight it affords into the actual rational working processes of science that now must be examined more closely.

## 2. *The Logic of Existence Assertions in Physics*

In the preceding section we have seen that in physics there are cases in which a distinction is made, on scientific grounds, between the way (or ways) in which an entity can or cannot exist, and the way (or ways) in which, for the sake of dealing with certain scientific problems, it is possible and convenient to "idealize" that entity. This distinction may now be put in a more general way as follows. On the one hand, we have assertions that certain entities do, or do not, or might, or might not, exist; or, putting the point linguistically, we have terms that can occur in such contexts as ". . . exist(s)," ". . . do (does) not exist," ". . . might exist," ". . . might not exist." (It should be emphasized that what are of relevance are not the *terms* involved, but rather their uses; thus, in a problem in which we treat the electron as a point charge, we may refer to the point charge *itself* by the term "electron." The context of usage, however, will indicate that the term is in such a case concerned with the idealization and not directly with electrons as they really are.) There are *clear* cases of such terms, or of such uses of terms. Sometimes they have to do with entities that are presumed to exist ("electron"), although they might not exist (or might not have existed). Others have to do with (purported) entities that, although they have at some time been claimed to exist, do not ("ether," "phlogiston"). And, finally, still others refer to (purported) entities whose existence is claimed (on presumably good grounds) by some good theorists, but whose existence or non-existence has not yet been established ("quark"). It should be noted that this class of terms includes also many terms (uses) which have commonly been classified as observational—"table," "planet"—as well as terms (uses) usually classified as theoretical.

On the other hand, we have expressions like "point particle" (in the Lorentz theory) and "classical rigid body" (from the viewpoint of special relativity), which do not designate (purported) entities, although they are *related,* in the ways discussed in the preceding section, to terms (or uses of terms) which do. Thus their reference to existing things, for example, is indirect. To put the point in another way, such terms or expressions cannot occur in the context ". . . exist(s)," *except* in a derivative sense, namely, that they have to do with "idealizations" of entities that do exist. And although these terms can occur in contexts like ". . . do (does) not exist," the sense in which they can so occur is stronger than, for example, the sense in which we can say "Vulcan does not exist" or "Nebulium does not exist"; for, in the cases of Lorentzian point particles or classical rigid bodies (according to special relativity), it is physically impossible that they should exist.

The first type of terms or uses may be called "existence terms"; or, inasmuch as, in the cases we have considered, what are alleged to exist are certain kinds of entities, they may be called "entity terms." But not all terms in science which refer to something "non-idealized" are naturally classed as entity terms. Many of them are more naturally referred to as having to do with "properties," or with "processes," or with "behavior of entities," for example. Furthermore, there are many borderline cases that are not easily brought under any of these headings. Finally, there are differences between the uses so classified which might lead us, for certain purposes, to distinguish between those cases or to classify them very differently. But all these points only bring out the fact that it is not the terminology of "entities," "idealizations," etc., that is important; what is important is the logic of the scientific usage of the terms so classified. Such expressions as "entity term" must therefore be understood as terms of convenience employed in order to call attention to certain features of the cases discussed—features which, although important and real, nevertheless are not the bases of some ultimate and final classification that excludes all others.

For similar reasons, these classifications should not be considered as providing a new "metascientific" vocabulary—a set of concepts which will serve for unambiguous discussion of *any* scientific work—which can (or even must) be employed in *any* attempt (past, present, or future) to characterize the scientific endeavor. Unless we stretch the meanings of the words beyond utility, the concept of "entity," for example, which is so naturally applied in some cases in science, becomes more and more inapplicable in others.

All these remarks apply also to the second type of uses of terms which, we suggested earlier, may be referred to as "idealization terms." In particular, as was noted at the end of section 1, there are differences between the cases discussed there which might lead us, for some purposes, to distinguish between them. Also those types of cases do not exhaust the kinds that, in science, are to be contrasted with "existence terms." There are many other kinds of cases in physics, not entirely like the cases we have examined and called "idealizations," in which we might feel it more appropriate to talk of, say, "abstractions" (e.g., considering a system of entities as isolated from the rest of the universe), or "approximations" (e.g., calculation only within a certain range of accuracy), or "simplifications" (e.g., considering electronic orbits to be circular rather than elliptical). For, though all these terms—"idealization," "abstraction," "approximation," "simplification"—are used loosely and often interchangeably in talk about science, there are nevertheless significant differences between cases clearly not having to do with things as they really are, which might make distinct employment of those terms useful. (Often, even usually, in science these logically different procedures are combined in attacking a problem, and this fact has encouraged writers to ignore the logical differences between them.)

The relations between "entities" and their "idealizations" may also be different from the cases considered in section 8. For example, it may be that the entities concerned are held not to exist in the way they are treated, not because their existence in such a form would contradict some basic theoretical principle, but rather because such entities have not been found (Vulcan, fifth force of nature), or because the reasons for supposing them to exist have, for one reason or another, been abandoned (ether, nebulium). And yet in some such cases, even after what we might call the "existence claim" of such concepts has been abandoned, there may still remain a *use* for the concepts. Thus, in contrast to the cases of impetus, phlogiston, caloric, and ether, where the conclusion that the (type of) entity did not exist was accompanied by a loss of utility of the concept, the nineteenth-century abandonment of the view that light consists of the transmission of particles, and the adoption of the wave theory, did not preclude the usefulness and possibility of employment of ray optics (which was associated with the particle theory) under certain circumstances.

> . . . We are thus led to recognize two kinds of optics. In the first, called "geometric optics," or "ray-optics," light is propagated along rays (straight or curved); it thus behaves much as beams formed of corpuscles might be expected to do. In the second form of optics, called "wave-optics," the wave nature of light becomes conspicuous. Of course the distinction we are here making is artificial. In all truth, there is only one kind of optics, namely, wave-optics; as for ray-optics, it is a mere ideal limiting case never rigorously realized in practice. The present situation is very similar to the one we have met with time and again. Under certain limiting conditions, a theory or a manner of interpreting things, though wrong in the last analysis, is nevertheless so nearly correct in its anticipations that we are often justified in accepting it, even when we recognize that it is a mere approximation. An example was mentioned in connection with classical mechanics: We have every reason to suppose that the theory of relativity is more nearly correct than classical mechanics; and yet, since under the limiting conditions of low velocities and weak gravitational fields, the predictions of the theory of relativity tend to become indistinguishable from those of the classical theory, we are justified in retaining classical mechanics in many practical applications. Likewise, when the energy values are high and the energy transitions are small, we may neglect the refinements of the quantum theory and base our deductions on the classical laws of mechanics and of radiation.
>
> The relationship between wave-optics and the less rigorous ray-optics is of the same type. Whenever the irregularities or inhomogeneities of the medium in which light is transmitted are insignificant over extensions of the order of the wave length of the light, the conclusions derived

from wave-optics tend to coincide with those obtained from the application of ray-optics. In such cases, ray-optics may advantageously be applied because of its greater simplicity.[30]

This example is of further interest for at least three reasons. First, it shows how such notions as "ideal," "limiting case," and "approximation" can be used interchangeably in talk about science. Second, the example brings out the fact that entire theories (sets of concepts and propositions), as well as individual concepts, can be spoken of as "idealizations" in the sense that, although it is known, for scientific reasons, that things are not really as they are alleged to be in the theory, it is nevertheless often convenient and possible to treat them in that way. But, again, ray optics is not an "idealization" because *all* scientific theories are "idealizations"; it is an idealization because it is contrasted with another theory which is *not* an idealization. And, finally, the example—and indeed all our examples—brings out the way in which concepts can *change* status—from entity term to idealization term—over the history of science. Such shifts are only to be expected, in view of the fact, brought out in earlier discussions, that the distinction between entities as they are or are not (or can or cannot be) and the ways in which they are treated ("idealized") is based on physical grounds rather than, say, on logical, metaphysical, or linguistic grounds.

Noting, as we have, that there are clear cases in which a distinction is made, on scientific grounds, between existence concepts and idealization concepts does not, of course, constitute an analysis of what is involved in a claim that something "exists." And, one might argue, the history of philosophy has shown that that concept is too opaque to offer any reasonable hope of illuminating the scientific process or any part of it. In reply to this objection, three points must be emphasized. *First*: the examples presented here show that talk about the "existence" (and the same holds for the "real properties") of electrons is not some metaphysician's talk about a subject (physics) which itself offers no assertions about existence; rather, they indicate that the distinction between the way certain things are (or are not) and the way or ways we can and do treat them in certain problem situations is one which is made *within* science, for scientific rather than for metaphysical or otherwise extrascientific reasons. And, in view of this fact, we might paraphrase Kant by saying that the question is not *whether* a realistic interpretation of at least some science is possible, but rather, how we do it—how we manage to make and justify existence-claims, and what the implications of the fact that we do, on occasion, are for the interpretation of science.

*Second*: it will be recalled that one of the major purposes of Part I of this paper was to show that many—although perhaps not all—of the puzzling features of such statements as "Electrons exist," and much of the apparent gravity of the so-called problem of the ontological status of theoretical entities, can be attributed merely to phrasing the problem, and demanding a solution of it, in terms of the empiricist-positivist distinction between "theoretical" and "observational" terms. If this contention is right, then we need not shy away from taking talk about the existence of electrons seriously just because the question of existence currently *seems* to be so overwhelmingly perplexing and difficult. On the contrary, as was suggested earlier, we may actually be encouraged by the multitude of very serious weaknesses which has been exposed in the theoretical-observational distinction itself. And, if one looks at cases, one finds, as also was noted earlier, that there are at least as many clear cases in which it is natural to use scientific terms in the context ". . . exists," etc., as there are cases in which the term is naturally classified as "theoretical" or "observational."

*Third*: There are some things that we can say about the role played by existence-claims in physical reasoning. To say that "A exists" implies (among other things, surely) at least the following:

1. A can interact with other things that exist. Particles that exist can interact with other particles that exist, and, derivatively, can have effects on macroscopic objects and be affected by them. "Convenient fictions" or "constructs" or "abstractions" or "idealizations" cannot do this, at least not in any ordinary sense.

This feature of existence-claims brings out an important error in one common positivistic view of science: namely, the interpretation of so-called correspondence rules, co-ordinative definitions, or rules of interpretation. We saw earlier that, modeling so much of their approach to the philosophy of science on mathematical logic, positivists talked about the linkages correlating "theoretical" with "observational" terms as being analogous to the interpretation of a formal system in mathematics and logic. Such linkages in logic clearly are unlike the kinds of connections that can be asserted to hold between existent things (e.g., causal interaction). (And we noted earlier how so many of the puzzling features of statements like "Electrons exist" arise here out of a bad analogy and a bad distinction.) But talk of electrons as existing enables us to consider assertions of linkages between electrons and, for example, scintillations or clicks (to say nothing of positrons) as assertions of causal actions or interactions (i.e., the particles involved cause the scintillations or clicks that we "observe"). Note, incidentally, that what counts as an "interaction" is also specified on scientific grounds.[31]

2. To say that "A exists" implies that A may have properties which are not manifesting themselves, and which have not yet been discovered; and, contrariwise, it is to say that some properties currently so ascribed may be incorrectly so attributed. We may be wrong in saying that a certain property of an entity has a certain quantitative value. Or we may be wrong in thinking that that property is fundamental—it may be, to use Leibniz' colorful phrase, a "well-founded phenomenon," being a manifestation of some deeper reality (e.g., as Wheeler claims that many properties of particles may be explainable as mere manifestations of an underlying "geometrodynamic field"). Or, again—although these kinds of cases are rarer, especially in more sophisticated stages of scientific development—we may be wrong in thinking that the entity has the property at all. Finally, we may be wrong in thinking that we have exhausted all the properties of the entity, and may discover wholly new ones (spin, strangeness).

These features are all hard to understand if electrons, for example, are mere "convenient fictions." (Note that what counts as a "property" is also specified on scientific grounds.)

3. To say that "A exists" is to say that A is something about which we can have different and competing theories. From the theoretical work of Ampère and Weber to that of Lorentz, from the experimental work of Faraday on electrolysis to Millikan's oil-drop experiment, there was an accumulation of reasons for holding that electricity comes in discrete units. The notion of the electron thus acquired what amounts to a theory-transcendent status: it was an entity about which theories—theories *of* the electron—were constructed. It is indeed ironic that the term "electron"—often taken as a paradigm case, in the philosophical literature of the positivistic tradition, of a "theoretical" term—should have this status; for the comparability of different, competing theories is now seen to be, not (at least not solely) their sharing of a common "observational vocabulary," but rather their being about the same kind of entity. The erstwhile "theoretical term" is now seen to be the source of what is perhaps the most important aspect of the "comparability" of competing theories: for electrons are what those theories are in competition about.

### 3. *Applications and Extensions of the Analysis*

The forgoing investigations have shown that there is, within at least some cases in science itself, ground for distinguishing between the ways in which entities can or cannot (do or do not) exist and the ways in which, for the sake of dealing with certain problems, it is possible and convenient to treat them, despite the fact that we know, on physical grounds, that they could not really be as they are treated. Thus, the present paper may be said to constitute, in part, a defense of a "realistic" interpretation of at least some scientific concepts.

But the significance of the analyses made in Part II does not lie simply in the support they offer to "scientific realism." For, by looking at science from the perspective of the approach taken in the preceding two sections, we find that new light is thrown on a number of classical problems of interest to the philosopher of science, while others are revealed which hitherto have been slighted. We saw at the end of section 2, for instance, that an unexpected twist has been given to the "comparability problem" by looking at science in terms of the entity-idealization distinction: an important aspect of the comparability of competing theories lies in their dealing with the same entities (formerly called "theoretical") rather than merely in some shared "observational vocabulary."

We have also seen, in our examination of cases in science, something of the rationale by which scientific theories at various stages are considered to be "incomplete" or "non-fundamental." These notions, which have largely been ignored by philosophers of science, have an important bearing on the attempt to analyze the concept of explanation in science. For, what we saw in section 1, in reviewing the fortunes of the Lorentz electron theory, may be outlined as follows: Against the background of a certain body of knowledge (both "theoretical" and "factual"), certain questions arose which required answers. At that stage, the directions in which it was plausible to look for solutions—at least *some* of the possible alternative paths along which it was reasonable to look for solutions—could be discerned, not only with hindsight, but even by the participants in the enterprise themselves. I would suggest that an understanding of what scientific explanation consists in may be gained by filling in the details of this outline: by analyzing the ways in which a background body of knowledge generates such questions, how alternative pathways of research are delineated for answering the problem, and how it is decided when the problem is resolved.[32] Note, incidentally, that—as our examination of cases has already revealed—what is involved here is the notion that there is a rationale, a logic, to the development of science.

Illumination of the notions of "theory" and "observation" ("evidence," "data," "facts"), as well as of the difficulties into which that distinction has fallen, may also be obtained through the present approach. For one thing, the distinction made here cuts across the lines of the old theoretical-observational distinction, for we have found reasons to classify electrons (formerly "theoretical") with tables and rocks. On the other hand, the class of "theoretical terms" now has been broken up into various categories of terms having very different functions: existence terms, idealization terms, and many other (rough) groupings of terms whose functions have only been touched on here. Thus we can see that it was futile to try to find some relatively simple form of relationship between theoretical and observational terms: the class of "theoretical" terms covered many kinds of concepts, having distinct sorts of relationships to one another, and, in particular, in the case of such types of concepts as idealizations, being related only indirectly (through the associated entity terms or concepts) to "observation" or "evidence." It is no wonder that both "realistic" and "instrumentalist" interpretations of science faltered with regard to their analyses of some theoretical term or other. For at least those areas of science where such terms as "existence" are appropriate, the problem may now be seen in a different way: to delineate the relationships of those terms (rather, uses) which do not have to do directly with entities as they actually exist, to those terms (or uses) which do; and to analyze the reasons for accepting existence-claims.

Just as the positivistic tradition—as we said in Part I—used the theoretical-observational distinction as an analytic tool for framing certain problems and dealing with certain topics, so the notion that science makes existence-claims, and the contrast between such claims on the one hand and idealizations (and like types of concepts) on the other, now has become a focus, a tool, in the effort to understand science. We can expect difficulties to accumulate regarding this approach, just as they did over the years for the theoretical-observational distinction. And, just as the latter distinction exaggerated or distorted certain problems and features of science while ignoring or slighting others, so also we need not be disappointed if similar limitations are found to exist in this approach. On the contrary, they should be expected and even welcomed, because they might suggest further approaches, which in turn might expose further problems and features of science to be dealt with—just as the positivistic failure to deal with the problem of ontological status in terms of an objection-riddled theoretical-observational distinction made clear the need for a re-examination of that problem independent of the theoretical-observational distinction. More generally, even if such difficulties and limitations are found in the present approach, it can nevertheless have important *heuristic* value if it enables us to phrase certain problems and to discuss certain features of science in new and at least partially illuminating ways which are free of doctrines that have proved seriously defective, and if it brings to light further problems and features of science which have been obscured by previous approaches. And this much, if the arguments of this paper are sound, we have already seen it do.

Viewed in the light of these remarks, the contributions of the positivistic tradition deserve greater appreciation than we might give it if we considered only the deficiencies of so many of its contentions. For, by looking at science carefully in certain ways, it also raised a great many problems regarding the interpretation of science; its answers to those problems, while perhaps far from being all that one would have wished, still provided considerable illumination and insight; and, finally, its very defects, as this paper contends, pointed the way toward a new approach.[33]

## NOTES

1. A. Einstein, "On the Electrodynamics of Moving Bodies," *Annalen der Physik*, 1905; reprinted in translation in *The Principle of Relativity* (New York, Dover), pp. 35–65. Einstein declares that "the theory to be developed is based—like all electrodynamics—on the kinematics of the rigid body, since the assertions of any such theory have to do with the relationships between rigid bodies (systems of co-ordinates), clocks, and electromagnetic processes. Insufficient consideration of this circumstance lies at the root of the difficulties which the electrodynamics of moving bodies at present encounters" (p. 38).

Some thinkers (e.g., E. A. Milne) maintain that the concept of a rigid body need not be taken as playing a fundamental role in special relativity. I will not consider their arguments here, for I am presently concerned only with whatever rationale there might be for employing the concept, in any sense, in discussions of relativity. Einstein himself appears to have vacillated somewhat in his opinion of the centrality of the concept of rigid bodies to his theory; but in general he seems to have held that the concept is central and fundamental. This attitude is revealed in the following comment:

> One is struck [by the fact] that the theory (except for the four-dimensional space) introduces two kinds of physical things, i.e., (1) measuring rods and clocks, (2) all other things, e.g., the electro-magnetic field, the material point, etc. This, in a certain sense, is inconsistent; strictly speaking measuring rods and clocks would have to be represented as solutions of the basic equations (objects consisting of moving atomic configurations), not, as it were, as theoretically self-sufficient entities. However, the procedure justifies itself because it was clear from the very beginning that the postulates of the theory are not strong enough to deduce from them sufficiently complete equations to base upon such a foundation a theory of measuring rods and clocks.

(Einstein, "Autobiographical Notes," in P. A. Schilpp, ed., *Albert Einstein: Philosopher-Scientist* [Evanston, Ill.: Library of Living Philosophers, 1949], p. 59.)

The *kinematic* use to which Einstein puts the notion of a rigid rod in his 1905 paper raises the question of the relevance of the *dynamic* difficulty noted in this paper; indeed, one writer, discussing these points, declares that "the concept of an infinitesimal rigid rod is *kinematically admissible* in the relativistic scheme, but there is a *dynamical* difficulty" (J. L. Synge, *Relativity: The Special Theory* [Amsterdam: North-Holland, 1965], p. 32, italics his). For our purposes, we need not be concerned with Synge's use of the word "infinitesimal" here. However, full conceptual clarity would certainly seem to demand an explanation of how a concept can be used in one context (kinematic) when the introduction of dynamical considerations (simply exerting a force on the object) leads immediately to a drastic contradiction of a fundamental principle of the theory concerned. In any case, reference to rigid reference frames (systems of co-ordinates) cannot evade the difficulty, because the bodies whose behavior is being referred to such reference frames will exert forces on them. Thus, the distinction between the kinematics and dynamics of rigid bodies does not obviate the need for an understanding of the sense in which a concept which contradicts a fundamental principle of a theory can nevertheless be usefully employed in expositions of that theory.

2. Apart from the view held by Milne and others (see note 14) that "rigid body" is to be considered a derivative and not a fundamental concept in the theory of relativity.

3. "Definition: *Rigid bodies are solid bodies which are not affected by differential forces, or concerning which the influence of differential forces has been eliminated by corrections; universal forces are disregarded*" (H. Reichenbach, *The Philosophy of Space and Time* [New York: Dover, 1958], p. 22, italics his). The reference to "bodies which are not affected by differential

forces" corresponds to our R1; the remainder of the definition corresponds to our R2. Grünbaum's primary use of the expression "rigid rod" is stipulated implicitly in one of the opening questions of his book: "What is the warrant for the claim that a solid rod remains rigid or self-congruent under transport in a spatial region free from inhomogeneous thermal, elastic, electromagnetic and other 'deforming' or 'perturbational' influences?" (A. Grünbaum, *Philosophical Problems of Space and Time* [New York: Knopf, 1963], p. 3).

4. Or, in Grünbaum's terms, all bodies are either equally rigid or equally non-rigid.

5. V. Fock, *The Theory of Space, Time, and Gravitation* (New York: Pergamon, 1964), p. 106.

6. For a discussion of Born's definition and the criticisms thereof, see W. Pauli, *Theory of Relativity* (New York: Pergamon, 1958), pp. 130–32; see also Synge, *Relativity: The Special Theory,* p. 36, and *idem, Relativity: The General Theory* (Amsterdam: North-Holland, 1964), pp. 114ff.

7. W. H. McCrea, *Sci. Proc. R. Dublin Soc.,* 26 (1952): 27, italics his; quoted in W. G. V. Rosser, *An Introduction to the Theory of Relativity* (London: Butterworths, 1964), p. 239.

8. Synge, *Relativity: The Special Theory,* p. 32.

9. *Ibid.* For discussion of Synge's expression "kinematically admissible," see note 14 above.

10. [Note added,1982] It might be supposed that after the introduction of the Lorentz contraction, even classical elementary particles should be deformed at high velocities. However, it must be remembered that Lorentz himself supposed that the contraction should ultimately be explained in terms of interparticle ("intermolecular") forces. The unalterability of the elementary constituents of matter had been a supposition of philosophy as well as of science since Parmenides; for classical science, see J. B. Stallo, *The Concepts and Theories of Modern Physics,* Cambridge, Harvard University Press, 1960. (Original publication was in 1881.)

11. The case of the shift from the classical to the McCrea definition is relevant to the problem of the comparability of scientific theories. For, although we have here two different "definitions," the concepts concerned are not, as Kuhn and Feyerabend would have them be, "incommensurable." On the contrary, we see here physical reasons why an old definition proved inadequate, and how it was possible to continue to use that concept in discussion of the new theory even though the old concept contradicted a fundamental principle of the newer theory. We also see that the two concepts, although they appear on the surface to be very different, are nevertheless quite comparable, the common element being expressible in the statement "A rigid body is one through which signals are transmitted at the maximum possible velocity." The disagreement concerns what that maximum possible velocity is. The classical definition has an *implication*—that signals must be transmitted instantaneously through a rigid body—which is contradicted by special relativity. McCrea's definition concentrates on this implication (rather than directly on the old definition—hence the appearance of radical dissimilarity) and modifies it to meet the relativistic requirement.

12. H. A. Lorentz, *The Theory of Electrons* (New York: Dover, 1952), esp. pp. 213–14.

13. Technical reviews of the more modern form of the problem are found in J. Schwinger, ed., *Quantum Electrodynamics* (New York: Dover, 1958); R. Stoops, ed., *The Quantum Theory of Fields* (New York: Interscience, 1961), esp. the papers by W. Heitler, "Physical Aspects of Quantum-Field Theory," and R. P. Feynman, "The Present Status of Quantum Electrodynamics"; see also the excellent survey of the problem in H. M. Schwartz, *Introduction to Special Relativity* (New York: McGraw-Hill, 1968), sec. 7–5, pp. 271ff. A non-technical discussion is given in L. de Broglie, *New Perspectives in Physics* (New York: Basic Books, 1962), pp. 41–50.

14. Application of the Einstein formula is admissible only if we allow "the Lorentz theory" to have two "versions": Lorentz' and the relativistic reformulation of his "theory." The ambiguity here, of course, arises from the lack of precision, in usual discussions, as to what is to count as "(part of) a theory." There is, however, no difficulty in the present discussion so long as we are clear as to the issues involved.

Note that when the word "theory" has been used in discussions in this section, no technical meaning has been presupposed; on the contrary, we have referred to specific theories whose contents are specifiable in any particular context. Thus, in the context of the first example of section 8, the expression "special theory of relativity" refers to two propositions (the principle of relativity and the principle of the independence of the velocity of light from its source) and their consequences.

15. There seems to be some disagreement as to whether this conclusion is based on experimental fact alone or on some stronger ("theoretical"? "logical"?) consideration. De Broglie, for example, remarks that, "if the charge is assumed to be a point charge, . . . the interaction between the particle and the electromagnetic field results in the energy of the particle at rest, and hence its mass (according to the principle of the inertia of energy), having an infinite value, *which is inadmissible*" (*New Perspectives in Physics,* p. 45, italics mine). Landau and Lifshitz refer to the "physical absurdity of this result" (L. Landau and Lifshitz, *The Classical Theory of Fields* [Reading, Mass.: Addison-Wesley, 1962], p. 102).

16. W. K. H. Panofsky and M. Phillips, *Classical Electricity and Magnetism* (Reading, Mass.: Addison-Wesley, 1962), p. 13. Panofsky and Phillips later declare that ". . . in classical electrodynamics the only thing known about the electron is that it has a certain total charge, and any calculation of its radiation field cannot involve details of how this charge may be distributed geometrically within the electron. On the other hand, it is impossible to assume that the charge has zero physical extent without introducing various mathematical divergences. But *certain features of the radiation field are actually independent of the radius of the electron, provided only that it is small compared with the other dimensions of the radiation field*" (*ibid.*, p. 341, italics mine). This, of course, constitutes a (physical) reason for being able to *ignore* the question of the radius rather than specifically for *idealizing* it as being of

zero extent; but it can serve as a supplementary rationale (in addition to the kinds of reasons adduced in the passage quoted in the body of the present paper) for one's being able to consider the electron as having some specific radius, including zero. Indeed, Panofsky and Phillips continue, "In our discussion of the electron and its behavior we shall assume that it has a finite radius, but we shall ascribe physical significance only to those properties which are independent of the magnitude of the radius" (*ibid.*).

17. H. A. Lorentz, *Problems of Modern Physics* (New York: Dover, 1967), pp. 125–26.

18. H. Poincaré, "Sur la dynamique de l'électron," *Rend. Palermo,* 21 (1906): 129; cf. W. Pauli, *Theory of Relativity* (New York: Pergamon, 1958), esp. pp. 184–86.

19. Cf. H. Weyl, *Space-Time-Matter* (New York: Dover, 1952), pp. 203–6; Pauli, *Theory of Relativity,* pp. 184–86; A. Sommerfeld, *Electrodynamics* (New York: Academic Press, 1964), pp. 236, 278.

20. The program of developing a "purely electromagnetic world picture" based on the Maxwell-Lorentz theory was carried on by Gustav Mie, whose views will be discussed shortly.

21. H. Yilmaz, *Introduction to the Theory of Relativity and the Principles of Modern Physics* (New York: Blaisdell, 1965), p. 51.

22. Landau and Lifshitz, *Classical Theory of Fields,* p. 102. Our treatment of this case illustrates one difference between the philosophy of science and the history of science: for the philosopher is interested not only in what *did* happen but also (among other things) in what *could* have happened—in a precisely specifiable sense of "could" (in this case no new theoretical or mathematical techniques, other than those available at the time, were required for one to have seen the contradiction).

23. Pauli, *Theory of Relativity,* pp. 188ff; see also Weyl, *Space-Time-Matter,* pp. 206ff. Along the same lines as Mie's work is the theory advanced by Born and Infeld, which showed how the problems concerning the radius and energy of the electron could be overcome within classical theory of the Maxwell-Lorentz-Mie type. However, the non-linearity of the equations of these theories, together with the fact that they were not incorporated within the highly successful quantum theory, led to their not being widely accepted (M. Born, *Proc. Roy. Soc. London,* A143 [1934]: 410; *idem, Ann. Inst. Henri Poincaré,* 7 [1937]: 425; M. Born and L. Infeld, *Proc. Roy. Soc. London,* A144 [1935]: 425; Schwartz, *Introduction to Special Relativity,* pp. 273–74).

24. Pauli, *Theory of Relativity,* p. 185.

25. Of the two fundamental principles postulated in the special theory of relativity, only the principle of the constancy of the velocity of light is held in common with classical (Maxwell-Lorentz) electrodynamics. No hypotheses are necessary concerning the nature or causes of light or, more generally, of electromagnetic phenomena. In particular, it is unnecessary to assume that electricity or light is particulate in nature.

26. What is of importance in the cases examined here is their relative *in*completeness or *non*-fundamentality. The question naturally arises as to

whether these notions imply or presuppose criteria of absolute completeness or fundamentality of theories—whether there could be an absolutely complete (or an absolutely fundamental) theory. (The precise relationships between completeness and fundamentality would also need to be examined.)

It is sometimes alleged that no explanation could ever be truly final (fundamental), because any explanation must always begin with certain postulates, and those postulates could in principle be deduced from more fundamental ones. (The notion of explanation as deduction plays a crucial role here.) Similarly, it is sometimes alleged that no explanation could ever be complete, because there are always phenomena that are not covered by the supposed complete explanation (or, the notion of absolute completeness presupposes the possibility of a definite and final listing of types of phenomena—a list which, it is maintained, cannot be given). The possibility that such arguments, while they have a point, do not tell the whole story about explanation, is suggested by the following considerations. Suppose that we had a theory which over a period of six thousand years was successful in answering all questions that were posed to it. *Despite* the correctness of the above-mentioned logical point, I expect that people would begin to suppose that they had a "complete," and perhaps even a "fundamental," theory.

27. I. Newton, *Mathematical Principles of Natural Philosophy (Principia)* (Berkeley: University of California Press, 1946), p. 398.

28. *Ibid.*, p. 399.

29. It is interesting that Boscovich, on the basis of a similar "principle of induction," argued that elementary particles have *no* extension—that they are point particles:

> Taking it for granted, then, that the elements are simple and non-composite, there can be no doubt as to whether they are also non-extended or whether, although simple, they have an extension of the kind that is termed virtual extension by the Scholastics. For there were some, especially among the Peripatetics, who admitted elements that were simple, lacking in all parts, and from their very nature perfectly indivisible; but, for all that, so extended through divisible space that some occupied more room than others; and such that in the position once occupied by one of them, if that one were removed, two or even more others might be placed at the same time.
>
> Since then we never find this virtual extension in magnitudes that fall within the range of our senses, nay rather, in innumerable cases we perceive the contrary; the matter certainly ought to be transferred by the principle of induction, as explained above, to any of the smallest particles of matter as well; so that not even they are admitted to have such virtual extension.

(R. Boscovich, *A Theory of Natural Philosophy* [Cambridge, Mass.: M. I. T. Press, 1966], p. 44, pars. 83–84.)

30. A. d'Abro, *The Rise of the New Physics* (New York: Dover, 1951), 1: 278–79. In connection with his remark that "there is only one kind of optics, namely, wave-optics," d'Abro says in a footnote that "in this chapter we are explaining the situation as it appeared to the physicists of the nineteenth century, and we are not taking into consideration the more recent discoveries in the quantum theory."

31. My colleague Kenneth Schaffner has given an analysis of the relations

between theoretical and observational terms as causal connections in his forthcoming paper "Correspondence Rules."

32. This suggestion will be developed in detail in my "Explanation and Scientific Progress," to be presented at the Boston Colloquium for the Philosophy of Science, and to appear in a forthcoming volume of the Boston Studies in the Philosophy of Science.

33. These remarks should not be taken as implying that we are forever condemned in the philosophy of science to viewing science only from different and irreconcilable, though perhaps complementary, perspectives—each of which is limited and objectionable, but each of whose limitations and objections are perhaps covered by one or another alternative perspective. There is no reason, at least not in any of the present discussion, why a unified interpretation of science cannot be achieved; but one way of getting at such an interpretation may be to begin with different approaches, as suggested here.

CHAPTER 18

# REASON, REFERENCE, AND THE QUEST FOR KNOWLEDGE

This paper examines the "causal theory of reference", according to which science aims at the discovery of "essences" which are the objects of reference of natural kind terms (among others). This theory has been advanced as an alternative to traditional views of "meaning", on which a number of philosophical accounts of science have relied, and which have been criticized earlier by the present author. However, this newer theory of reference is shown to be equally subject to fatal internal difficulties, and to be incompatible with actual science as well. Indeed, it rests on assumptions which it shares with the purportedly opposing theory of meaning. Behind those common assumptions is the supposition that the nature of science can be illuminated by an examination of alleged necessities of language which are independent of the results and methods of scientific inquiry. An alternative view of science is proposed, according to which the goals and language of science develop as integral parts of the process of demarcating science from non-science, a process in which the notion of a "reason" gradually assumes a decisive role. On this view, the comparability, competition, and development of scientific ideas are understood without reliance on either common "meanings" or common "references" as fundamental tools of analysis.

## I

There is a traditional theory of meaning, or, more exactly, a class of theories of meaning, which has been much discussed in recent years both within and outside philosophy of science. Briefly, the idea is the following: we choose certain features as defining a type of thing, as giving the "meaning" of the term we choose to use for that type of thing. From then on, those features will serve as "criteria", in the sense of necessary and, or perhaps or, sufficient conditions for anything's being that sort of thing, for deserving to be referred to by that term. If the criteria are not satisfied (or, for that version known as the "cluster" theory, if "enough" of them are not satisfied), we will not apply the term in question to it. According to such theories, nothing we find out empirically about the things in question can be relevant to changing those criteria; for if a thing did not satisfy them, it would not even be that sort of thing, and hence could not be a counterexample affecting the criteria. One aspect of this type of theory that has been the focus of much attention among philosophers of science is that it, if anything, was supposed to account for continuity of discussion between different, and particularly successive, theories: theories would be "talking about the same thing", and thus really in agreement or competition with one another, if and only if the terms used in those theories have the same "meaning".

Saul Kripke and Hilary Putnam have raised two major objections against this traditional theory of meaning, in any of its forms. First, the properties used originally in identifying a substance or natural kind (or, more generally, specified as constituting the meaning of the term in question) need not belong to the substance or kind; they may—even all of them—be found not to be true of that substance or kind. And second, other substances or things may be found which have all those properties and yet are not that kind of substance or thing. Hence we must reject the view that the properties initially assigned to kinds of things and substances give the "meanings" of the terms we use to refer to those kinds and substances, in the sense either of a set of necessary and/or sufficient conditions for applying the term or a set (cluster) of conditions "enough" of which must be satisfied. "*A priori*", Kripke asserts, "all we can say is that it is an empirical matter whether the characteristics originally associated with the kind apply to its members universally, or even ever, and whether they are in fact jointly sufficient for membership in the kind" (Kripke 1980, p. 137; *cf.*, also, Kripke 1977; Putnam 1977, 1978, 1979).[1]

It should be clear from my previous writings that I agree fundamentally with these criticisms. I have argued in a number of papers that, for science, there are no conditions governing the application of a term which are immune from revision in the light of further experience. And I have diagnosed a number of difficulties and unacceptable positions in the interpretation of science—difficulties such as those revolving around the subject of "meaning change", and views such as that of the alleged "incommensurability" of at least some scientific theories—as due to confusions engendered in considerable part by reliance on traditional doctrines of meaning. I have urged that such doctrines should be abandoned as inadequate tools for the attempt to forge a satisfactory interpretation of science and scientific change. (Shapere 1964, 1966, 1971, 1974b, 1977, 1980, 1981). In the remainder of this paper, I will provide still more support for these views. However, I will argue that the alternative view which Kripke and Putnam offer in place of those doctrines must also be rejected, at least as it applies to science. My purpose will nevertheless go far beyond mere criticism of that specific view; for I will try to show that both the Kripke-Putnam approach to the interpretation of science in terms of reference, and the opposing approach in terms of meaning, share certain fundamental assumptions, and that, despite their deep differences, those approaches stem from the same tradition. And I shall argue that those assumptions constitute a violation of the spirit of science, and that the tradition of philosophical analysis from which the two views flow must be banished from the philosophy of science if we are to make true progress in understanding the scientific enterprise. But more importantly, as far as I am able within the confines of this brief paper, I will use my discussion of these matters to bring out, in ways that further what I have said in previous writings, some central features of the approach and views which I have come to believe are appropriate and correct.

The alternative with which Kripke and Putnam would replace the traditional theory of meaning makes reference, rather than meaning, the guarantor that we are "talking about the same thing" despite any changes in the descriptive criteria associated with the thing. The properties by which we originally identify cats or gold or water do not establish conditions which anything must satisfy in order to count as a cat or gold or water. Cats and gold and water are what cats and gold and water *really are*, and the properties by which we initially identify them may have nothing to do with what they really are; those properties do not constitute the irrevocable "meaning" of the term, but only serve, as Kripke says, to "fix the reference". It is the latter, not the meaning, which remains "rigid"—that is, which, in being passed down historically from a hypothetical initial baptism, remains unaltered throughout any vicissitudes, no matter how radical, in what we believe about and attribute to that thing or kind of thing. What we are talking about—referring to—all along, throughout the history of our use of the term after the initial baptism, is what the object or kind really is; and our aim in science is to discover what the thing or kind of thing really is, its very nature, its essence. According to Kripke, therefore, "In general, science attempts, by investigating basic structural traits, to find the nature, and thus the essence (in the philosophical sense) of the kind." (1980, p. 138) The discovery of that essence would be an *a posteriori* one. Of course we might be wrong in thinking that we have arrived at that discovery—that a certain property is indeed an essential property of the substance or kind in question. But *if* we are right—if we have examined a certain substance and found what it "really is"—then from that point on, we will call by the name of that substance or kind all and only those things, in this or any possible world, that have that essential property. If we encounter something in some hitherto-unexamined region of the universe that resembles that substance in every respect except that of possessing what we have found to be the "very nature" of the substance, we will not call that new thing by the name of that substance. Our referential practice—what we say or would say when we discover what a substance or kind really is, and what we are talking about pending that discovery—thus conforms to the existence of metaphysically necessary truths, of essences; and it is in view of the existence of those metaphysical truths and the linguistic practices which manifest them that the so-called "causal theory of reference" claims that science is a search for "essences" in the "philosophical sense". My argument against this thesis will fall into two parts. First I shall argue that the alleged linguistic practice of calling something by the name of a substance (once we know what that substance really is) if and only if it has the essential property of that substance neither would nor should nor does take place. Having so argued, I shall then ask what aspects of science could remain to be illuminated by the doctrine that there are metaphysical truths about essences which it is the object of science to discover. (It is at this point that I shall explore the view that we refer to the essence of a thing or kind even before we have discovered what that essence is.) And I shall conclude that none remain; for an understanding of science, we have no need of that doctrine.

There are certain difficulties in the formulation of the Kripke-Putnam thesis which I will not discuss here. I will simply mention one of these, because it lurks in the background of some things I will talk about. Nothing satisfactory is said about how we are to decide what is to count as an "essential" property. In particular cases, this is by no means an easy matter. There are properties which are true of a substance which Kripke and Putnam would not want to consider "essential" to it; and on the other hand, an essential property apparently need not be a fundamental one (being an element of atomic number 79 is not a "fundamental" property of gold, at least in the sense that it is explainable in deeper quantum-mechanical terms). Talking about "what a substance really is" is thus quite ambiguous in the absence of any discussion of criteria of essentiality. Much could and should be said about this point, but I shall have to pass over it with only one brief comment: we must not suppose that universality is a *criterion* of essentiality—that is, that in order to decide whether a property is an essential property of a certain kind or substance, we must first ascertain (perhaps in addition to other things) whether it belongs to all instances of that kind or substance. On the contrary, for Kripke and Putnam, we discover, from an examination of things of that kind in our spatiotemporal region, what the essence is, and *from then on* refuse to consider anything to be that kind of thing unless it has that property.

## II

With regard to the case of gold, then, the thesis before us is the following. *Given* that we have found gold to be an element of atomic number 79, we will thenceforth call something gold if and only if it has that property. In this paper, I will examine just the "only if" portion of this thesis. But that will, I think, be enough to establish the points I want to make about the Kripke-Putnam thesis.

Let us grant, then, uncritically for the moment, that scientists have found, in their examination of our spatiotemporal region of the universe, that gold has certain fundamental or essential properties: it is an element of atomic number 79. And let us assume—still leaving aside any "epistemic" questions of how they find out—that they are right: the claim is true, and furthermore they know it. But later they find some other region in which there is no substance which is an element of atomic number 79; let us even suppose that in that region there are no "elements", and nothing corresponding to "atomic numbers". Is it clear that, if we were to come across such a region, we would not call something in it gold because that substance is not an element of atomic number 79—that we would call a substance gold *only if* it had that characteristic, and that, since nothing in that region had that characteristic, nothing in it would be (*i.e.*, be called by us) gold? Certainly it is not clear; in fact, I shall argue that if certain conditions prevail in that region, precisely the contrary will be in accord with the scientific spirit.

Suppose there is a substance in that region having all the characteristics of gold except of being composed of an atomic nucleus of atomic number 79. We might come to understand this in terms like the following: there exists in that region a peculiar field which smears out the discrete particles of the nucleus and the characteristics of it relevant to its atomic number, but leaves the more "superficial" physical and chemical properties intact; the field also alters the ratio of strength of the electromagnetic to the strong force, leaving that substance easily splittable into two other substances. If the field were removed (or the substance removed from the region), the nuclear characteristics and the ratio we know between the two forces would be restored. (Perhaps, on the other hand, we might conclude on various grounds that *ours* is an exceptional region in which a peculiar field warps substances into the form of elemental nuclei.) Our physics has such patterns of explanation within it; and should we find such an explanation of the situation, it seems clear that the existence of that explanation, together with the similarities between our gold and the particular substance in question, would lead us to call that substance gold *despite* its not having the characteristics Kripke says it must have if we are to call it gold. Even if we had only a range of alternative applicable explanations, without knowing which if any was correct, we would still undoubtedly call the substance gold on the ground that we would understand how the situation *could* arise. Nor need we wait until the substance has been transported out of that region (or put into our own peculiar field, if that is the way things are) before deciding to call it gold; the reasons for accepting the distorting-field explanation might be quite convincing in themselves. Even if we were unable to remove the substance from that region or to bring it into our own (suppose such transport always resulted in its destruction), we might still have abundant reason for calling it gold.

But the case is stronger yet. For although having such an explanation (or such a range of possible explanations) might give us grounds for calling the substance gold, the existence of such an explanation, or of any other explanation, is not a precondition of our deciding to call it gold. Extending our usage of the term 'gold' is not dependent on a presupposed explanation or even a presupposed set of alternative possible explanations. It is not necessary to assume that *understanding* of the situation always be present as a condition of extending our use of the term to the stuff in the new region, as long as other considerations are sufficiently compelling. For example, if *all*, or even *many*, of the substances in the new region had all or most of the same characteristics as certain corresponding substances in our region, but lacked in each case the property found to be "essential" in our region, our suspicions would, to say the least, be aroused. Our hypothesis would be that the new substances were counterparts of substances in our region (there is absolutely no problem here about what are the counterparts), despite the fact that they do not have the characteristics we have found (truly, as we have assumed for the sake of argument) to belong to those substances in our region, and also despite the fact that we have no explanation or even possible explanation of that fact as yet; and we would *then* begin looking for explanations of that difference. That is, our question would be, Why is gold here not an element of atomic number 79? If the total evidence were somewhat weaker—if, for example, not so many, though still several, substances in the new region shared all or most properties of their earthly counterparts except for the essential property—the question would be correspondingly more tentatively stated: Is this substance perhaps gold, and if so, why is it not an element of atomic number 79? It is only if the total evidence were quite weak that we would withhold the term 'gold' from the substance in the new region—though reasons might accumulate later which would strengthen the case that it is gold. And on the other hand, no matter how strong the evidence might be, our attribution of the term 'gold' to the counterpart substance is defeasible: we might decide, in the light of our investigation, that despite the initial evidence, the substance was not gold after all.

Finally, note that a return to our earthly laboratories to re-examine our prototypical instances of gold would be irrelevant in the example I have given. For, since they are in our field, those instances would show the same nature we had already found for them, and would shed no light on the behavior of the corresponding substances in the other region. Nor would we need to transport our prototypical instances to the other region to see what would happen to them there (again, suppose even that we *could* not so transport them). Under the circumstances I have envisaged, our reasons for calling the new substance gold would be strong enough, independent of such investigations (though such investigations might, of course, strengthen them yet further).

The upshot of our discussion so far is, of course, that *it is not just one property or set of properties—the "essential" ones—that determines or affects how scientists will apply terms in new situations; all the (true) properties may, as in this example, play a role, and furthermore, the properties and behavior of other entities (substances, etc.) may also play a role—as is again the case in this example*. No doubt, too, the possession of certain properties (more fundamental ones) will play a more important role than the possession of others; but importance will be balanced against other factors, such as, in the example given, sheer numbers. Still further, the availability in current science of applicable explanatory patterns, or the actual existence of such an explanation, may also play a role, though, as I have argued, it need not.

I have focused on Kripke's arguments; but Putnam has advanced very similar ones. In a particularly imaginative and influential discussion, he asks us to consider a "Twin Earth" in which is found a substance indistinguishable from water at normal temperatures and pressures, but whose chemical formula is not $H_2O$, but something else, XYZ. Then, he claims, that liquid is not water; *we* would not call it water, even if the inhabitants of Twin Earth did. His conclusion is the same as Kripke's: ". . . once we have discovered the nature of water, nothing counts as a possible world in which water doesn't have that nature. Once we have discovered that water (in the actual world) is $H_2O$, *nothing counts as a possible world in which water isn't* $H_2O$." (Putnam 1977, p. 130; 1979, p. 233; italics his) But the reply to Putnam's example is the same as to Kripke's: if the substance had all, or even a great many, of the properties of our water except that of being $H_2O$, and especially if other substances on Twin Earth similarly resembled substances on Earth, scientists would suspect strongly that it *was* water under circumstances that, from our point of view, were extraordinary.

The Kripke-Putnam reply to my argument would undoubtedly be that our encounter with the substance in the new region only shows that we were *mistaken* in our belief that gold or water was "essentially" what we had supposed; all that is shown by my example, they would say, is that being an element of atomic number 79 turned out not to be the "very nature" of gold, as we had previously thought it to be. And in general, according to this response, *any* particular claim as to what is true—or rather, essentially true—of something may turn out to be false. But, the response concludes, the point is that *if* the claim is true, and also satisfies further conditions of essentiality, then certain consequences will follow and will be reflected in the linguistic practices of scientists.

The first problem with this reply is that it seems impossible to see how, on the Kripke-Putnam view, scientists could ever come to the conclusion that they were mistaken, once they have come to know, or even to suppose they know, that gold is essentially an element of atomic number 79. For in consequence of coming to that knowledge they will, according to them, thenceforth refuse to call anything gold unless it has that property, and so *cannot* come to the conclusion that the substance in the other region was gold, and that therefore they were mistaken in thinking gold was, in its very nature, an element of atomic number 79. Whether they are right or not, once they commit themselves at any stage—*whether at an initial baptism or after long investigation*—to supposing they have found the essence of gold, they are thenceforth stuck with that conclusion, and it can never transpire that they will find themselves wrong; there can *be* no counterexamples. The gold-like substance in the new region would be *dismissed out of hand* as not being gold, just as, on the old theory of meaning, one would dismiss out of hand the offer of this married man as a counterexample to the thesis that all bachelors are unmarried. The common ground of the old theory of meaning and the new theory of reference begins to be revealed.

Putnam is aware of the difficulty, but only in passing. He admits that ". . . we can perfectly well imagine having experiences that would convince us (and that could make it rational to believe that) water *isn't* $H_2O$. In that sense, it is conceivable that water isn't $H_2O$. It is conceivable but it isn't possible! Conceivability is no proof of possibility." (1977, p. 130; 1979, p. 233) With that remark, however, Putnam simply changes the subject, leaving his and Kripke's theory with a paradox that should have been its epitaph. It is impossible to hold *both* that scientists commit themselves in the way Kripke and Putnam suggest, *and* that they might yet find themselves to have been mistaken. Given the doctrine of rigid reference, merely to *assert* that we might be mistaken is not enough. For, having found what we suppose to be the essence, if we then engage in the practice of rigid reference, there is no way to discover that we might have been wrong; but if we admit that we *could* be wrong, we are not practicing rigid reference.

Suppose, though, that we waive this objection; suppose, that is, that the Kripke-Putnam view has been successfully supplemented by a theory that shows how we can be mistaken as to the essence of a substance despite our rigidly refusing to consider anything an instance of that substance that does not have that essential property. Then if one of the Kripke-Putnam examples is shown not to support the thesis of rigid reference, it would be open to a defender of the thesis to say that it just wasn't a good example, that the property previously held to be essential turned out not to be essential after all. Even then there would remain a serious flaw in the proposed reply to my argument. For the real import of this possibility lies in the fact that *no examples at all* would ever be admitted as having the credentials required to bring about the freezing of usage that Kripke and Putnam suggest does or might take place. If nothing will be admitted as a telling counterexample, it is equally clear that neither are there, nor even could there be, any supporting instances. And if examples can throw no light on the theory, the theory can likewise throw no light on the examples. The proposed reply, by divorcing the

Kripke-Putnam position from its own examples, only succeeds in alienating that position still further from actual science.

Behind these two difficulties of the proposed reply lies the root of the failure of the Kripke-Putnam position as regards science, at least as I have discussed it so far. For it is not only that (as I have argued) scientists would not, when they have found what a substance truly ("essentially") is, apply the name of that substance "rigidly" to all and only those things having that essential property. It is also that, if they did, they would be justified in doing so only if they had knowledge of an extent that we do not have; at best (*i.e.*, granting the certainty of the conclusions we draw from what we have observed), the knowledge that we do have does not countenance the commitment Kripke and Putnam envision. At best—questions about the adequacy of *that* claimed knowledge aside—we know only the things we have observed thus far. This "epistemic" consideration *cannot* be dismissed as irrelevant to the Kripke-Putnam thesis about linguistic commitment; for what it brings out is that that thesis is irrelevant to real science, which does not, even at best, obtain the sort of knowledge required for such commitment. To base rigid commitment on the knowledge that we do have would be sheer dogmatism, incompatible with the spirit of science which is reflected in what scientists clearly would do in hypothetical cases like those of gold-like and water-like substances in a new region of the universe. And thus, not only is it the case that scientists *would not* observe the kind of linguistic commitment that Kripke and Putnam allege they do or might; as I have now argued, they *should not,* because to do so would violate the spirit of science.

## III

But it is also the case that science, at least when it is most characteristically scientific, *does not* observe such linguistic commitment. The examples given by Kripke and Putnam are science fiction stories, and what I have done is simply to turn those examples against them. Science fiction stories have the disadvantage that they often permit the construction of arguments too vague and general to be critically evaluated. But what is important about the use to which I have put the Kripke-Putnam stories is that the kind of *attitude* exemplified in my version, far from being an artificial construction, is characteristic of the scientific approach.

For we found in my revision of their examples the following four ingredients: the attempt to find relationships and differences; the use of those relationships as bases for classification; the shaping of vocabulary in the light of the relationships and differences we have found; and the refusal to be bound irrevocably to some prior categorization, or to some prior vocabulary or concepts, however well-founded. These four features are typically found in the dynamics of forming and revising the subject-matter investigated by a scientific field—in what I have elsewhere referred to as the formation and change of scientific domains. They can be given a generalized systematic treatment, but it is not necessary to do that here. What is pertinent here is the way those four features operate in the case of actual linguistic practice in science.

For what happens in the sorts of cases relevant to the present issue is the following. We find reason to believe that things are of a certain sort. Particularly when those beliefs approach or achieve the ideal of being successful and free from specific doubt, we accept them and use them as bases for our further thinking and talking about the world. These processes are well illustrated by the chemical revolution of the eighteenth century. There we find the language in terms of which we talk, scientifically at least, about the world or certain aspects of it, reformulated, and on a very broad scale. The reformulation was based on an incorporation of new beliefs, based on reasons, into the vocabulary for naming and describing the objects of chemical study. In particular, in that case, the beliefs so incorporated were a specific version of the idea that material substances are to be understood in terms of their constituents—what they are made of. When a more mature chemistry added two further ideas—that material substances are to be understood not only in terms of their constituents, but also in terms of the arrangement of those constituents and the forces holding them together—the view amounted to what may be called the *compositional* approach to the understanding of material substances. (Shapere 1974a) Here I will use that term in reference to the first of these three ideas.

Whatever one might suppose, the view that material things can be understood in terms of their constituents is by no means an *a priori* or necessary truth. While it was present from early times, it was rivalled by, among others, a view according to which material entities—at least "earthy" entities—are all of one or at most a few basic types, whose similarities and differences are to be understood in terms of their various degrees of fulfillment or perfection of that one type, and not, or at least not simply, in terms of the intermingling of more basic constituents in them. And laboratory manipulation of those substances was to consist of bringing the substances to their proper perfection. That trend of thought, so foreign to the compositional approach, was more powerful than the latter in the alchemical tradition. In the sixteenth through eighteenth centuries, when it underwent a revival, the compositional approach was confronted with a serious objection. What, precisely, does fire (then the primary means of operating on material substances) do to the substance on which it acts? To the compositionalist, fire merely separates substances into their constituent parts, while not altering those parts in any way. But how could we know that the fire did not act on the substance so as to *alter* it, so as to *produce* the breakdown products which did not exist as constituents of the substance prior to the application of the fire? Things might have been that way. But the compositional approach gradually gained credence through the discovery that a set of substances could be found which were breakdown products in a great many reactions, which themselves could not be broken down by any means at the disposal of the chemistry of the time, and which could be reassembled to form the original substance. But in order to develop that compositional view in a consistent way, more had to be involved in the view than that. The concept of an element had to be given a new interpretation; further, a positive (at first, positive or zero) weight had to be shown to be of central relevance chemically; and it had to be argued that ''airs'' (gases) were true constituents of material substances, *i.e.*, that they participated in chemical reactions, despite the fact that many, including many compositionalists, had believed that they did not. The totality of such considerations was not sufficient to establish the compositional approach in any final way, or even anything near that; it still had to prove its worth in a fuller way. But those circumstances in which Lavoisier's version of a compositional approach was successful were coupled with another, largely independent consideration to justify a radical revision of scientific linguistic practice. For in the process of discovering new substances and new techniques for handling them, the chemists of the seventeenth and eighteenth centuries had come more and more to the view that sensory qualities, which had hitherto served as the primary bases for classifying and naming material substances, were inadequate for that purpose. On their basis, substances by then known to be different had been confused; and the same substance, if produced by different methods, was often considered to be two or more different substances. The group centered around Lavoisier therefore proposed a revision of the vocabulary of material substances to reflect a, and more specifically his, compositional understanding of those substances: the names of compound substances (like sodium chloride) embodied the theory in terms of which they were understood. Had the compositional view, in a form at least descended from Lavoisier, proved fruitless, no doubt that reform of language would have been rescinded, those beliefs peeled away from the vocabulary in which they had been incorporated—as, indeed, the vocabulary of the phlogiston theory was so discarded, and as, in a parallel case, classifications of stars and consequent naming of types of stars in the light of theories of stellar evolution were withdrawn when those evolutionary theories had to be discarded. (Shapere 1977) As it was, however, the reform of chemical nomenclature was vindicated in its essentials in succeeding decades; and the new vocabulary contributed significantly to suggesting new problems and lines of research.

Though not ordinarily so sweeping, similar extensions and shifts of vocabulary are characteristic of many stages of the scientific enterprise. The development of science consists, in part, of such a shifting of associations of items into new classifications, and of a constant redescribing, and often a renaming, of the items with which it deals. Change is by no means limited merely to renaming of previously-conceived kinds; the pattern of division into kinds is also altered, old kinds being split or united, and new ones introduced. Classifications into kinds of things or substances is not something given, there for us to understand, brought to science to investigate but not to alter, whose essences we are trying to ascertain. Rather, our classifications are shaped and reshaped in the light of what we learn. And the objects to be classified, too, are not simply given; what objects there are is something to be found out, if indeed we find out that there are objects (and kinds) at all in any fundamental sense. Finally, however apt it may be in some cases to talk about the shifts as wholesale, kaleidoscopic, revolutionary, they are not, in the most characteristically scientific cases at least, capricious. There are cases where the considerations on which the revisions are based are abundant and compelling; where they are less strong, as in the case of the reform of chemical nomenclature, fuller justification may have to await further developments. That we may have to await such after-the-fact justification should not be surprising, especially when one considers relatively early cases of scientific thought. For as science is, among other things, a search for better ways of thinking and talking about the world, and for an understanding of the relations between what is talked about, it should be clear that at stages when its objects and relations are but little understood, there will be correspondingly little guidance. Then it may have to rely on ill-grounded hypotheses, obtained, sometimes, from what are judged, then or later, to be extraneous considerations. With the accumulation of beliefs that have proved successful and relatively free from reasons for specific doubt—beliefs in the form of both vocabulary and understanding of relationships—the reasoning by which science advances becomes more autonomous, more self-generating. It becomes more able to generate problems, lines of research, hypotheses, and so forth, from its own structure.

But not only do shifts of the sorts we are considering take place *as a result* of beliefs that have been found successful and free from doubt (and perhaps satisfy other conditions as well); as we have seen, science also *incorporates* such beliefs into its conception of the subject-matter of its inquiry. Such utilization of prior well-grounded belief for determining the strategies of science, and the incorporation of such beliefs into the structure and language of science, is an instance of a more general process that has come to characterize science, a process which we may call the internalizing of relevant considerations. It is a process of gradual reformulation, in the light of what we have learned, of the scientific enterprise—its goals, problems, patterns of explanation, and indeed all its aspects, in addition to its subject-matter. The aim is to make all these so tightly bound by considerations of relevance, and so comprehensive of such relations, that all reasoning about a subject-matter and its problems can be based solely on a consideration of that subject-matter and the well-grounded information known to be relevant to it. It is in that respect that we may say that, in science, we not only learn, but also learn how to learn; and part of that learning consists in learning how to talk and think about the world. Further, this internalizing of relevant considerations exemplifies a third factor, in addition to "success" and "freedom from doubt", that must go into any analysis of what a "scientific reason" is: namely, the intuition that in any argument concerning a subject-matter, those considerations will count as reasons which have to do with that specific subject-matter. (Shapere 1981)

Viewed from another perspective, that process through which scientific reasoning becomes more "internal" is essentially one of gradually distinguishing the scientifically relevant from the irrelevant, of gradually separating what will be counted as internal and what as external to science, of gradually demarcating science from nonscience.[2] It is an ideal we have learned to seek; but it is far from being fully achieved, simply because we do not know everything. It should therefore again not be surprising that "internal considerations" are not yet always adequate for formulating or dealing with scientific problems. Nevertheless, the distinction between scientifically relevant and scientifically irrelevant considerations is far better delineated now than it was in the days, say, of Kepler or Lavoisier.

The process of internalization includes, as we have seen, the building of well-founded information into our descriptive vocabulary. But here as elsewhere, science is constantly open to the possibility that doubt may (though it need not) arise, that our present views, including the ways we "conceptualize" objects and kinds, and name and describe them, may have to be revised or rejected and replaced. Despite the fact that compositionalist views were built into the very language in terms of which we name and describe material substances, that view may even yet have to be rejected; today, indeed, it again faces potential crisis, at least where fundamental theory is concerned. For if quark confinement is accepted, the notion of an independent elementary particle may come to be viewed as a high-energy, short-range approximation, and a truly fundamental physical theory might no longer be characterizable in terms of the particles it postulates. Our reasons for belief—and that includes naming, describing, and classifying—are not in any case conclusive; it is always possible, at least as far as our present understanding is concerned, that reason for doubt may arise.

Even in that majority of cases of scientific change where reform of language does not take place, the properties or criteria associated with the application of a term may be altered. We may indeed *call* by 'gold' only those things which satisfy our latest or most fundamental well-grounded beliefs about nature and about a specific kind of substance in particular; to this extent Kripke and Putnam are right. But we are also always prepared to find reasons for changing what we attribute to (or call) gold no matter how fundamental and well-established that attribution may be. Kripke-Putnam commitment never comes.

## IV

In the account I have given, our scientific linguistic practice has been described wholly within the framework of a conception in which, in the light of new findings, we arrive at new beliefs and incorporate them into the ways we think and talk about nature. Is there any remaining place for the concept of "essence", of "metaphysically necessary truth"? As we have seen, Kripke and Putnam hold that if scientists were to discover what a substance in our region "really is", they would thenceforth call something by the name of that substance only if (and if) it has that property. Against this doctrine, I have argued that scientists would not so behave were they to make such discoveries, and furthermore that they should not and do not so behave. The examples adduced by Kripke and Putnam were turned against them, and it was shown that in fact no specific examples of scientific reasoning can support the Kripke-Putnam thesis. If there is to be anything to their claims, then—at least with regard to science—the point cannot be exhibited in the behavior of scientists subsequent to any discovery, even of the sort they have in mind. Can it then be manifested in the behavior of scientists—in the way they talk—in the very process of seeking discoveries? More specifically, can we say that the discovery of essences is the *aim* of science, an aim which is reflected in our (here, scientists') alleged referential practice of "talking about" what gold, water, and so forth "really are", no matter what specific beliefs we may hold regarding those substances, and whether or not we do or even can ever achieve that aim?

The trouble with this view is that, unless the notion of "essence", of the "very nature" of a substance or kind, is expanded to the point of emptiness, it cannot describe the aim of science without limiting the options of science in an unacceptable way. The problem is not just that Kripke and Putnam sometimes talk in ways that suggest that understanding the "very natures" of substances would be understanding those substances in a compositional way; the validity of the latter approach, if indeed it is valid, is contingent. But the notion of "essence" could, I imagine, be expanded to cover alternatives to compositional understanding. Nor is the problem merely that there is no guarantee that there are well-circumscribed boundaries between substances or kinds, and well-defined sets of essential properties for them, after the fashion of Aristotelian essences; perhaps the difference between kinds is rather like the difference between metals and non-metals. That, too, is a contingent matter; but no doubt the concept of essence could be expanded to take in that contingency too. Nor is the difficulty simply that fundamental understanding may have nothing to do with distinct kinds, which might ultimately be understood in far more general terms; that possibility, I am willing to grant, might be covered by some appropriate distinction between knowledge of essential truth and knowledge of fundamental truth. The real difficulty is that there are such a variety of possibilities, and it is folly to suppose that we are aware of all of them that might arise in future science. Even to cover all those of which we are aware, we would have to include the possibility that we might arrive at the conclusion that we cannot understand what things are "really like", and even that the very notion of things being "really like" anything may be inapt. One might think that the Copenhagen Interpretation of quantum mechanics is wrong; but can it be ruled out as impossible? Yet in at least some versions, it suggests that certain properties (*e.g.*, position and momentum) tell us nothing about "entities" themselves, but only about the system consisting of object plus instrument. True, some properties remain attributable to entities (*e.g.*, mass and charge); but is it impossible that a scientific theory might involve—for definite reasons, as in the case of position and momentum—the conclusion that *no* properties are so ascribable? If that view, or anything like it, were to prevail, we might say that we had found a kind of truth; but in that case, it would be only in a very stretched sense that we would have found out something about the essences of kinds and substances. Yet we would certainly not want to conclude that such a theory had failed to fulfill the aims of science, any more than we would want to say that about quantum theory. Furthermore, in such a case, there might even be reason, as some have argued, for concluding that there are no such things as essences.

The aim of science, then, cannot be restricted in any way in advance of the investigation of nature. Earlier, we found that the Kripke-Putnam thesis about the linguistic commitment of scientists after their discovery of essences can receive no support from any specific scientific belief; rather, it must be compatible with any outcome of scientific investigation, with any scientific belief. It is hopeless to try to view this compatibility as an advantage of the thesis—as showing the thesis to be independent of the mere contingencies of scientific belief. For now, in considering whether the discovery of essences can be the aim of science, we see that such compatibility can be purchased only at the cost of interpreting the concept of "essence" so broadly that it is without content. Like its ancestors in the history of philosophy, the new doctrine of essences can be true only if it is empty; to whatever extent it is specific, it is false.

But we must be clear as to just where that falsity lies. Note that I am not objecting to the idea that specific sense might be made of the notion of "essence"; nor do I even object to the idea that someday, somehow, we might be in a position to obtain knowledge which is certain, for which there is not even an in-principle possibility of doubt, and which satisfies whatever sensible conditions of essentiality philosophers might lay down. What I have been calling the "spirit of science" is, after all, itself not an *a priori* or necessary condition of science, but is based on a particular view about the way things are and the way we can find out about them that we have developed through encounter with the world. Roughly speaking, we have come to accept the possibility that specific doubt may always arise because we have come (for good reasons) to believe that space and time are of incomparably vaster extent, and the entities existing therein of incomparably vaster number and internal complexity, than we can examine adequately in our experience, and also that the only way to learn about the way things are is to examine the things existing in space and time intensively and serially. And though we have no specific reason to doubt that view, there may still be profound surprises in store regarding it. In short, given a better analysis of essentiality than Kripke and Putnam have provided, I am willing to admit that the discovery of essences, in some specific sense, might be an eventual achievement of science. But whatever may be the case in non-scientific contexts, in science we have learned not to impose inviolable restrictions on possible outcomes of our investigations: we do not search exclusively for "essences" in any specific (*i.e.*, restrictive) sense of that term; we search for whatever we find. If—or to the extent that—scientists talk as though they seek to find out what things are "really (essentially) like", they do so because, in grappling with experience, they have found reason to talk that way, and not because such talk is imposed on them as a necessary consequence of the nature of referential language. And thus—my earlier objections aside—what I am objecting to is the claim that science *must* always aim at discovery of "essences", *where that claim is supposed to have any specific content*. What we strive for in science are views which will be as successful and free of specific doubt as possible, if any such views can be arrived at; and even that is something we have learned to seek. But nothing can be specified, in advance of the actual study of nature, about the particular sort of view we may arrive at. If the concept of essence is understood broadly enough to cover *any* outcome of scientific investigation, well and good—though saying that science is a search for essences then becomes highly misleading. The trouble is that the Kripke-Putnam view *seems* to claim more: that it constitutes a profound comment on the aims of science. This is what I object to.

Everything I have said is compatible with "realism" to at least the following degree: that we *may* eventually arrive at a view which is fully successful in all respects; regarding which no specific reason for doubt ever arises which is not removed satisfactorily; and to which no equally successful alternative view is ever found. My account does not even rule out the possibility that we may arrive at beliefs that we could know fulfilled those conditions. But such eventualities are by no means guaranteed: whatever one says about "truth", the "truth" which science seeks must be compatible with *either* "realism" *or* the discovery that realism cannot be accepted. The Kripke-Putnam view, at least if it is interpreted as having some specific content, limits science to seeking a "realist" account. What I have argued is that that is too limiting. And in any case, it cannot be their theory of reference that explains our putative success in achieving knowledge; for that view is, or at any rate must be, compatible with *any* beliefs we may hold. Nothing else is needed, nor is anything else available, to explain the success of science than the reasons we have for accepting the beliefs we do; and nothing else is available or needed for providing truth-conditions for claims about gold and so forth except the beliefs we have best reason to accept.[3]

The issue here is a deeply important one for the attempt to understand science. For on the Kripke-Putnam view, the aim of science is specified in a way that is independent of the content of science. In this respect, that view is a descendant of the idea that science must make "metaphysical presuppositions", though that old idea is now given a new twist in terms of the philosophy of language. Or from another, broader perspective, their view, like the old theory of meaning, offers a bifurcated account of the scientific enterprise. On the one hand, we are told, there is the content of scientific belief, established on the basis of "epistemic" considerations; and on the other hand, there is the aim of science, which is independent of that content, and is established on grounds independent of "epistemic" considerations. And so our question is this: Can the scientific enterprise be understood without appealing to concepts which transcend the content of scientific belief and the methods by which we arrive at it? Can we provide an account of science in which such concepts as reference, and as the aims of science, are integrated with an account of the content of scientific belief itself, or must that account be bifurcated in the way I have described? My answer should by now be clear: I have argued that we neither would, nor should, nor do, nor need, employ the concepts of "essence" and "metaphysical necessity" in the ways Kripke and Putnam allege we do, unless those concepts are understood so broadly as to be empty. More positively, my suggestion is that the process of "internalization" which characterizes science is in part a process directed toward making scientific reasoning autonomous, making it grounded fully in the investigation of the experience it tries to understand; and that included in that process is the shaping—the internalization—of its own goals in the light of what we learn.

If a realistic account is indicated by the successes of present science (and in many aspects of science it is), that account is a conclusion from our experience, not the product of an *a priori* or transcendental argument in the philosophy of language. Even more than I object to the idea that science aims at the discovery of essences, I protest against this latter view—the view, namely, that what science must aim at can be established by an examination of the nature of language. Here my criticisms of the Kripke-Putnam view flow together, and join also with my objections to the traditional theory of meaning. I remarked earlier that the arguments raised by Kripke and Putnam against the traditional theory of meaning were sound and correct: any properties (criteria) we originally use in identifying some thing or kind may later be found not to hold of that thing or kind. My agreement can now be seen to be simply a manifestation of my view that *all* of what we say about the thing or kind may, in principle, be subject to doubt and revision or rejection (barring certain sorts of new discoveries about how to learn, discoveries that we have no specific reason whatever to expect). But from this point of view, my disagreement with Kripke and Putnam is a manifestation of the very same point. For in saying that there *is* something at whose discovery we might arrive (and to which we have been referring all along) and whose discovery will lead to practices which are in principle (because of the nature of reference) immune to revision, they have abandoned the very point which they so incisively raised against the traditional theory of meaning. Their view remains in agreement with that theory on one fundamental point: that there is something about science that is in principle immune to revision in the face of experience. The difference between them is merely that, for the traditional theory of meaning, that which is immune to revision is something that is established at an inception, while for Kripke and Putnam it is something that is aimed at and perhaps arrived at after long empirical investigation. I reject the fundamental assumption common to both views.

Both the traditional theory of meaning and the Kripke-Putnam theory of reference stem from the same source: the idea that, by an analysis of features of language, we should be able to lay down inviolable conditions on the knowledge-seeking enterprise. Having rejected the principle of inviolability with regard to *any* aspect of that enterprise, I reject it in this instance also. Neither meaning nor reference, at least as understood by the views considered in this paper, can serve as tools in the attempt to understand science.[4]

## V

Yet though we may thus reject the ideas of meaning and reference which are the technical artifacts of philosophers, those notions have their source in homely observations for which place must be made, if not as tools, then as products. After all, we do talk about things, and we do attribute meanings to what we say. But the work that needs to be done with regard to those observations can be done in the context of the picture of science and its changes that I have drawn. Consider, for example, the notorious problem of explaining how (or whether) continuity of discussion is possible between successive theories in science. To account for such continuity, it is unnecessary to assume *either* a common "meaning" *or* a common ("rigid") reference of terms. That Stoney, Thomson, and Feynman were, in certain aspects of their work, all "talking about" electrons is not guaranteed by a shared set of necessary and sufficient conditions for applying the term 'electron'; there *may* be shared ascriptions[5] (though not in the sense of necessary and sufficient conditions), but there is nothing to prevent the ultimate abandonment of all of them; they are not irrevocable. To assume that there *must* be shared and irrevocable "criteria" is simply a Platonic fallacy. Nor is continuity guaranteed by a baptismal act by Stoney and a "causal" handing-down of the term; that view, as I hope I have shown, results in the assumption of a something-I-know-not-what that does even less explanatory work, and certainly more mischief, than its Lockean ancestor. In particular (and in addition to my earlier objections), that view is no help at all in resolving the problem of continuity or comparability of scientific theories: for to say that continuity is guaranteed by the fact that we are talking about (referring to) the same "essence", where we do not or cannot know what that essence is, is merely to give a name to the bald assertion of continuity. Nor can it help to look at the "causal" connections between successive usages: as adherents of incommensurability might well argue by pointing to methods of education in science, a new generation may be conditioned—caused—to *think* it is referring to the same thing by 'electron' as its predecessors.

An emphasis on *causal* connections is thus futile for escaping the usual incommensurability arguments. (Here, at least, the old theory of meaning has the edge over its rival: for, its other flaws aside, it *could* answer those arguments *if* there were "meanings" in the sense it claimed, and *if* those meanings were indeed shared by successive members of a scientific tradition. But those conditions are not satisfied, and in any case the other objections to the theory make it unable to do any better than its rival in accounting for scientific change.) Rather than either common meanings or common references, what serve, and what alone can serve, in science at least, as providers of continuity are the *reasons* connecting the successor idea to its ancestors. Where it occurs in science, continuity is achieved by what may be called a "chain-of-reasoning connection". An example is the chain which led in successive stages from Stoney's views about the electron, through Thomson's, and on to Feynman's, there having been reasons at each stage for the dropping or modification of some things that were said about electrons and the introduction of others. The meaning of the term 'electron', in the only sense that can be made or is needed, then, is this: a family of criteria related by a chain-of-reasoning connection. This is sufficient warrant for the claim that the theories concerned were saying competing things about electrons.

Thus, what Feynman and his associates "meant" in the homely sense in discussing electrons is understood in terms of what Feynman and his associates *said* electrons were—the set of properties they attributed to electrons.[6] What they were talking *about* is similarly determined by those properties, or, if you prefer, criteria. But those properties or criteria were in turn related to Thomson's and Stoney's by specifiable connections. The considerations forming those connections had been determined to count as "reasons" as a result of the process of "internalization" described in Part III, above. We thus need not say that Feynman and the others were all really referring to a common something-I-know-not-what; all the work that concept tries and fails to do can be done simply through this analysis, together with the recognition, as part of our analysis, that Feynman's or the others' criteria—what they said about electrons, how they used the term 'electron'—might become subject to doubt or replacement. Electrons can thus be understood as "transtheoretical", something about which we can have competing theories, without assuming that there is either a common meaning or a common reference of the term 'electron'.[7] A somewhat more subtle analysis must be given for the competition of theories at least one of which is rejected and thus does not belong to a chain of theories each of which (or at least the later ones in the series) was *accepted* in turn. But even there, there will be found, on proper analysis, a common subject-matter, common problems, and usually at least some common set of other relevant information from the scientific storehouse.

As I hope I have shown in this and other papers, the technical concepts of meaning and reference stemming from the philosophy of language have failed to clarify the scientific enterprise. On the contrary, they have only succeeded either in introducing hopeless confusion or in contradicting some of the most fundamental aspects and achievements of that enterprise. Their vagaries, confusions, and paradoxes, their arbitrary presuppositions and apriorisms, their epistemological relativisms and metaphysical absolutes, must all be avoided. The only way of doing this is to abandon those technical concepts themselves, as philosophers and others have understood them, and to exorcise completely the error of supposing that scientific reasoning is subservient to certain alleged necessities of language, and that the study of the latter is therefore deeper than the study of the former. The situation, I have argued, is rather the reverse. I have tried to show how this is so, and how its being so can be recognized in a more adequate understanding of the scientific enterprise.

## REFERENCES


Kripke, S. (1977), "Identity and Necessity", in Schwartz, S. (ed.), *Naming, Necessity, and Natural Kinds*: 66–101. Ithaca: Cornell University Press.

Kripke, S. (1980), *Naming and Necessity*. Cambridge, Mass.: Harvard University Press.

Margalit, A. (1979), "Sense and Science", in Saarinen, E., *et. al.* (eds.), *Essays in Honour of Jaakko Hintikka*: 17–47. Dordrecht: Reidel.

Putnam, H. (1973), "Explanation and Reference", in Pearce, G., and Maynard, P. (eds.), *Conceptual Change*: 199–221. Dordrecht: Reidel.

Putnam, H. (1977), "Meaning and Reference", in Schwartz, S. (ed.), *Naming, Necessity, and Natural Kinds*: 119–132. Ithaca: Cornell University Press.

Putnam, H. (1978), *Meaning and the Moral Sciences*. London: Routledge and Kegan Paul.

Putnam, H. (1979), *Philosophical Papers*, Vol. II: *Mind, Language, and Reality*. Cambridge, Eng.: Cambridge University Press.

Shapere, D. (1964), "The Structure of Scientific Revolutions", *Philosophical Review 73*: 383–394.

Shapere, D. (1966), "Meaning and Scientific Change", in Colodny, R. (ed.), *Mind and Cosmos*: 41–85. Pittsburgh: University of Pittsburgh Press.

Shapere, D. (1969), "Notes Toward a Post-Positivistic Interpretation of Science", in Achinstein, P., and Barker, S. (eds.), *The Legacy of Logical Positivism*: 115–160. Baltimore: Johns Hopkins University Press.

Shapere, D. (1971), "The Paradigm Concept", *Science 172*: 706–709.

Shapere, D. (1974a), "On the Relations Between Compositional and Evolutionary Theories", in Ayala, F., and Dobzhansky, T. (eds.), *Studies in the Philosophy of Biology*: 187–202.

Shapere, D. (1974b), "Scientific Theories and Their Domains", in Suppe, F. (ed.), *The Structure of Scientific Theories*: 518–565. Urbana: University of Illinois Press.

Shapere, D. (1977), "The Influence of Knowledge on the Description of Facts", in Suppe, F., and Asquith, P. (eds.) *PSA 1976*: 281–298. East Lansing: Philosophy of Science Association.

Shapere, D. (1980), "The Character of Scientific Change", in Nickles, T. (ed.), *Scientific Discovery, Logic, and Rationality*: 61–116. Dordrecht: Reidel.

Shapere, D. (1981), "The Scope and Limits of Scientific Change", in Cohen, L. J., *et. al.* (eds.), *Logic, Methodology and Philosophy of Science VI*, forthcoming. Amsterdam: North-Holland.

Shapere, D. (forthcoming), "The Concept of Observation in Science and Philosophy."

## NOTES

[1]In speaking of Putnam in this paper, I refer only to the views he expressed in the above-mentioned works (with the exception of the final chapter of [1978]). More recently, he has abandoned some but not all of those ideas, and for this reason I have focused here more on Kripke's than on Putnam's views. But I have included discussion of some of the latter, both because they remain influential, and because they often throw additional light on the topics at issue. But there are many aspects even of Putnam's earlier views that I do not discuss here.

[2]Note that the idea of an "internalizing of relevant considerations" provides a resolution of the debate, among philosophers and historians of science, as to whether scientific change is governed primarily by "internal" or "external" factors.

[3]Margalit has put the arguments I am rejecting in a way that exhibits my point glaringly: "In determining the extension of 'salt', say—which is part of the task of determining the reference conditions of the sentence constituents, which is, in turn, part of the task of determining its truth conditions—the chemical structure of salt is crucial. The reasoning here is clear. Our ability to say something true about the world depends not just on what we *believe* about the world, but also on its actual structure. In order to explain the success of our language in saying truthful things about salt we need information about the constitution of salt . . . The chemical structure of salt is, therefore, essential for determining the truth conditions of sentences where 'salt' occurs in a non-empty way. . . . . one of the enterprises undertaken by the theory of language is to explain the connection between language and the world. The explanation of this connection requires that we understand both sides of it, *i.e.*, language and the world. The world, however, is what *science* tells us that is the world." (Margalit 1979, p. 22) But the contrast made here between "beliefs" and the "actual structure" of the world plays no role in determining the truth-conditions *we use*. Scientific information is not something distinct from "beliefs", but itself consists of (well-grounded) beliefs. It is the final sentence in this quotation from Margalit, not the preceding discussion, that carries the real weight.

[4]In this paper, I have not addressed the contention that Kripke's views on reference and essences stem from certain technical features of (current) modal logic. If those views follow necessarily from the apparatus of modal logic, then if my criticisms are justified, that

apparatus does not do justice to actual science. But it seems to me questionable whether his views of reference and essences *are* so required by modal logic. If they do not so follow, then of course it remains an open question whether, and to what extent and in what ways, modal logic can illuminate the structure and reasoning of science. In that case, as with all formal systems, its applicability cannot be taken for granted, but is something that must be established.

[5]We must of course distinguish between "criteria for (properties and behavior used in) identifying something as an *x*" and "criteria for being an *x*"; but my not doing so in the present context is unimportant. It should be remembered that, in the usual situation in mature areas of science, the connection between the two is tight, since the former—*e.g.*, shapes of tracks of electrons—are explained in terms of the properties ascribed to *x*'s, together with a knowledge of the ways *x*'s interact with other entities and a knowledge of the particular circumstances of observation.

[6]We thus need not suppose that there is a "concept" of "electron" consisting of a set of necessary and sufficient conditions remaining unchanged in all uses of the term 'electron'. If the ascription of some particular subset of properties happens to be common to all uses of the term 'electron', that is a contingent fact: for people to be "talking about the same thing" when they use that term, there need be no such set, the "concept of electron" being simply the family (ancestors and descendants) of reason-related criteria each "generation" of which consists of what is attributed to electrons at a given stage in the succession.

When combined with the earlier discussion of the "internalizing of relevant considerations" (in the final four paragraphs of Section III, above), this concept of 'concept' suggests an analysis of what it is to be a "reason" in science (and in other areas of human thought as well). As I remarked in that earlier section, for something to be a "reason" is for it to be (among other things) *relevant*, in a sense broader than is capturable by formal logic, to that for which it is alleged to be a reason. This implies that, in the process of discovering what is and what is not relevant to given claims—that is, in delineating the distinction between "internal" and "external" considerations, and increasing its ability to arrive at and test ideas on the basis of internal considerations alone—science gradually clarifies what is to count as a scientific reason in a given domain of inquiry. The discussion given now, in the present section, implies that such clarification of what *counts* as a reason constitutes at the same time a clarification of the *concept* of "(a) reason" itself, in the only sense of 'concept' that can ultimately make sense. This suggested way of understanding 'reason' as inextricably bound to what is "internal" to a subject points the way toward understanding how it is possible for science to progress on the basis of considerations that are "objective" and "rational". These remarks are developed in detail in a forthcoming work, "The Concept of Observation in Science and Philosophy", where it is shown that even allegedly "metascientific concepts" undergo a process of revision in response to new scientific findings.

[7]For the notion of a "transtheoretical term", see Shapere (1969, Part II). Putnam (1973) has appealed to this idea in developing his own theory of reference; as should be clear, his use of it was not what I had in mind.

CHAPTER 19

# MODERN SCIENCE AND THE PHILOSOPHICAL TRADITION

One of the recurrent and most profound clusters of problems of our intellectual tradition has been that of trying to understand the process of seeking and acquiring knowledge, or at least what we take to be knowledge. What, for example, are the proper methods for seeking knowledge or well-grounded belief? What are the characteristic features by which we recognize a belief as being knowledge, or at least as being well-grounded?

The major traditional approaches to these questions, as exemplified in such thinkers as Plato, Kant, the early Wittgenstein, and the logical empiricists, have held a common assumption: namely, that there is something which is presupposed by, or essential to, or necessary to, the knowledge-seeking enterprise. That presupposed or essential or necessary ingredient is alleged to be immune to revision or rejection in the light of any new knowledge or beliefs at which we might arrive. Further, it has often been held, that ingredient (or set of ingredients, for there need not be just one) is supposed to constitute the characteristic, defining feature of science. Let me give two examples of this general viewpoint. First, it has sometimes been argued that there is a method, the scientific method, by application of which we obtain knowledge or well-grounded belief about the world, but which, once discovered, is not subject to revision or alteration in the light of any beliefs at which we might arrive by applying that method. The classic version of this view is the claim that, at a certain point in history, Galileo discovered (or invented) the scientific method, which was thenceforth *applied* in the discovery of further and further truth about nature, though all the while the method itself remained unaltered, inviolate, in the light of the knowledge acquired by its means. Furthermore, it is held, this method is what is characteristic of, definitory of, science. This might be called the "methodological" version of the general thesis that science presupposes something. A second kind of example, which can be called a "presuppositionist" version, holds that science makes certain *presuppositions* which could not be rejected as long as we are to seek and possibly to find knowledge or well-founded belief. There have been many versions of this "presuppositionist" sort of view. For example, Kant held that knowledge, and the seeking of it, could not exist without what he called "forms of intuition" (space and time) and "forms of thought" or "categories" (among which he included the idea that every event must have a cause, and the idea of substance, that is, of independent, permanently existing objects), and furthermore, that these forms of intuition and categories *could* not be abandoned or, presumably, even modified, no matter what we might find in our experience. Other philosophers proposed other "necessary presuppositions" of knowledge and the enterprise of seeking it; I will only give a rough statement of some of the most famous: the principle of the uniformity of nature – that nature always "acts in the same way"; the principle of the simplicity of nature – that nature ultimately is "simple"; the principle of the unity of nature – that nature is open to being understood through a unified theory; John Maynard Keynes' "Principle of Limited Independent Variety," the idea that there are only a finite number of different kinds of things in nature; the "Principle of Induction," often stated in some form such as "The future will be like (obey the same laws as) the past." One of the most recent versions of the "presuppositionist" viewpoint, differing in important ways from those I have just mentioned, is that of logical empiricism (often called logical positivism) in its mature form. According to that view, there are what its supporters called "metascientific" concepts, like 'evidence,' 'observation,' 'theory,' 'explanation,' 'confirmation,' which are used in talking *about* scientific concepts, claims, or arguments, which must have meanings which are wholly independent of the specific content of ongoing science, and whose meanings constitute the identifying characteristics of anything that is to count as a piece of evidence, an observation, a theory, an explanation, and so forth – that is, whose meanings constitute *what it is to be* evidence, observation, and so forth. And when you have understood each of these and put them all together, you have defined *what it is to be science* or scientific – the "essence" of science, to use a word they despised but which nevertheless fits. To do something that didn't fit those definitions would be to do something that, by definition, was not science, so that you couldn't violate these definitions as long as you did science.

Now what all these examples, both methodological and presuppositional, have in common is, to repeat, the *general* view that there is *something*, whatever its precise character, which is presupposed by, or is essential to, or is necessary to, the knowledge-seeking enterprise, and in particular to science, and that that presupposed or essential or necessary ingredient cannot be revised or rejected in the light of any new scientific knowledge or beliefs at which we might arrive. I will refer to this general view as the *Principle of Inviolability*. A second view which, as I have said, is usually associated with the Principle of Inviolability is the view that that inviolable ingredient is supposed to constitute the defining characteristic of science. Since these two views are usually held together, and since they are held together by the views I have been and will be talking about, I will therefore refer to both of them together as "The Inviolability Thesis."

We are all familiar with the fact that such approaches, holding one or another version of the Inviolability Thesis, while by no means extinct, are no longer as popular as they once were. I will mention three sources of challenge to them. Within philosophy itself, the attack on the doctrines of essentialism and necessary truth, and the defense of the view that science is an ever-changing, ever-evolving enterprise, was launched by the American pragmatists, especially Peirce and Dewey. Their attack has been sustained and advanced by such writers as Morton White and W. V. Quine, who have denied that there are essential or necessary truths and defended the view that science is a corporate body of beliefs no one of which is immune to revision or rejection. Entirely outside the pragmatic movement (and rarely mentioning science), the later Wittgenstein put forward the thesis that what we take to be *necessarily true* is a function of the "language-game" we play, the "form of life" in which we participate; thus there are no necessary or essential truths apart from such language-games or forms of life, and when we suppose some proposition to be such a truth, we must be suspicious of "the hardness of the logical must" and seek to expose the linguistic bases of our mistaken supposition.

A second source of the reaction against the Inviolability Thesis has come from within the science of the last century or so. For there we have seen before our eyes the downfall, one after the other, of what had previously been considered to be necessary truths, beliefs (or, in some cases or aspects, standards or methods) that, it was thought, neither would nor could ever be rejected or even revised. The idea that all ultimate explanation must be in terms of interactions of matter in motion; the idea of the inviolability of Euclidean geometry; the view that there is a unique set of events in the universe simultaneous with a given event; the thesis that the fundamental laws of nature must be deterministic; the idea that space and time are Parmenidean, or rather Democritean, "non-being," background arenas unaffected by the events taking place in them and not affecting them in turn – all these are only the most famous examples of alleged "necessary truths" that have been rejected as not even being true, much less necessarily so. And on the positive side, new ideas have been introduced into science that go so far beyond anything that might have been anticipated as to seem utterly bizarre.

The very distinction between matter and the space-time in which it is located is blurred and threatened with obliteration; we dare now to speak of holes in the fabric of space and time. And at the level of the very small, the very notions of space, time, and matter, as traditionally conceived, verge on the inadequate. Elementary particles have properties so far from traditional and macroscopic expectations that they – all of them – fully deserve to be called "strange"; and even the terms 'elementary' and 'particle' are not wholly appropriate in terms of the older senses of those words, while the term 'interaction' has moved so far in its applications that it has come (for good reasons) to cover cases where only one particle is involved. Surely in all this and more, it seems, there is reason to agree that anything we think necessary and inviolable must be suspect, that scientific thought cannot be constrained by any limits on its thought.

These indications coming from recent philosophy and from modern science that the Inviolability Thesis can no longer be maintained have been reinforced from a third area, the study of the history of science, which has only in recent decades become a truly professional discipline, written not by scientists dividing the past into their own anticipators and their own obstructors, but by highly trained historians of science who look deeply into the background of ideas in terms of which scientists of a period conceived their problems, their alternatives, their methods, their standards. And the upshot of these ever-broadening historical investigations has been the powerful suggestion that it is not just in recent times that science has undergone radical and profound change. Rather, such changes have been characteristic of it throughout its history, and go far deeper than mere discovery of new facts and simple successive alterations in the body of substantive beliefs about the world held at a given time, but extend also to the methodology employed (contrary to the adherents of an inviolable scientific method), the distinction between a legitimate and an illegitimate scientific problem (contrary to Karl Popper as well as adherents of the verifiability theory of meaning), what counts as a possible or a correct explanation (contrary to Carl Hempel and others), what counts as an observation or observational evidence), and even the standards and goals of inquiry (contrary to practically all positivistic philosophers). The suggestion is that there is nothing in science that is sacrosanct, that is immune in principle to revision or rejection: neither alleged fact, nor theory, nor concept, nor problem, nor method, nor way of thinking, nor even the definition of "science" as embodied in some allegedly immune set of "metascientific" concepts.

And yet we must be philosophically severe with ourselves: the revelation that science has undergone profound changes, that *many* ideas which were formerly argued to be necessary truths turned out not to be; that *many* alleged methods or concepts or standards which were formerly argued to be requirements of the very possibility of scientific inquiry turned out not to be; that *many* characteristics of science which were formerly argued to be essential to, definitory of, science, turned out not to be – none of this *proves* that there are *no* necessary truths, necessary requirements, or eternal defining characteristics of science. That *certain* ideas have been violated does not establish that *all* ideas are open to violation, even though it sharpens our suspicions of any alleged inviolability. That is, the second and third reasons I have discussed for challenging the Inviolability Thesis – the results of modern science, and the revelations of the history of science – while they are enormously compelling, are not conclusive refutations of that thesis. Indeed, for someone suspicious of inviolability to *demand a proof* that anything in science can be violated would be for that person to betray his own principles: for would not such a *proof* make it a necessary, inviolable truth that there can be no necessary, inviolable truths?

There is yet a further point which we must remember soberly when we challenge the Inviolability Thesis, and that is the ultimate motivation behind it, throughout its history, whatever specific form the thesis took in particular versions of it. That motivation has been the desire to preserve the possibility of knowledge, to avoid the Scylla of relativism according to which what we take to be knowledge is only a function of our arbitrary presuppositions, and the Charybdis of skepticism according to which we can have no knowledge. The motivation has always been so: it was thus with Plato against the relativism of the Sophists, for whom man was the measure, the determiner, of all things, of the true that it is true as well as of the good that it is good; so, Plato thought, there must be Absolutes. It was thus, too, with Kant against the skepticism into which Hume's philosophy collapsed, leaving us not only with the inability to offer the slightest justification for any belief whatever, but, worse, with the inability to determine that what we are saying or thinking is not utter gibberish, so that in the end we could only grunt with animal faith; to save us from that, Kant maintained, we must have a set of presupposed, necessary categories and forms of intuition. And it was thus with Mill, with his Principle of the Uniformity of Nature, Keynes with his Principle of Limited Independent Variety, the mature Russell (in *Human Knowledge*) with his Postulates of Scientific Inference, and a host of others, against the epistemological holocaust threatened by the Problem of Induction; so, they all maintained, there must be inviolable presuppositions of the kind each of those thinkers proposed.

The perils of relativism have certainly appeared real enough in recent times; for the radical, fundamental changes in science brought out by historians have led many of them in precisely that direction. Thus in the highly influential work of Thomas Kuhn, his *The Structure of Scientific Revolutions*, science periodically undergoes drastic "revolutions," after which the entire fabric of science has been changed, the "paradigm" before and the paradigm after the revolution being "incommensurable" with respect to what each accepts as observations, facts, problems, methods, standards, and so forth; and, for Kuhn (at least in that book), there are no trans-paradigmatic standards in terms of which we could choose between paradigms. The views of numerous other historians of science and philosophers influenced by developments in the history of science (most notably, Paul Feyerabend) have also tended toward relativism. And what, after all, is the ultimate difference between relativism and skepticism? For both amount to a denial of the possibility of knowledge.

In my discussion of the three motivations for rejecting the Inviolability Principle earlier, you will remember that the first consisted of suggestions coming from within philosophy, notably the pragmatists and the later Wittgenstein. But just as the attacks on the Inviolability Principle coming from modern science and the history of science fail to provide a conclusive refutation of that Principle, so also the work of these philosophers fails to provide all we would wish for in this connection. Insofar as Wittgenstein's contributions are not merely negative – refuting allegations of necessary truth (not that such refutations are unimportant), – his positive views are all too easily seen as compatible, and have been taken as being compatible, either with relativism (the idea that there are different "language-games" or "forms of life," any one of which is as good as any other) or with absolutism (one can never legitimately violate "ordinary language"). Pragmatism of the Peirce-Dewey variety has often been accused of being relativism in sheep's clothing, while the similarities between Quine's views and Kuhn's are, for many readers, too close for comfort. (Consider, for example, Quine's claim that "Any hypothesis can be defended come what may," or his views about ontological relativity or the indeterminacy of translation.) The perils of rejecting inviolability seem manifest in these writers; one feels the pressures that led traditional philosophers to suppose that the only alternative to relativism and its close kin, skepticism, was to defend some version of the Inviolability Principle; there has seemed to be neither a third nor a middle ground.

Besides, none of these philosophers made much of an attempt to go deeply into the analysis of science (or at least *modern* science) and its history (or at least *modern* studies of the history of science). And yet, as I have tried to suggest, it is in those areas that our problems have arisen; is it too much to suppose, then, that the issues about human knowledge which I have been raising might find a resolution through a deeper, philosophically sensitive examination of the knowledge-seeking enterprise as it is engaged in science?

Let me, then, try to summarize the issues that I have been developing. On the one hand, we have a long tradition claiming that there is – and must be – something inviolable about or necessarily presupposed by science. There have been powerful suggestions from within philosophy, science, and the study of the history of science, however, that that Principle is at least questionable. But *can* it be questioned without forcing us into relativism or skepticism?

One solution would be simply to accept one of the two alternatives: either that there must be something inviolable, or that there is no "knowledge" that does not rest ultimately on the wholly arbitrary, and is therefore not what we would expect "knowledge" to be. Both these alternatives seem unappealing. Candidates for inviolability have repeatedly proved either unable to guarantee the possibility of knowledge or else not to be inviolable after all. Relativism-skepticism seems to deny the evident fact that we have, through science, learned a great deal about the way things are; the problem, as Kant would have said, is not *whether* science has gained knowledge; it is rather, *how* science has been able to do so.

The real Scylla and Charybdis, then, are not relativism and skepticism; those two are really, in their most important implication, the same: the denial of the possibility of knowledge. The real problem, the real perils between which we must try to steer if possible, are relativism-skepticism on the one side, and the Principle of Inviolability on the other. In other words, the central issue, in trying to understand the enterprise of knowledge-seeking and knowledge-acquiring, has now emerged in the following form: Given that we cannot demand a *proof* that the Inviolability Principle is incorrect (for that would be to accept a version of the thesis itself), can we at least develop a coherent view which will show that both of the following are simultaneously satisfiable?

> (1) That *all* aspects of science – including, for example, all those which have hitherto been distinguished as matters of observation, fact, theory, method, standards, defining concepts ("metascientific" concepts), and so on – are in principle open to revision, though of course they need not in fact ever actually *be* revised.

(2) That those revisions can be made in the light of *reasons*, reasons based on what we *learn* in the course of engaging in the scientific enterprise.

To satisfy the first of these conditions would be to dispense with the Principle of Inviolability; to satisfy the second would be to avoid relativism and skepticism. And to satify both simultaneously would be to do what so much of our philosophical tradition has believed cannot be done: namely, to show how it is possible (at least) for us to gain knowledge – to avoid relativism and skepticism – without having to accept any absolutes whatever.

As I have formulated it, the issue comes to a focus in the concept of a "reason"; for the question really comes down to the following: Is it possible to make intelligible the idea that scientific change can come about *for reasons* without having a science-transcending criterion of what it is to be "a reason"; what it is for a change to be "reasonable," without having a prior, independent criterion of what it is to be a "reason," or a "reasonable change"? Can we make coherent the idea that criteria of rationality can themselves evolve rationally, as part and parcel of the scientific enterprise, without supposing that there are higher-level standards or criteria of rationality, themselves immune to alteration, in terms of which changes of lower-level criteria of rationality could be judged to be rational? For if we can show this, we will have gone far toward showing the *possibility* of knowledge without having given any guarantees that knowledge can be obtained, or even that there is knowledge to be found. We would, in other words, have taken an important step toward showing that knowledge, or at least justified rational belief, is possible without assuming anything inviolable, and in particular, inviolable criteria of rationality.

A full development of such a view is imperative in the light of the developments I have been discussing; at the very least, if the effort is made through a close study of science and its evolution, it might clarify the issue of what might be inviolate about science, if there is anything; and it might show that nothing is. It is a large order, certainly not satisfiable in one short lecture, if it can be satisfied at all. But focussing on the problem of what is to count as a reason makes it possible at least to suggest a direction in which, I believe, such an investigation should be directed. For what is it for something to be a "reason" for an activity or a belief? A valuable hint comes from the idea that, in order to count as a *reason* for or against some belief or activity, the alleged reason must be *relevant* to the belief or activity. "Stick to what's relevant"; "Keep to the relevant facts"; "That's not a reason; that's irrelevant."

Such are important moves in any argument. Perhaps a clue to understanding what a "reason" is in science, and how we can develop criteria of rationality *as part of the enterprise of science rather than as transcending and presupposed by it* might be found in the process, in the history of science, of coming to understand what is relevant to what.

But is the following not an intuitively plausible view of the development of science: that science is precisely the discipline, *par excellence*, of determining relations of relevance and irrelevance? The development of science consists, in large part, of beginning with suppositions about what is and is not relevant to a certain claim, and of gradually refining the claim and our understanding of what is relevant to it and what is not: of shedding certain beliefs as irrelevant; of introducing new ones which we find to be more so; of refining our modes of conceiving and describing the world around us so as to bring out more clearly and firmly the ways in which things are related to one another; of *finding out* what is a relevant consideration for the acceptance or rejection of specific beliefs, and even of finding out *that* we must stick to the relevant.

According to this picture of science, we do not have in advance unalterable criteria of relevance, of what is to count as a reason. We arrive at new ideas, which enable us to do new things; and we elevate the considerations which led us to those ideas to the status of reasons, and the more general character of that reasoning to the status of criteria of reasoning, to be tested further in terms of their success. As for success, the situation is similar: far from having or needing transcendent and unalterable criteria of what is to count as successful, we find that we can do things with certain approaches (things we may not even have thought of beforehand), and *then* elevate certain general aspects of those new approaches to the status of criteria of success and ways of looking for further successes. We *found*, in the sixteenth and seventeenth centuries, that we could understand material substances in terms of what they are made of rather than in terms of their perfectability; and looking for constituents then *became* a standard method and criterion of success in understanding material substances.

The process of developing such standards or criteria is far more complex than past and prevailing philosophies of science have assumed. It must consist partly in a revision of our descriptive language to bring out better what we have learned; it must consist further of revisions in our ways of learning about nature; and it must consist of much else as well. But to put it as simply as possible, it must consist not only of our coming to know about the world (the traditional focus of the theory of knowledge); it must also consist of learning how to learn, to think, and to talk about nature; it must, in short, consist of gradually coming to understand how to understand – of gradually reasoning out what it is to reason – of learning what it is to learn. We may, of course, always turn out to have been wrong or misguided or confused in the beliefs we have arrived at, or the standards of rationality we have forged, or the understanding we have developed of what it is to understand: to abandon the Principle of Inviolability is to abandon the offer of guarantees; the best we can have in the way of what we take to be knowledge and reasons are, simply, the best we have arrived at in our searches; and even what we take to be "best" is a hypothesis which may be debatable and may be rejected later in the light of what we then come to accept as reasons for doubt. The possibility of doubt and error is always open; but we do have, or at least develop, better and worse hypotheses about what are better and worse hypotheses, methods, descriptions, and so on.

But is the loss of guarantees something to be mourned? After all, we can have no confidence that we ever have or ever will be given any guarantees, since the promised guarantees have so often failed. Would it not be more to the credit of man that, born in a world he does not initially understand, and does not even initially understand how to understand, he could come to understand, and to understand how to understand – that he can learn, and learn what it is to learn? Would it not be more to his credit if he could avoid the arrogance of assuming that he can know in advance how the world is, or at least what he must accept, before ever examining it? Would it not be more to his credit to shed the embarrassment of finding his most certain expectations about nature and knowledge repeatedly shattered? For to show that such a view of man and his knowledge can be made coherent would simply be to recognize that there may be more in heaven and earth than is dreamt of in our philosophies, but also to hope that, by our efforts, we might come to learn something about it.

# LIST OF PUBLICATIONS

*of Dudley Shapere*

## I. BOOKS

1. *Philosophical Problems of Natural Science*, New York, Macmillan, 1965.
2. *Galileo: A Philosophical Study*, Chicago, University of Chicago Press, 1974.
3. *The Concept of Observation in Science and Philosophy*, New York, Oxford, forthcoming.

## II. ARTICLES

4. "Philosophy and the Analysis of Language," *Inquiry* III (1960), 29–48; reprinted in R. Rorty (ed.), *The Linguistic Turn: Recent Essays in Philosophical Method*, Chicago, University of Chicago Press, 1967, 271–283.
5. "Mathematical Ideas and Metaphysical Concepts," *Philosophical Review* **LXIX** (1960), 376–385.
6. "Descartes and Plato," *Journal of the History of Ideas*, October–December, 1963, 572–576; also in *Ithaca: Proceedings of the Tenth International Congress of the History of Science (1962)*, Paris, Hermann, 1964, Vol. **I**, 275–278.
7. "Space, Time, and Language," in B. Baumrin (ed.), *Philosophy of Science: The Delaware Seminar*, Vol. **II**, New York, Wiley, 1963, 139–170.
8. "The Causal Efficacy of Space," *Philosophy of Science*, April, 1964, 111–121.
9. "The Structure of Scientific Revolutions," *Philosophical Review* **LXXIII** (1964), 383–394; reprinted (in French translation) in F. Jacob (ed.), *De Vienne à Cambridge*, Paris, Gallimard, 1979; reprinted in G. Gutting (ed.), *The Kuhnian Revolution: Applications and Appraisals of Thomas Kuhn's Philosophy of Science*, South Bend, University of Notre Dame Press, 1980, 27–38.
10. "Newton," in *Encyclopedia of Philosophy*.
11. "Newtonian Mechanics and the Nature of Mechanical Explanation," in *Encyclopedia of Philosophy*.
12. "Meaning and Scientific Change," in R. Colodny (ed.), *Mind and Cosmos: Explorations in the Philosophy of Science*, Pittsburgh, University of Pittsburgh Press, 1966, 41–85; reprinted in I. Hacking (ed.), *Scientific Revolutions*, Oxford, Oxford University Press, 1981, 28–59.
13. "Plausibility and Justification in the Development of Science," *Journal of Philosophy*, October 27, 1966, 611–621.
14. "The Philosophical Significance of Newton's Science," *The Texas Quarterly*, Autumn, 1967, 201–217; reprinted in R. Palter (ed.), *1666: The Annus Mirabilis of Sir Isaac Newton*, Cambridge, M.I.T. Press, 1970, 285–299.
15. "Biology and the Unity of Science," *Journal of the History of Biology*, Spring, 1969, 3–18.

16. "Notes Toward a Post-Positivistic Interpretation of Science," in P. Achinstein and S. Barker (eds.), *The Legacy of Logical Positivism*, Baltimore, Johns Hopkins University Press, 1969, 115–160.
17. "The Paradigm Concept," *Science*, Vol. 172 (14 May 1971), 706–709; reprinted in M. Marx and F. Goodson (eds.), *Theories in Contemporary Psychology*, New York, Macmillan, 1976, 53–61; reprinted (in Spanish translation) in A. Schneider (ed.), *Paradigmas en Economía: Análisis y Applicaciones*, Mexico City, 1979.
18. "On the Relations between Compositional and Evolutionary Theories," in F. Ayala and T. Dobzhansky (eds.), *Studies in the Philosophy of Biology*, London, Macmillan, 1974, 187–202.
19. "A Note on the Concept of Selection" (with Gerald Edelman), in F. Ayala and T. Dobzhansky (eds.), *Studies in the Philosophy of Biology*, London, Macmillan, 1974, 202–204.
20. "Discussion with Charles Birch and Bernhard Rensch," in F. Ayala and T. Dobzhansky (eds.), *Studies in the Philosophy of Biology*, London, Macmillan, 1974, 256–258.
21. "Natural Science and the Future of Metaphysics," in R. Cohen and M. Wartofsky (eds.), *Methodological and Historical Essays in the Natural and Social Sciences*, Dordrecht, D. Reidel, 1974, 161–171.
22. "Scientific Revolutions," *New Catholic Encyclopedia: Supplementary Volume*, 1974.
23. "Discovery, Rationality, and Progress in Science: A Perspective in the Philosophy of Science," in K. Schaffner and R. Cohen (eds.), *PSA 1972*, Dordrecht, D. Reidel, 1974, 407–419.
24. "Scientific Theories and Their Domains," in F. Suppe (ed.), *The Structure of Scientific Theories*, Urbana, University of Illinois Press, 1974, 518–565; Spanish translation, "Las Teorias Cientificas y sus Dominios," in Suppe, *La Estructura de las Teorias Cientificas*, Madrid, Editora Nacional, 1979, 570–618.
25. "Copernicanism as a Scientific Revolution," in A. Beer and K. Strand (eds.), *Copernicanism Yesterday and Today*, New York, Pergamon, 1975, 97–104.
26. "Scientific Theories and Social Values," *Science: An American Bicentennial View*, Academy Forum, Washington, D.C., National Academy of Sciences, 1977, 36–38. (Publication of a nationally-broadcast panel discussion at the National Academy of Sciences, November 12, 1975.)
27. "The Influence of Knowledge on the Description of Facts," in F. Suppe and P. Asquith (eds.), *PSA 1976*, East Lansing, Philosophy of Science Association, 1977, 281–298.
28. "Unification and Fractionation in Science: Significance and Prospects," *The Search for Absolute Values: Harmony Among the Sciences, Proceedings of the Vth International Conference on the Unity of the Sciences*, New York, 1977, 867–880.
29. "What Can the Theory of Knowledge Learn from the History of Knowledge?" *The Monist*, October, 1977, 488–508.
30. "Interpreting the Meaning of 'Evolutionary Synthesis'," in E. Mayr and W. Provine (eds.), *The Evolutionary Synthesis*, Cambridge, Harvard University Press, 1980, 388–398.
31. "Scientific Thought: A One-Year Course in History and Philosophy of Science," *Teaching Philosophy*, Vol. II, No. 2, 130–134.

32. "Progress and the Limits of Science," *Proceedings of the VIth International Conference on the Unity of the Sciences*, New York, 1978, 989–1001.
33. "The Character of Scientific Change," in T. Nickles (ed.), *Scientific Discovery, Logic, and Radionality*, Dordrecht, D. Reidel, 1980, 61–116.
34. "The Scope and Limits of Scientific Change," in L. J. Cohen, H. Pfeiffer, K. P. Podewski, and J. Los (eds.), *Logic, Methodology and Philosophy of Science VI*, Amsterdam, North-Holland, 1982, 449–459. Also to appear in *Theorema* in Spanish translation.
35. "Questions on the Development of Particle Physics," in L. Brown and L. Hoddeson, *The Birth of Particle Physics*, New York, Cambridge University Press, forthcoming. (Proceedings of conference held at Fermi National Accelerator Laboratory.)
36. "Reference, Necessity, and the Role of Reasoning in Scientific Change," to appear in E. Villanueva (ed.), *Proceedings of the First International Conference on Philosophy, Queretaro, Mexico*. (In Spanish translation.)
37. "Observation and Knowledge," to appear in E. Villanueva (ed.), *Proceedings of the Second International Conference on Philosophy, Oaxaca, Mexico*. (In Spanish translation.)
38. "Reason, Reference, and the Quest for Knowledge," *Philosophy of Science*, March, 1982, 1–23.
39. "The Concept of Observation in Science and Philosophy," *Philosophy of Science*, December, 1982, 485–525.
40. "Modern Science and the Philosophical Tradition," forthcoming in *Theorema* (in Spanish translation).
41. "The Role of Observation in Science," forthcoming in *Theorema* (in Spanish translation).

## III. REVIEWS

42. Review of William Dray, *Laws and Explanation in History*, in *University of Toronto Quarterly*, July, 1958, 471–473.
43. Review of L. W. H. Hull, *History and Philosophy of Science*, in *Philosophy of Science*, April, 1960, 218–220.
44. Review of W. A. Wallace, *The Scientific Methodology of Theodoric of Freiberg*, in *Philosophy of Science*, January, 1962, 101–102.
45. Review of *Classics in Science: A Course of Readings by Authorities*, in *Philosophy of Science*, July, 1962, 326–328.
46. Review of R. Merton, *The Sociology of Science: Theoretical and Empirical Investigations*, in *Physics Today*, Vol. 27, No. 8 (August, 1974), 52–53.

# INDEX OF NAMES

# BOSTON STUDIES IN THE PHILOSOPHY OF SCIENCE



1. Marx W. Wartofsky (ed.), *Proceedings of the Boston Colloquium for the Philosophy of Science 1961-1962.* 1963.
2. Robert S. Cohen and Marx W. Wartofsky (eds.), *In Honor of Philipp Frank.* 1965.
3. Robert S. Cohen and Marx W. Wartofsky (eds.), *Proceedings of the Boston Colloquium for the Philosophy of Science 1964-1966. In Memory of Norwood Russell Hanson.* 1967.
4. Robert S. Cohen and Marx W. Wartofsky (eds.), *Proceedings of the Boston Colloquium for the Philosophy of Science 1966-1968.* 1969.
5. Robert S. Cohen and Marx W. Wartofsky (eds.), *Proceedings of the Boston Colloquium for the Philosophy of Science 1966-1968.* 1969.
6. Robert S. Cohen and Raymond J. Seeger (eds.), *Ernst Mach: Physicist and Philosopher.* 1970.
7. Milic Capek, *Bergson and Modern Physics.* 1971.
8. Roger C. Buck and Robert S. Cohen (eds.), *PSA 1970. In Memory of Rudolf Carnap.* 1971.
9. A. A. Zinov'ev, *Foundations of the Logical Theory of Scientific Knowledge (Complex Logic).* (Revised and enlarged English edition with an appendix by G. A. Smirnov, E. A. Sidorenka, A. M. Fedina, and L. A. Bobrova.) 1973.
10. Ladislav Tondl, *Scientific Procedures.* 1973.
11. R. J. Seeger and Robert S Cohen (eds.), *Philosophical Foundations of Science.* 1974.
12. Adolf Grünbaum, *Philosophical Problems of Space and Time.* (Second, enlarged edition.) 1973.
13. Robert S. Cohen and Marx W. Wartofsky (eds.), *Logical and Epistemological Studies in Contemporary Physics.* 1973.
14. Robert S. Cohen and Marx W. Wartofsky (eds.), *Methodological and Historical Essays in the Natural and Social Sciences. Proceedings of the Boston Colloquium for the Philosophy of Science 1969-1972.* 1974.
15. Robert S. Cohen, J. J. Stachel and Marx W. Wartofsky (eds.), *For Dirk Struik. Scientific, Historical and Political Essays in Honor of Dirk Struik.* 1974.
16. Norman Geschwind, *Selected Papers on Language and the Brain.* 1974.
18. Peter Mittelstaedt, *Philosophical Problems of Modern Physics.* 1976.
19. Henry Mehlberg, *Time, Causality, and the Quantum Theory* (2 vols.). 1980.
20. Kenneth F. Schaffner and Robert S. Cohen (eds.), *Proceedings of the 1972 Biennial Meeting, Philosophy of Science Association.* 1974.
21. R. S. Cohen and J. J. Stachel (eds.), *Selected Papers of Léon Rosenfeld.* 1978.
22. Milic Capek (ed.), *The Concepts of Space and Time. Their Structure and Their Development.* 1976.
23. Marjorie Grene, *The Understanding of Nature. Essays in the Philosophy of Biology.* 1974.

24. Don Ihde, *Technics and Praxis. A Philosophy of Technology.* 1978.
25. Jaakko Hintikka and Unto Remes, *The Method of Analysis. Its Geometrical Origin and Its General Significance.* 1974.
26. John Emery Murdoch and Edith Dudley Sylla, *The Cultural Context of Medieval Learning.* 1975.
27. Marjorie Grene and Everett Mendelsohn (eds.), *Topics in the Philosophy of Biology.* 1976.
28. Joseph Agassi, *Science in Flux.* 1975.
29. Jerzy J. Wiatr (ed.), *Polish Essays in the Methodology of the Social Sciences.* 1979.
32. R. S. Cohen, C. A. Hooker, A. C. Michalos, and J. W. van Evra (eds.), *PSA 1974: Proceedings of the 1974 Biennial Meeting of the Philosophy of Science Association.* 1976.
33. Gerald Holton and William Blanpied (eds.), *Science and Its Public: The Changing Relationship.* 1976.
34. Mirko D. Grmek (ed.), *On Scientific Discovery.* 1980.
35. Stefan Amsterdamski, *Between Experience and Metaphysics. Philosophical Problems of the Evolution of Science.* 1975.
36. Mihailo Marković and Gajo Petrović (eds.), *Praxis. Yugoslav Essays in the Philosophy and Methodology of the Social Sciences.* 1979.
37. Hermann von Helmholtz: *Epistemological Writings. The Paul Hertz/Moritz Schlick Centenary Edition of 1921 with Notes and Commentary by the Editors.* (Newly translated by Malcolm F. Lowe Edited, with an Introduction and Bibliography, by Robert S. Cohen and Yehuda Elkana.) 1977.
38. R. M. Martin, *Pragmatics, Truth, and Language.* 1979.
39. R. S. Cohen, P. K. Feyerabend, and M. W. Wartofsky (eds.), *Essays in Memory of Imre Lakatos.* 1976.
42. Humberto R. Maturana and Francisco J. Varela, *Autopoiesis and Cognition. The Realization of the Living.* 1980.
43. A. Kasher (ed.), *Language in Focus: Foundations, Methods and Systems. Essays Dedicated to Yehoshua Bar-Hillel.* 1976.
46. Peter L. Kapitza, *Experiment, Theory, Practice.* 1980.
47. Maria L. Dalla Chiara (ed.), *Italian Studies in the Philosophy of Science.* 1980.
48. Marx W. Wartofsky, *Models: Representation and the Scientific Understanding.* 1979.
50. Yehuda Fried and Joseph Agassi, *Paranoia: A Study in Diagnosis.* 1976.
51. Kurt H. Wolff, *Surrender and Catch: Experience and Inquiry Today.* 1976.
52. Karel Kosík, *Dialectics of the Concrete.* 1976.
53. Nelson Goodman, *The Structure of Appearance.* (Third edition.) 1977.
54. Herbert A. Simon, *Models of Discovery and Other Topics in the Methods of Science.* 1977.
55. Morris Lazerowitz, *The Language of Philosophy. Freud and Wittgenstein.* 1977.
56. Thomas Nickles (ed.), *Scientific Discovery, Logic, and Rationality.* 1980.
57. Joseph Margolis, *Persons and Minds. The Prospects of Nonreductive Materialism.* 1977.
58. Gerard Radnitzky and Gunnar Andersson (eds.), *Progress and Rationality in Science.* 1978.

59. Gerard Radnitzky and Gunnar Andersson (eds.), *The Structure and Development of Science*. 1979.
60. Thomas Nickles (ed.), *Scientific Discovery: Case Studies*. 1980.
61. Maurice A. Finocchiaro, *Galileo and the Art of Reasoning*. 1980.
62. William A. Wallace, *Prelude to Galileo*. 1981.
63. Friedrich Rapp, *Analytical Philosophy of Technology*. 1981.
64. Robert S. Cohen and Marx W. Wartofsky (eds.), *Hegel and the Sciences*. (Forthcoming).
65. Joseph Agassi, *Science and Society*. 1981.
66. Ladislav Tondl, *Problems of Semantics*. 1981.
67. Joseph Agassi and Robert S. Cohen (eds.), *Scientific Philosophy Today*. 1982.
68. Władysław Krajewski (ed.), *Polish Essays in the Philosophy of the Natural Sciences*. 1982.
69. James H. Fetzer, *Scientific Knowledge*. 1981.
70. Stephen Grossberg, *Studies of Mind and Brain*. 1982.
71. Robert S. Cohen and Marx W. Wartofsky (eds.), *Epistemology, Methodology, and the Social Sciences*. 1983.
72. Karel Berka, *Measurement*. 1983.
73. G. L. Pandit, *The Structure and Growth of Scientific Knowledge*. 1983.
74. A. A. Zinov'ev, *Logical Physics*. (Forthcoming).
75. Gilles-Gaston Granger, *Formal Thought and the Sciences of Man*. 1983.
76. R. S. Cohen and L. Laudan (eds.), *Physics, Philosophy and Psychoanalysis*. 1983.
77. G. Böhme et al., *Finalization in Science*, ed. by W. Schäfer. 1983.